Bones of Contention

Bones of Contention

A Creationist Assessment of Human Fossils

Marvin L. Lubenow

Baker Books

A Division of Baker Book House Co
Grand Rapids, Michigan 49516

Published by Baker Books
a division of Baker Book House Company
P.O. Box 6287, Grand Rapids, MI 49516-6287

Seventh printing, February 1998

Printed in the United States of America

Library of Congress Cataloging-in-Publication Data

Lubenow, Marvin L.
    Bones of contention : a creationist assessment of the human fossils / Marvin L. Lubenow.
      p. cm.
    Includes bibliographical references and indexes.
    ISBN 0-8010-5677-2
    1. Fossil man. 2. Creationism. I. Title
GN282.5.L82   1992
573.3—dc20   92-20925

For information about academic books, resources for Christian leaders, and all new releases available from Baker Book House, visit our web site:
                    http://www.bakerbooks.com/

Contents

List of Charts, Illustrations, and Definitions

Charts

Illustrations

Definitions

Preface

THE HUMAN FOSSIL RECORD is strongly supportive of the concept of Special Creation. On the other hand, the fossil evidence is so contrary to human evolution as to effectively falsify the idea that humans evolved. This is not the message we hear from a hundred different voices coming at us from a dozen different directions. But the human fossils themselves tell the real story.

The field of paleoanthropology has been, and will continue to be, an area of intense controversy. It necessarily must be. Its concepts go far beyond the data of science alone. They strike at the very heart of human origins, human nature, and human destiny. No questions are more basic to us—or more controversial. Hence, the raw data of paleoanthropology will always be "bones of contention." The title of this book came from the Oxford anthropologist, Sir Wilfred Le Gros Clark, who used the term when he delivered the Huxley Memorial Lecture in 1958. Others, including science writer Roger Lewin, have also used it.

This book is the fruit of twenty-five years of research. Because its conclusions are different from what is almost universally believed, it is heavily documented from the most recent scientific sources. The human fossil charts found in chapters 6, 7, 12, and 16 form the backbone of my argument. They are among the most comprehensive human fossil charts to be found anywhere in the scientific literature. Because of what these charts reveal, this type of chart is difficult to find in evolutionist writings.

Although the dates on these charts speak in terms of millions of years, I do not accept those dates. I march to the tune of a different drummer based upon the data set forth in the early chapters of Genesis. The purpose of this book is to demonstrate that even when the human fossils are placed on time charts according to the evolutionist's dates for these fossils, the results do not support human evolution but conflict with it. Because of the nature of my approach, the evidence presented here should be of value to old earth creationists as well as to my fellow young earth creationists.

The fossil category known as *Homo erectus*, which includes Java Man and Peking Man, is the key to the proper understanding of the human fossil material. For this reason I have devoted a number of chapters to the discovery of Java Man and to the fossil material in the *Homo erectus* category. The conclusions I draw from the *Homo erectus* material were evident to the unprejudiced mind one hundred years ago when Eugene Dubois first discovered Java Man.

The fossil humans with a somewhat different skull shape (morphology) are actually products of the post-Flood Ice Age. The Ice Age, in turn, is inseparably linked in terms of cause and effect to the worldwide Genesis Flood. The scientific community, rejecting the Genesis Flood, has been singularly unsuccessful in developing an adequate explanation for the Ice Age. We who believe in a worldwide Genesis Flood do our Christian brothers who believe in a local Noachic Flood a disservice in not insisting that they explain the Ice Age in the absence of a global flood.

Because the creationist literature on the human fossil evidence is limited, I have intended that this book serve both as an interpretation of the human fossil material and as an introduction to the entire subject. Hence, I have defined terms and concepts used by workers in this field but unfamiliar to the general public. Also, the first place I mention the name of a particular authority on fossils I include in parentheses the name of the institution with which that person is (or was) affiliated. Indexes include all of the persons mentioned in the text and all of the fossils referred to in the text and in the charts. A number of fossils come from what was formerly known as the Union of Soviet Socialist Republics. That area is now in political turmoil. At this writing no new designation has been determined for that geographic area of the world, so I have used the older designation (U.S.S.R.) in the text and in the charts, even though it is now obsolete.

Whenever possible, the extensive documentation utilizes the most recent scientific literature so that the serious student can verify my conclusions as well as do independent study. For this reason I refer often to Richard G. Klein's *The Human Career*. Having used this work as a text in teaching a graduate-level course on paleoanthropology, I expect that Klein's work will become one of the standard texts on the subject in the 1990s, just as Milford Wolpoff's *Paleoanthropology* has been the standard work in the 1980s.

It is impossible to deal with the subject of human fossils without also dealing with the subject of the radiometric dating methods. Because the flaws in these dating methods have been adequately exposed by other creationists, I have chosen a different approach. The appendix of this book is a case study of the tortured ten-year attempt to date the famous fossil KNM-ER 1470 discovered in 1972 by Richard Leakey's team. This case study clearly reveals that the radiometric dating methods are not independent confirmations of evolution and an old earth, nor are the various dating methods independent of each other. These dating methods are, instead, "faithful and obedient servants" of evolution. Because the chapter on the dating methods is by its very nature more technical, it is included as an appendix.

I wish to thank Dr. Steven A. Austin and Dr. Gerald E. Aardsma, radiometric dating authorities at the Institute for Creation Research, for reading the chapter on the dating of KNM-ER 1470. I also appreciate the suggestions of Michael Oard (National Weather Service) on chapter 14, and Dr. Donald De Young (Grace College) on chapter 18. Obviously, responsibility for these chapters, as well as for the rest of the book, is mine alone. Dr. John Whitcomb and Dr. Duane Gish read the manuscript and offered valuable comments. I also want to thank Dianne Stark, Terri McKelvey, Joseph Pearson, and Dr. Henry Morris for reasons which each will readily understand.

I must express my appreciation to Dr. Michael Charney, professor of anthropology at Colorado State University and one of the nation's leading forensic anthropologists. I first met Mike when I enlisted him as a participant in a creation-evolution debate I organized at Colorado State. Later we team-taught on the creation-evolution issue. Although we differ in matters of human origins and human destiny, Mike is a dear friend. His challenges have contributed much to my intellectual growth.

Special appreciation must go to Wilbert H. Rusch, Sr., of Concordia College, Ann Arbor, Michigan. One of the founders of the Creation Research Society, he is a gracious Christian gentleman and an extraordinarily gifted teacher. He is the person who first got me "hooked" on this twenty-five year study of the human fossils. My wife will probably never forgive him.

Marvin L. Lubenow
Christian Heritage College
El Cajon, California
December 1992

1

The Family Gathering

PEOPLE STARE. As they approach the table lined with human skulls, they stare. The mood is one of silence and incredible wonder. When someone finally dares to break the silence, I know instinctively what the question will be. I have heard it hundreds of times. "Are they real?"

When I inform the questioner that the skulls are plaster casts of the original fossils, his mood changes to relief that he is not in the presence of death. However, even my assurance that they are accurate and expensive casts of the original fossil material doesn't restore the mystique that was obvious before the question was asked.

The very thought that a professor at a Christian liberal arts college would possess thirty *original* human fossils reveals the magnitude of the misconception that exists in the mind of the public regarding these fossils. It represents the first myth about human evolution that I want to discuss.

Although I have visited most of the major natural history museums in the United States and some overseas, I have never seen an original human fossil. Neither have most of the anthropologists who teach human evolution in our universities. Neither have you. In fact, you may not have seen even a picture of an original fossil. What you thought were pictures of original fossils may have been pictures of casts.

No prisoner on death row is under greater security than those ancient relics called human fossils. Most of the original fossils are sequestered inside vaults of concrete or stone, accessible only through massive steel doors—the type you would expect to see at First National Bank.

The former director of the National Museums of Kenya in Nairobi was Richard Leakey. Many of the fossils housed there were found by him and his teams of national workers. They are kept in the Hominid Room, which has reinforced concrete walls and is designed to withstand conventional bomb blasts. Leakey and one other trusted museum staff member were the only ones who had keys to the room. Inside the room are locked boxes with hinged lids containing the fossils that rest on form-fitted blocks of foam rubber.

Hominid

The word is used by the evolutionist community to mean "humans and their evolutionary ancestors." It includes the genus *Homo*, the genus *Australopithecus*, and all creatures in the family *Hominidae*. As an evolutionist term it is meaningless in a creationist worldview. The creationist counterpart would be the term *human*. I use the term *human* to refer to those who are descendants of Adam.

Most of the South African fossils reside at the Transvaal Museum in Pretoria. They are kept in a strong room known as the Red Cave because of the three-foot-thick walls that are painted red. This vault was originally designed to house valuable documents. The fossils rest on red velvet placed over foam-rubber lining.

Germany has built a two-story museum to celebrate the fossil skull known as Steinheim Man, discovered in 1933. Visitors, however, see only plastic replicas. The fossil itself is kept in a small safe several miles away. This safe is set into the thick stone wall of a 250-year-old military arsenal outside Stuttgart. The fossil's former home was a bank vault. The story is told that when scientists came to study the fossil they were blindfolded, driven to the bank, and unmasked only when safely inside so that they would not even know the location of the bank.

Why this incredible security? Whether we are creationists or evolutionists, these fossils—certainly the human ones—are the remains of our ancestors. They are priceless treasures of human history. Their discovery has been the result of hard work, great expense, and often incredible luck. They are irreplaceable. How would one replace a fossil that has been lost or damaged beyond repair? Where would one go to find another just like it? Since in paleoanthropology and archeology quantity makes for quality in the study of human variation, finding a similar fossil does not make up for the loss of the first one.

Furthermore, many of the fossils are extremely delicate. Sometimes their teeth will shatter at the slightest impact. Chunks of bone may flake off at the scratch of a fingernail, since some of them are not completely fossilized—that is, the organic material is not completely replaced by inorganic minerals. Even the air in these fossil rooms is maintained at a constant temperature and humidity to minimize contraction and expansion that could crack the fossils.

Unfortunately, some fossils have been lost, such as the original Peking Man fossils which were lost in 1941. Although we do have plaster casts of them made by Franz Weidenreich, their loss is still keenly felt by both evolutionists and creationists. Many other *Homo erectus* fossils (the present classification of Peking Man) have been subsequently discovered, but these new ones have not made up for the lost information on early human populations that the original Peking Man fossils would have provided.

Because of their incalculable value and fragile nature, the original human fossils are so protected that the total number of people who have access to them is actually fewer than the total number of heads of state—monarchs, presidents, prime ministers, dictators—in the world today. However, like Camelot, there was one brief, glorious moment when this condition did not exist.

In 1984, the American Museum of Natural History in New York sponsored its famous "Ancestors" exhibit in which more than forty of these original fossils were brought together for the first time ever for the public to view and for scholars to study. Obviously, security had top priority. Each fossil was accompanied by the curator of its home museum. Special agents met them at Kennedy International Airport and whisked them through a special section of customs without even opening the containers housing the fossils. Black Cadillac limousines with police escorts rushed them to the American Museum. When the fossils were put on public display, they were placed behind one-inch laminated acrylic panels in batter-proof, bullet-proof, electronically monitored exhibit cases. Even work on the subway line under the museum was halted until after the exhibition to protect the fossils from vibration.

Although many nations refused to send their fossils and expose them to risk—China, Australia, Tanzania, Kenya, and Ethiopia (where Lucy is kept)—the exhibit was considered a resounding success. Scholars from all over the world were able for the first time to study the originals side by side. One-half million people were able to view them. To everyone's relief, nothing was broken. But, because of the high risk involved, most authorities predict that such a "family gathering" will never take place again.

If the risk to the fossils was so great, why was this "family gathering" held even once? The idea of having the world's leading paleoanthropologists study these fossils was just an afterthought to the main purpose of allowing the *public* to view the original fossils. What situation could loom so large as to pry these fossils loose from the security of their shelters and expose them to public view? The answer: the rising threat of creationism![1]

Paleoanthropology

Anthropology is the Greek word for "the study of man." *Paleo* means "old." Paleoanthropology is the study of fossil humans. The term replaces the older term *human paleontology*.

Eric Delson, John Van Couvering, and Ian Tattersall, American Museum scientists who were largely responsible for the Ancestors exhibition, admit that the creationist assault on evolutionary biology was a matter of "great and growing concern" at the museum. They go on to say that the primary purpose of the exhibition was to show people— lay and professional—the evidence for evolution. They refrained from making any kind of political statement regarding human evolution lest they "dignify" the challenge of "creation science."[2]

Bernard Wood (University of Liverpool), writing in *Nature*, one of the most prestigious science journals in the world, states that the Ancestors exhibition was the response of the American Museum to creationist attempts to influence both public opinion and legislators in their "attack on the foundations of all scientific endeavor—namely reason and evidence."[3]

It is obvious that in spite of the decibels, communication is not taking place. The problem is not with the fossils. It is with the interpretation of the fossils. Delson makes this naive comment: "How can you be anti-evolution when you see so much tangible evidence of our own roots?"[4] Apparently, evolutionists believe that all one has to do is look at the fossils to experience a "born-again" conversion to evolution. They seem oblivious to the fact that the human fossils can be arranged another way, a better way. To show that way is the purpose of this book.

Except for that one glorious moment in the summer of 1984, the original hominid fossils are not generally available for study, even by paleoanthropologists. In fact, Milford Wolpoff (University of Michigan) is said to have seen more of the original hominid fossil material than any other paleoanthropologist, although even he has not seen all of it. On the other hand, Ian Tattersall and Niles Eldredge (American Museum of Natural History), who have written extensively on the human fossil record, confess that they have seen only a fraction of the available material. They go on to say that it is not comforting to realize that many of the statements made by others regarding human evolution "are similarly removed from the original data."[5] Even the Ancestors exhibit displayed only a tiny fraction of the total material that has been recovered.

One would assume that those who have the proper academic credentials and are able to travel to where the original fossils are housed would have access to them. However, this is not always the case. Science writer Roger Lewin quotes Donald Johanson, the discoverer of Lucy, as agreeing that sometimes ". . .only those in the inner circle get to see the fossils; only those who agree with the particular interpretation of a particular investigator are allowed to see the fossils."[6]

Johanson's comment is a bit humorous. Paleoanthropologist Adrienne Zihlman (University of California, Santa Cruz) tells of writing to Johanson when he was at the Cleveland Natural History Museum and asking permission to see the fossils (including Lucy) that he had discovered in Ethiopia. He replied that he would grant permission only if he were allowed to review any article she might write on them before she sent it to a journal. She interpreted this as his insisting that he must approve it. Since she felt that this was a form of censorship, she declined. The result was that she didn't get to study those fossils before they were sent back to Ethiopia where they are permanently housed.[7] Zihlman also suspects there were times she was denied access to fossil collections because she is a female worker in a male-dominated field.[8]

In spite of some obvious cases of injustice, I do not wish to imply that this lack of general access to the original fossil material is some sort of evolutionist plot. The problem rests with the basic nature of the material itself—its fragility and its irreplaceability. Most of the fossil material, especially the newer material, is housed in the particular country in which it was found. As ancestor remains these fossils are national treasures of incredible value. In some countries the protection of these fossils seems to be far more important than the study of them. However, this lack of access has important implications for the study of human origins. It means that paleoanthropology is in the strange situation of being a science in which many of its workers do not have general access to the material upon which their science is based. They are at least one step removed from the objects of their study.

What, then, do they work with? Reproductions made of plaster or some other material. This means that the authority of the statements paleontologists make regarding fossils depends on the quality of the casts upon which they work. Obviously, one cannot make a universal statement about the general quality of these casts. That quality would depend upon the accuracy of the molds used, the type of material used, the care in making the casts, and other factors.

It is possible to have fossil reproductions that are of excellent quality. The Peking Man casts are said to be of such quality. C. Loring Brace (University of Michigan) tells the story of a tiny piece of new Peking Man cranial material that was found many years after the other originals were lost. This new piece fit perfectly into the space on the cast of the original from which the new piece had come.

The classic illustration that casts can be far from ideal is the account of the fraudulent Piltdown Man fossils. Piltdown Man was a combination of a late-model human cranium and a piece of the lower jaw of an orangutan. The teeth of the orangutan mandible had been filed down to make them look human and to match those in the upper jaw of the cranium. Louis Leakey in his book *Adam's Ancestors* tells of several attempts to make a detailed study of the original Piltdown fossils. On each occasion when he visited the British Museum to do so he was given the original fossils for just a few moments. They were then taken away, and he was given casts to work on. The file marks on the orangutan teeth were visible on the originals, but they were not visible on the casts.[9]

Given the unavailability of the originals, casts are the next best medium of study. Yet, it is common knowledge that casts or reproductions, while giving a general impression of the original, often lack the detail of the original. Becky A. Sigmon (University of Toronto) says there is a general feeling among paleoanthropologists "that casts should not be used as resource material for a scientific paper."[10] However, there is another problem with the use of casts. Casts of only a small percentage of the total fossil material and of only about half of the most important fossil material are available for study. This in itself is a serious situation. It would seem to place a degree of contingency on all conclusions reached in the study of human evolution.

Descriptions of fossils in the scientific literature, although a poor substitute for casts, are probably the most common tools used in the study of the human fossil material. Since only the original fossils should be used in the writing of such papers, this would seem to place serious limitations on their preparation. Unfortunately, seldom do authors of such papers indicate what their sources were: the literature, casts, or the original fossil material. Milford Wolpoff, commenting on the value of the 1984 "Ancestors" exhibition, which allowed him and others to compare points of difference between fossils by seeing them side by side, says: "You can't do that properly through the literature."[11]

Perhaps the best example of the problem facing paleoanthropology is that many of the scholars who felt that casting technology was now able to provide copies as good as the originals, after studying the originals in the 1984 American Museum exhibition, admitted "that technology still has a long way to go."[12] The crowning blow came at the beginning of the public display. The precision mounts for the original fossils were carefully prepared on the basis of casts supplied in advance. When the original fossils were placed in those mounts, most of them did not fit. No better illustration could be found showing that "casts are no substitute for originals."[13]

Since the original fossils are virtually beyond access even to most who teach in the field of paleoanthropology, and only a few of them are available as reproductions, and reproductions are not recommended in the preparation of scientific papers, and those scientific papers themselves cannot adequately convey differences between fossils, paleoanthropology seems to have a problem.

The myth in the mind of the public is that the human fossil material is readily available and is thoroughly studied by all who teach and write on the subject. The truth is that paleoanthropology is in the awkward position of being a science that is at least one step removed from the very material upon which it claims to base its findings.

2

An Inexact Kind of Science

SPEAKING AT AN ANNUAL CONVENTION of the American Association for the Advancement of Science (AAAS) in San Francisco, astronomer Carl Sagan (Cornell University) explained how science works. "The most fundamental axioms and conclusions may be challenged" and the prevailing hypothesis "must survive confrontation with observation." "Appeals to authority," he said, "are impermissible," and "experiments must be reproducible."[1] This, of course, is the concept of science which the general public has.

Sagan's comments give us an insight into how he, and evolutionists like him, work. He gives lip service to the accepted methodology of science. However, when presenting his views on the evolution of everything, he gives the public a freewheeling fantasy in which one cannot separate science from science fiction. The result is that all of it is accepted as science. The undiscerning public permits this because it considers scientists to be some sort of high priests of our society, paragons of objectivity who have no philosophical axes to grind. Hence, the public is often fed a diet of philosophy under the guise of science.

Sagan then made a comment that is both true and profound. "Not all scientific statements have equal weight." He cited Newtonian dynamics, the first and second laws of thermodynamics, and the law of angular momentum as being on extremely sound footing because of the millions of experiments and observations that have been performed on their reliability. Sagan's remark about scientific statements having various weights based on the data backing them is obvious. But few people put it into practice.

Human evolution allegedly took place in the past over vast periods of time. Evolutionists readily admit that evolutionary processes work so slowly that they are not observable over the lifetime of one individual or even over the successive lifetimes of hundreds of generations. In other words, there are no *direct* observations or experiments that can confirm the process of human evolution.

On a scale from zero to ten, it is then possible to assign relative values to various scientific statements based on the number of direct experiments and observations involved. If, based on Sagan's statement, we assign a value of ten to Newtonian dynamics and the laws of thermodynamics because of the millions of confirming experiments and observations, what value can we assign to statements regarding human evolution when there is not one direct observation to back them up? The only value to assign to those statements is zero.

Evolutionists use several lines of evidence in promoting the concept of human evolution. One is studies of living animals, specifically primates. Another is the arrangement of the fossil material. A third is molecular data.

Studies on living primates—their behavior, genetic make-up, and anatomy—are used to support human evolution. All of these studies are fundamentally flawed. The flaw is known in logic as begging the question. In begging the question, you assume to be true the very thing you are trying to prove. Let me illustrate.

A man in Chicago was observed walking down the street snapping his fingers. Finally, someone was driven by curiosity to ask him why he repeatedly snapped his fingers. "It keeps the elephants away," the man replied. "Why, man, there aren't any elephants within ten thousand miles of this place!" responded his questioner. "Pretty effective, isn't it?" exclaimed the man. The man first assumed that his finger snapping kept elephants away. He then used the absence of elephants to prove the effectiveness of his actions. To presuppose the truth of what you are trying to prove is the illogical practice of begging the question.

Studies regarding human evolution and our relation to the other primates could cast light on human nature and behavior only if evolution were true. The evolutionist first assumes that humans and the other living primates are related. He then uses studies on the other primates to cast light on the alleged evolution of humans. If evolution is not true, studies on living primates, although valuable in their own right, are worthless in shedding light on human origins and human nature.

Related to this logical fallacy of begging the question is the seeming failure of evolutionists to understand the difference between scientific and historical evidence. Again let me illustrate. It is believed that in the American Revolution George Washington and his men crossed the Delaware River to attack the city of Trenton. How would one go about proving that event? If one used the scientific method, he would do research on boats, measure the width and flow of the river, do studies on the rowing of boats, and perhaps even row across the Delaware River himself. Would all of this data prove that Washington crossed the Delaware? No. Scientific evidence is not what is needed. Historical evidence, such as records of eyewitnesses or of persons closely associated with those who were involved, is what is needed. All the scientific method could prove is the *possibility* that Washington crossed the Delaware, not that he actually did so.

G. A. Kerkut (University of Southampton) discusses this problem of historical versus scientific evidence as it pertains to evolution. The problem concerns all phases of evolution, and if evolution in general did not take place, humans did not evolve either. Kerkut first lists seven basic assumptions in evolutionary theory, which, he claims, are seldom mentioned in discussions on the subject. He then states:

> The first point that I should like to make is that these seven assumptions by their nature are not capable of experimental verification. They assume that a certain series of events has occurred in the past. Thus though it may be possible to mimic some of these events under present-day conditions, this does not mean that these events must therefore have taken place in the past. All that it shows is that it is *possible* for such a change to take place. Thus to change a present-day reptile into a mammal, though of great interest, would not show the way in which the mammals did arise. Unfortunately we cannot bring about even this change; instead we have to depend upon limited circumstantial evidence for our assumptions. . . .[2]

All experiments performed with present-day animals, plants, or biological molecules are logically flawed. They cannot prove or even support the alleged evolutionary processes of the past. The extensive use of present-day experiments to try to demonstrate evolution reveals that evolutionists do not understand the difference between scientific and historical evidence.

Another major line of evidence used to support the concept of human evolution is the fossil record. We have all seen pictures of the impressive sequence allegedly leading to modern humans—those small, primitive, stooped creatures gradually evolving into big, beautiful you and me. What is not generally known is that this sequence, impressive as it seems, is a very artificial and arbitrary arrangement because (1) some fossils are selectively excluded if they do not fit well into the evolutionary scheme; (2) some human fossils are arbitrarily downgraded to make them appear to be evolutionary ancestors when they are in fact true humans; and (3) some nonhuman fossils are upgraded to make them appear to be human ancestors.

A major section of this book will consider the human fossil evidence. At this point I merely want to emphasize a phenomenon that seems almost universally unrecognized: *Any series of objects created by humans (or God) can be arranged in such a way as to make it look as if they had evolved when in fact they were created independently by an intelligent being.* The fact that objects can be arranged in an "evolutionary" sequence does not prove that they have a relationship or that any of them evolved from any of the others.

In a certain graduate course I took in paleontology at a state university, the professor attempted to teach us the concepts of taxonomy and the construction of those familiar evolutionary family trees. He handed each student a packet of about 150 metal objects such as nails, tacks, and paper clips. Utilizing the various rules of evolutionary taxonomy, such as small to large, simple to complex, and generalized to specialized, we were each expected to arrange these objects in evolutionary order. Starting with generalized nails, we went on to nails gradually increasing in size and then branching off into various specialized types of nails and tacks. Naturally, no two students in the class arranged their objects in exactly the same way although there was an overall similarity. When the project was finished, we all had created a beautiful series of phylogenetic trees showing the "evolution" of nails, tacks, and paper clips.

What I found fascinating about the project was that as we played with our object lesson, no one sensed that the illustration was totally invalid; it had no relationship to reality. Each of the objects that we had arranged in such a convincing evolutionary sequence had in fact been individually created for a specific purpose by humans. There was no actual evolutionary relationship between them. We were able to ar-

Taxonomy

Taxonomy is the science of the classification of living things. The common classification system used today involves classification according to structure. Humans are classified as follows:

Phylum	Humans are chordates
Class	Humans are mammals
Order	Humans are primates
Family	Humans are in the family *Hominidae*
Genus	Humans are in the genus *Homo*
Species	Humans are in the species *sapiens*
Subspecies	Humans are in the subspecies *sapiens*

Modern humans are classified as *Homo sapiens sapiens*. Only the genus term is capitalized. Since they are Latin terms, all are italicized.

Homo. Humans are the only living forms in the genus *Homo*. Biblically, there are no creatures past or present who would qualify for the genus *Homo* or "true humans," other than descendants of Adam.

sapiens. The first *sapiens* in the classification refers to the species level. The Latin term means "wise." The species level is the level of reproduction and of reality. The higher levels are constructs of the human mind to bring order out of the complex world in which we live. The scientific term *species* is a very involved concept which has yet to be defined with finality. The biblical word *kind* is not a synonym of *species* and should not be confused with it.

sapiens. The second *sapiens* in the classification refers to the subspecies level. The racial distinctions of humans are so slight that they are well below the subspecies level. All humans belong to the same subspecies, the same species, and the same genus. This amazing unity of the human family is in itself strong evidence for creation and against evolution.

range them in an "evolutionary" sequence even though none of them had evolved. That fact did not seem to dawn on anyone in the class, including the professor.

Fossils cannot reproduce, nor can they reveal their genetic relationships. These relationships are often in the mind of the arranger. Even if we were to find the fossilized bones of two individuals, father and son, there is no way from the fossils themselves that we could prove conclusively their relationship. That proof would have to come from historical records.

Because of the problems I have discussed in these chapters, creationists feel that paleoanthropology does not deserve the same status in science that is accorded fields like chemistry or physics. When we creationists make such statements, we are often accused of being antiscience. But we are merely saying what Carl Sagan said: "Not all scientific statements have equal weight."

The relative value of different scientific statements has a direct bearing on the origins controversy. The late Bernard Ramm criticized those who give Genesis a literal reading for what he saw as their inconsistency, calling them hyperorthodox:

> Hyperorthodoxy is inconsistent in actual practice for it will certainly use the *practical* achievements of modern science, e.g., radio, television, phones, cars, medicines, furnaces, glasses, artificial teeth, etc., etc., etc. It is not intellectually consistent to condemn science as satanic while having teeth repaired by scientific technicians, wearing glasses prescribed and ground by other scientists, covered with clothing produced by chemists and engineers, with a body saved from premature death by an appendectomy performed by a scientist, and with a mind trained in a school system working with methods provided by educational scientists.[3]

Ramm's comment is a rather shallow appraisal of the situation. There is a world of difference in the quality of the scientific information utilized by those who make glasses and work on teeth compared to the quality of information utilized by those who postulate the origin of the universe in the Big Bang about fifteen billion years ago.

It is refreshing to know that there are those within the paleoanthropological community who have been open-minded enough to acknowledge the quality-of-information problem in origins, especially human origins. In an AAAS interview, Richard Leakey said, "I think the study of early man (physical anthropology in a paleoanthropologi-cal sense) is a science that is just reaching its adolescence. I do not think the science has matured. I think we are still doing a great deal of guessing."

The interviewer then asked, "So, paleoanthropology is on its way to becoming a real science?"

Leakey replied: "I sincerely hope so. Some people will have heard me say that I often felt that paleoanthropology was more of an art than a science. I think it is now about to be a science, and I sincerely hope that it will be a science, because the minute it becomes a science we begin the possibility of really understanding what's going on."[4]

Leakey's description of paleoanthropology as an art rather than a science is a gracious way of saying that this field is the scene of much prejudice, subjectivity, and emotionalism in the interpretation of the human fossils and in the construction of phylogenetic trees. The pro-fessionalism and objectivity found in other areas of science have been conspicuously absent in this area.

For many years David Pilbeam (Harvard University) had convinced his fellow paleoanthropologists that a fossil form known as *Ramapith-ecus* was a hominid. This assessment was almost universally accepted even though it was based on the flimsiest of fossil evidence. Later, when Pilbeam found more abundant fossil evidence, it became obvi-ous that *Ramapithecus* had nothing to do with human origins. In ex-plaining where he and the paleoanthropological world had gone astray, Pilbeam's confession reads almost like a Shakespearean soliloquy:

Theory shapes the way we think about, even perceive, data . . . . We are unaware of many of our assumptions.[5]

Conflicting visions of these [evolutionary] human ancestors probably says more about our conflicting views of ourselves than about the actual fossil data.[6]

In the course of rethinking my ideas about human evolution, I have changed somewhat as a scientist. I am aware of the prevalence of implic-it assumptions and try harder to dig them out of my own thinking. . . . Theories have, in the past, clearly reflected our current ideologies in-stead of the actual data . . . . I am more sober than I once was about what the unwritten past can tell us.[7]

It is sobering to reflect that at the very time Leakey and Pilbeam were confessing the lack of scientific methodology in interpreting human fossils, the public was being convinced of the truth of human evolution based upon what other paleoanthropologists were saying about those very same fossils. Although both Leakey and Pilbeam hoped that brighter days lay ahead for paleoanthropology, the public had already been convinced of a certainty that is far from legitimate.

Perhaps the most outspoken researcher in the field of paleoanthropology is Lord (Solly) Zuckerman (University of Birmingham, England), noted authority on the australopithecines. In discussing the degrees of objectivity in the various branches of science, he said that the behavioral sciences are at the low end of the objectivity spectrum. He continued:

> We then move right off the register of objective truth into those fields of presumed biological science, like extrasensory perception or the interpretation of man's fossil history, where to the faithful anything is possible—and where the ardent believer is sometimes able to believe several contradictory things at the same time.[8]

Later, he added:

> As I have already implied, students of fossil primates have not been distinguished for caution when working within the logical constraints of their subject. The record is so astonishing that it is legitimate to ask whether much science is yet to be found in this field at all. The story of the Piltdown Man hoax provides a pretty good answer.[9]

Sir Peter Medawar, who won a Nobel Prize in medicine, spoke of paleoanthropology as "a comparatively humble and unexacting kind of science." Andrew Hill (Yale University), following up on Medawar's comment, added, "It has certainly been possible to get away with being an unexacting practitioner."[10]

As one studies evolutionist literature, one cannot help but notice in its practitioners both a lack of logic and an inability to weigh evidence properly. Legal experts have also noted this. Some years ago Harvard-trained lawyer Norman Macbeth wrote a book, *Darwin Retried*.[11] After studying evolution for many years Macbeth, who is not a creationist, concluded that there were serious gaps in the evidence for evolution and errors in the reasoning of evolutionists. He claimed that evolution itself had become a religion. The alleged evidence for evolution, he charged, was not of the quality that would stand up in a court of law.

A similar conclusion was reached in a recent (1991) book, *Darwin on Trial*,[12] by law professor Phillip E. Johnson (University of California, Berkeley). Johnson describes himself as a Christian and a creationist, but not a biblical literalist. His book may be the most significant one to appear on the evolution debate in decades.

Johnson concludes that (1) evolution is grounded not on scientific fact but on a philosophical belief called naturalism; (2) the belief that a large body of empirical evidence supports evolution is an illusion; (3) evolution is itself a religion; (4) if evolution were a scientific hypothesis based upon a rigorous study of the evidence, it would have been abandoned long ago; and (5) since atheism is a basic *supposition* in the evolutionary process, it cannot be drawn as a *conclusion* from it.

Evolutionists got the cart before the horse, Johnson states. They first accepted evolution uncritically as a fact. Then they scrambled for evidence to support it, without too much success. This analysis is certainly true of human evolution. As we will show later, the concept of human evolution was well in place before any of the human fossils had been discovered.

In discussing the lack of objectivity in the interpretation of the human fossils, Johnson refers to Roger Lewin's description of the 1984 "Ancestors" exhibit. Lewin said that the paleoanthropologists were in awe as they held their fossil ancestors in their hands. Lewin described how moving and emotional that experience was. Johnson comments:

> Lewin is absolutely correct, and I can't think of anything more likely to detract from the objectivity of one's judgment. Descriptions of fossils from people who yearn to cradle their ancestors in their hands ought to be scrutinized as carefully as a letter of recommendation from a job applicant's mother.[13]

Johnson continues:

> The story of human descent from apes is not merely a scientific hypothesis; it is the secular equivalent of the story of Adam and Eve, and a matter of immense cultural importance. . . . The needs of the public and the [secular scientific] profession ensure that confirming evidence will be found, but only an audit performed by persons not committed in ad-

vance to the hypothesis under investigation can tell us whether the evidence has any value as confirmation.[14]

A widely held myth is that paleoanthropologists are able to speak with the same authority as other scientists. The reality is that the quality of their information and their interpretation of that information are open to serious challenge. As Carl Sagan said, "Not all scientific statements have equal weight."

3

Dead Reckoning

HAVE YOU EVER WONDERED how many hominid fossils have been discovered to date? It's a very important question because when we deal with historical evidence, quantity makes for quality. The total number of hominid fossil individuals is universally assumed to be quite small, since paleoanthropologists have for years complained about the lack of fossil material.

The first surprise—if not shock—comes when we learn that the actual number of fossil individuals discovered to date is very difficult to obtain. There does not seem to be a central clearinghouse, a publication, or a data bank to which the interested person can go to get that information.

One publication attempted to give that information as of 1968-1976. It is the three-volume *Catalogue of Fossil Hominids* edited by Kenneth P. Oakley, Bernard G. Campbell, and Theya I. Molleson, published by the British Museum (Natural History). Volume 1 deals with fossil hominids from Africa (the second edition of that volume was published in 1977); volume 2 deals with Europe and the U.S.S.R., published in 1971; volume 3 covers the Americas, Asia, and Australasia, and was published in 1975. The *Catalogue* was intended to serve as a reference for information on the fossil hominids. Its scholarship and authority are beyond reproach. It answers the question of approximately how many hominid fossils had been discovered up to 1969-1976.

Unfortunately, nothing as complete as the *Catalogue* has been published since 1977. And this work is extremely difficult to find even in the largest university or city libraries. In their introduction the editors of the *Catalogue* stated that it was their intention to keep all of the volumes updated. Although Oakley died in 1981, other persons at the British Museum are adequate for the task. However, the British Museum (Natural History) recently informed me that they were not aware of plans to update the volumes. I find this surprising. This is an area where the *Catalogue* stands unique, interest in the hominid fossils is increasing, and recent discoveries have been unparalleled.

For some unexplained reason, the *Catalogue* does not tabulate the total number of fossil hominid individuals that it covers. (An individual may be represented by just a tooth or all the way up to a complete skeleton.) According to my personal count, the total number of hominid fossil individuals, including all categories of the genus *Homo*, the genus *Australopithecus*, and categories such as *Ramapithecus* (since then disqualified as a hominid) listed in the *Catalogue* are as follows:

Africa—1390 fossil individuals discovered through 1976.

Europe and the U.S.S.R.—1516 fossil individuals discovered through 1970 (the figure for France, one of the most prolific fossil areas, goes only through 1969).

The Americas, Asia, and Australasia—1092 fossil individuals discovered through 1974.

A grand total of 3998 (approximately 4000) fossil hominid individuals have been discovered as of 1969-1976. This is a surprisingly large amount of evidence dealing with human origins.

Several comments about this figure are in order. Because of the many different ways the hominid fossils have been catalogued, it is probable that in some cases different numbers have been assigned to different fossil bones of the same individual. In seeking to arrive at a "body count" of the hominid fossils, there is no way to avoid this inflation of the numbers. However, several factors would compensate for this inflation:

(1) Editors of the *Catalogue* recognized that there were gaps in their coverage of the fossil material. They appealed to their readers to help

them track down unreported fossil discoveries. Some legitimate fossils that would strengthen the creationist position were omitted.

(2) In my tabulations, I always used the minimum figures in cases where exact fossil populations were unknown. For instance, for Peking Man the Lower Cave material indicates a minimum of forty individuals and a maximum of forty-five individuals. At Krapina, Yugoslavia, the fossil material indicates a minimum of fourteen individuals and a maximum of twenty-eight individuals. I have always used the minimum figures, although the probability is that in many cases a larger number of individuals was actually present.

(3) In cases where miscellaneous human fragments are mentioned, I have always counted these as just one individual unless it is obvious from the remains that more than one individual is involved. For instance, three femora (thigh bones) would indicate at least two individuals, and possibly three individuals.

(4) In evaluating fossil remains, when there is no duplication of bones it is usually assumed that just one individual is involved. I have tallied it as such. Although this is a logical inference, it cannot be proven. In some cases, more individuals could be involved.

I believe that my conservative approach in tallying the fossils listed in the *Catalogue* compensates for the possible inflation that I spoke of, and that the figure of approximately four thousand fossil individuals is a realistic appraisal of the total number discovered worldwide 1969–76. This is an immense amount of material.

By 1976 over two hundred individuals were classified as Neandertals, and over one hundred were classified as *Homo erectus*. This means that in over three hundred of these fossil individuals, enough material had been recovered to be diagnostic of these categories.

Much of this fossil material is only thousands of years old. Going back in time, the amount of fossil material drops off significantly. Evolutionists say that because evolutionary change occurs so slowly, the older fossils are more significant, and these are the ones that are in short supply. Yet, it is the more recent fossils that effectively falsify the concept of human evolution, specifically recent *Homo erectus*, our alleged evolutionary ancestor, who may have been alive and well just a few hundred years ago (the Cossack skull).

In light of the richness of the hominid fossil record, it is difficult to understand why we continue to read statements from authorities about the small number of hominid fossils that have been discovered.

About forty original fossils were brought to the American Museum exhibition in 1984. Ian Tattersall, one of the organizers of the exhibition and a curator of the museum, is quoted as having said, "When this exhibition is assembled, we'll have more than half of the most complete specimens in the human fossil record under this roof."[1] Because Tattersall is a responsible scientist, I suspect that he was misquoted. As it stands, the statement is so patently wrong as to be absurd. Even if he had said "more than half of the most important fossils," the statement would still be false. A large number of very important fossils were not brought to the exhibit.

Boyce Rensberger was a senior editor of the journal *Science 84*, a publication of the American Association for the Advancement of Science, one of the most respected scientific organizations in our nation. Directly under the title of his article, "Bones of Our Ancestors," is the caption, "In all the world there are only a few dozen. But these rare fossils attest to the evolutionary odyssey that created the human species."[2]

This type of comment has been issuing from even the best-known names in paleoanthropology. At the very time the *Catalogue* was being published, documenting approximately four thousand individual fossil hominids, Richard Leakey was excusing the many mistakes that had been made in interpreting the human fossil record: "I think this was inevitable by virtue of the fact we had so little material."[3]

We continually read comments like these—in *Time*: "Scientists concede that even their most cherished theories are based on embarrassingly few fossil fragments";[4] in *Scientific American*: "The human fossil record is short and scant";[5] in *New Scientist*: "The entire hominid collection known today would barely cover a billiard table";[6] and in *Science*: "The primary scientific evidence is a pitifully small array of bones from which to construct man's evolutionary history. One anthropologist has compared the task to that of reconstructing the plot of *War and Peace* with 13 randomly selected pages."[7]

In light of statements such as these, it is not surprising that the perception is far different from the reality. The public is unaware of the rich harvest of hominid fossils we now possess. Although some of the myths I discuss in this book are not the fault of the evolutionists, this one clearly is. It is because they have gone to their public wailing wall and lamented the tragic lack of human fossils.

Since every paleoanthropologist worth his salt knows about the *Catalogue* published by the British Museum as well as the many fossils discovered since its publication, the question arises, Is he lying? No. While he is certainly not telling the truth, in his own mind he is not lying. He is speaking a different language, and the public has not learned the translation. His professional colleagues understand the language. They have made the translation. To comprehend the field of paleoanthropology and its literature, you, too, must learn to translate. When a worker in this field speaks of the scarcity of the human fossils, he is actually saying, "Although there is an abundance of hominid fossils, the bulk of them are either too modern to help me or they do not fit well into the evolutionary scheme. Since we all know that humans evolved, what is so perplexing is the difficulty we are having in finding the fossils that would clearly demonstrate that fact."

A very common myth today is that not many hominid fossils have been discovered. The reality is that by 1969-1976, approximately four thousand hominid fossil individuals had already been unearthed. The period since that time has seen the most intensive and successful search for hominid fossils in the history of paleoanthropology. No one knows exactly how many have been found to date. However, by my own "dead reckoning," a conservative estimate, *the total number of hominid fossil individuals discovered to date exceeds six thousand.*

4

Monkey Business in the Family Tree

CARL SAGAN IS A MAN of deep faith, faith in the ability of science to lead us into truth. Speaking at an AAAS convention in San Francisco, he said:

> Scientists, like other human beings, have their hopes and fears, their passions and despondencies, and their strong emotions, which may sometimes interrupt the course of clear thinking and sound practice, but science is also self-correcting. . . . The history of science is full of cases where previously accepted theories and hypotheses have been overthrown to be replaced by new ideas which more adequately explain the data. . . . This self-questioning and error-correcting aspect of science is its most striking property, and sets it off from other areas of human endeavor such as . . . theology.[1]

Anthropologist Vincent Sarich (University of California, Berkeley) states that he is a humanist. He also has deep faith in the ability of science to lead us into truth. In one of his debates with creationist Duane Gish, he repeated a concept that he had stated many times:

We have a faith game and a science game. Really, these are both faith
games. Both are human constructs which are made to make sense out
of a very complicated world. Science is a very peculiar kind of new faith,
which assumes that faith itself can be self-correcting. Other faiths do not
have any kind of self-correcting mechanism built into them. They are
dogmatic. People doing science can be dogmatic, but they forget what
science really is. It is a continuously self-testing and self-correcting faith.
The faith comes in the idea that it is a way in which we can generate increasing understanding of the world around us.[2]

There is no question that this self-correcting aspect of science is
hailed by those who would make science a superior worldview over
biblical Christian faith with its belief in the inerrancy of the Word of
God and its "Thus saith the Lord!" Christ claims that he is "the way,
the truth, and the life" (John 14:6). The Bible claims to be the Word
of God in an ultimate sense. These claims are offensive to some. They
either are not interested in ultimate truth, do not like the biblical
teaching of ultimate truth, or do not believe that it is possible to know
ultimate truth.

We could ask the questions, How does one know that there is no ultimate truth? How does one determine that Christ's words are not ultimate truth? What criteria does one use to evaluate his words? The
Christian has such a criterion. He believes that Jesus Christ by his resurrection from the dead validated all his claims. Anyone who has power over death commands ultimate power and traffics in ultimate truth.

It is not necessary to make science and biblical faith an either-or situation, but Sagan and Sarich have chosen to do so. There is something
romantic about the thought of a scientist searching for truth. However,
like the steamship company slogan that says "getting there is half the
fun," many scientists say that they are not interested in finding truth.
They enjoy the search. They don't particularly care if they arrive or
not. It is not unusual to hear scientists say that the Bible would destroy
science, because there's no point in doing science if you already have
the truth—probably one of the most absurd statements ever uttered as
an excuse for rejecting God's Word.

One question, a basic question, needs to be asked, but I have never
heard anyone ask it. To justify science as a superior worldview, Sagan
cites situations in the history of science where the self-correcting
mechanism has worked. However, the question is not whether this
self-correcting mechanism has worked once, twice, a thousand times,

or a million times. The basic question is, How efficient is this self-cor-recting mechanism? or What is the batting average of science in this area? Out of the total number of mistakes made in science, how many have been corrected?

When we put the question this way, it is obvious that there is no way of knowing the total number of mistakes made in the history of sci-ence. Nor do we know how many uncorrected errors exist in science today. If we knew about them, they would be corrected. Hence, it is impossible to know how efficient this self-correcting element in sci-ence is. But if there is no way to determine its effectiveness, then we can never know if trusting science to lead us into truth is a very wise worldview or a very foolish one. We all agree that according to its methods, science could be somewhat self-correcting. But we are not living in an ideal world.

Behind the self-correcting aspect of science is the idea that when a scientist feels he has discovered something unique or innovative, he must publish both his results and his methodology. His work is not only subject to a process known as peer review, but it is eventually ex-posed to the entire scientific community for evaluation. This sounds like a very healthy and purifying process. Ideally, it is. However, in ac-tual practice it breaks down. Scientists simply do not have the time nor the money to check up on the research of other scientists. Scientists in the academic community are busy with their teaching assignments, graduate student supervision, and their own research programs. They are driven by the publish-or-perish attitude prevalent today. It simply does not benefit them in any way—no fame or fortune—to confirm or falsify the work of someone else. Scientists in industry have a bottom-line mentality. They must be productive in the areas in which their company specializes. They have no time to check out the work of an-other scientist just for the fun of checking him out or to prove that sci-ence really is self-correcting. There are exceptions, but in practice this is normally the case.

There is a touch of irony in the fact that science works in the very opposite way that people think it works. The self-correcting aspect of science implies a self-policing action by scientists, a checking up on one another. In actuality, scientists have demonstrated an incredible faith and trust in the work of their fellow scientists. They tend to ac-cept that work at face value without much investigation at all. A num-ber of recent scandals have developed where fraudulent medical research was incorporated into medical practice and the fraud discovered far down the line. The exposure of error usually occurs only when the effects are very obvious. If the error is not obvious, it can be perpetuated almost indefinitely. One of the most amazing illustrations of faith is the trust that scientists put in their fellow scientists who work outside their own fields of expertise.

The question still is, Is science really self-correcting? One of the areas of science with which I am familiar is paleoanthropology. I can testify that in this area the track record for self-correcting by scientists is very poor. Many illustrations could be used, but I will confine myself to two of the most notorious: the faulty reconstruction of the Neandertal skeleton from La Chapelle-aux-Saints, and the Piltdown Man fraud.

The Faulty Neandertal Reconstruction

In 1908, a rather complete skeleton of a Neandertal-type[3] individual was found buried ritualistically in the floor of a small cave near the village of La Chapelle-aux-Saints, France. The site is in the famous Dordogne River Valley, carved into a limestone plateau in southwestern France about 120 miles east of Bordeaux. The area is known for the rock shelters and caves the groundwater has carved into the limestone. These caves and rock shelters have yielded other Neandertal fossils—Le Moustier and La Ferrassie—as well as the Cro-Magnon fossils from Les Eyzies.

Marcellin Boule, the famous paleontologist at the National Museum of Natural History in Paris, was called upon to reconstruct the man from La Chapelle-aux-Saints, the most complete Neandertal skeleton found in Western Europe up to that time. The world could then see what a Neandertaler looked like. Not for a moment did Boule believe that the Neandertalers, with their low, wide skulls and their sloping foreheads, deserved a place in the direct ancestry of humans—certainly not in the direct ancestry of the French. He felt that the Neandertalers were a withered side branch of the family tree or a backward evolutionary group that became extinct leaving no issue. Piltdown Man was one of Boule's proofs that modern humans with their high-domed skulls dated much further back than the Neandertalers. In Boule's mind, the tall, high-domed Cro-Magnon people, the ones responsible for the beautiful cave paintings in France and Spain, were the true ancestors of modern humans.

Boule, who made the first detailed description of the bones of the Neandertalers, emphasized what he felt were simian (apish) features in the La Chapelle-aux-Saints skeleton rather than human features—based on his preconceived ideas of evolution. Although there was evidence that the vertebrae were severely deformed because of arthritis and rickets, Boule ignored the pathological evidence. He claimed that the spine lacked the curves that enable modern humans to walk erect. He placed the head in an unbalanced position on the neck, thrust far forward, so that the individual probably would have sprained his neck had he looked at the sky.

Boule also decided that this man could not extend his legs fully, but walked with a bent-knee gait. He made the foot only slightly arched, resting on its outer edge, with toes pointing in. Hence the man would have walked like an ape, pigeon-toed. Boule formed a wide separation between the big toe and the other toes, making the big toe like an opposable thumb—such as monkeys and apes have. Under these conditions, if Neandertal Man walked at all, he would have looked like a shuffling hunchback. His center of gravity was located so far forward of his center of support that he probably would have fallen flat on his face.

Using casts of the inside of the La Chapelle-aux-Saints skull, Boule felt that the brain of Neandertal, although larger than the average brain size of modern man (1620 vs. 1450 cc), resembled the brain of the great apes in organization. Boule concluded that Neandertal was closer to apes than humans in brainpower, had only a trace of a psychic nature, and had only the most rudimentary language ability—possibly not much more than a series of grunts.

The stone tools found in the same cave confirmed to Boule this primitive condition. Boule wrote:

> . . .the uniformity, simplicity and rudeness of his stone implements, and the probable absence of all traces of any pre-occupation of an aesthetic or of a moral kind, are quite in agreement with the brutish appearance of this energetic and clumsy body, of the heavy-jawed skull, which itself still declares the predominance of functions of a purely vegetative or bestial kind over the functions of the mind. . . .[4]

Boule's report on the Neandertalers was published between 1911 and 1913 in a series of articles in three volumes of the *Annals de Paleontologie*. This twisted view of the Neandertalers dominated the world for forty-four years until William L. Straus (Johns Hopkins Medical College) and A. J. E. Cave (St. Bartholomew's Hospital Medical College, London) published their paper in 1957 on the re-examination of the Neandertals. Attending an anatomy conference in Paris in 1955, Straus and Cave decided to take a look at the La Chapelle-aux-Saints skeleton. They immediately recognized that there were some very serious problems with the reconstruction. Their study revealed that the Neandertals, when healthy, stood erect and walked normally as modern humans do. Exit *Homo neanderthalensis*. Enter *Homo sapiens neanderthalensis*.

Thanks to Boule, the Neandertalers have had a bad reputation from the start, one they did not deserve. It is now admitted that their differences from modern humans are rather superficial. Their low, wide cranium and heavy brow ridges caused people to think of them as "savage," even though there is nothing in the anatomy of a person to indicate his morality, behavior, or degree of culture. Since the average cranial capacity of the Neandertalers is almost 200 cc higher than the average for modern humans, that should have helped their image. But thanks to Boule's prejudice, it did not.

We wish we could say that Boule was rather unsophisticated and that he made sincere mistakes. That would be the kinder explanation. However, he was world renowned for his abilities. It is difficult to escape the conclusion that his errors were deliberate. Yet, an evolutionist could argue that: (1) the entire evolutionist establishment should not be held responsible for the misguided mistakes of one man, and (2) since the errors were discovered and corrected, does that not prove that science really is self-correcting?

The answer is not quite that simple. During the time that the mistakes went undetected, the "savage-caveman" idea was being used worldwide as strong evidence for human evolution. The word *Neandertal* is still virtually synonymous with *brute*. Until recently, it would have been easy to find a children's book in almost any schoolroom where a picture of Neandertal was displayed as one of the major evidences for human evolution. We have the right to ask, Did no evolutionist look at this reconstruction for forty-four years? Whether the answer is yes or no, the implications for evolutionist scholarship are not complimentary. If evolutionists did look at the reconstruction and did not notice the obvious errors for forty-four years, that is hardly a mark of sound scholarship. If they did not look at it for forty-four years while using it as evidence for evolution, that is hardly an illustration of the self-correcting nature of science. It is not just a matter of mistakes by one man, Boule. It is a case where the entire evolutionist community allowed those mistakes to persist for forty-four years, failing to correct them. When it takes scientists forty-four years to correct very obvious mistakes, it is hardly fair to call that a successful case of self-correction.

However, the situation does not end there. Not only did it take forty-four years for the original mistakes regarding Neandertal to be corrected, it took the Field Museum of Natural History in Chicago, one of the great natural history museums of the world, another twenty years to correct their own Neandertal display. How long it took other museums I do not know.

During that period I visited the Field Museum perhaps fifteen times. It was not until the mid 1970s that the Field Museum removed their old display of the apish Neandertals and replaced it with the tall, erect Neandertals that are there today. What did they do with the old display? Did they throw it on the trash heap where it belonged? No. They moved the old display to the second floor and placed it right next to the huge Brontosaurus (Apatosaurus) dinosaur skeleton where more people than ever—especially children—would see it. They labeled it "An alternate view of Neandertal." It was not an alternate view. It was a *wrong* view. So much for the self-correcting mechanism in science as far as Neandertal is concerned.

The Piltdown Man Hoax

The second illustration of where paleoanthropology was notorious for its failure to be self-correcting is the famous Piltdown Man hoax. This is probably the best-known "whodunit" in all of science, and the mystery of "whodunit" is still unsolved.

Piltdown, England, is the least likely place in the universe where one would expect something momentous to happen. Forty miles south of downtown London and about twenty miles west of the site where the Battle of Hastings took place in 1066, Piltdown is a tiny hamlet in east Sussex in a once-forested area known as the Weald. Here, at a small gravel pit used for the repair of roads, Charles Dawson "discovered" Piltdown Man.

Dawson was a solicitor—a lawyer who represents the Crown in matters of the public interest. Quite knowledgeable of geology, he had made some significant fossil discoveries, was an honorary collector for the British Museum, and had been invited to be a member of the prestigious Geological Society. For all of this, he was still regarded as an amateur by professional geologists.

In 1908, a workman at the Piltdown gravel pit handed Dawson a portion of a human skull (parietal) which the workman said looked to him like a coconut. In late 1911, Dawson found several more pieces of the skull in the gravel refuse heaps beside the pit.

In early 1912, Dawson took his finds to Arthur Smith Woodward at the British Museum. His finds, Dawson believed, would rival Heidelberg Man (Mauer Mandible) which had been discovered in Germany in 1907. Smith Woodward was interested. A number of human fossils had been discovered on the Continent, and the British were embarrassed by their lack of respectable fossil ancestors. The time was ripe for Piltdown.

Studying at nearby Hastings was the thirty-one-year-old French Roman Catholic paleontologist-priest, Pierre Teilhard de Chardin, destined to become famous for his relationship both with Piltdown Man and Peking Man. Later he would become known as the formulator of a pantheistic evolutionary theology which was first condemned by the Roman Catholic Church but is now embraced by many Roman Catholics. Teilhard's "theology" is just now coming into its own under the banner of the New Age Movement.

In early June of 1912, Dawson, Woodward, and Teilhard began a series of digs at the Piltdown pit. More pieces of the skull were found, and Teilhard discovered a fragment of an elephant molar. Then, the lower jaw of Piltdown Man was discovered, together with more fossils of mammals. In December, Dawson and Woodward officially announced to the Geological Society of London the discovery of the earliest Englishman, *Eoanthropus dawsoni*, affectionately known as Piltdown Man.

The skull of Piltdown Man was quite large and quite "modern" in shape (morphology). However, the lower jaw was very "primitive" or ape like. Some doubted the association of the two fossils. Paleoanthropologists said that what was needed to resolve the strange association was the canine tooth. The following August the canine tooth was "discovered" at the feet of Teilhard de Chardin as he sat on a gravel refuse pile beside the pit.

The Piltdown pit also produced fossil bones of elephant, mastodon, rhinoceros, hippopotamus, beaver, and deer. Primitive tools and crudely flaked flint stones (eoliths) were found. The last fossil ever to come out of the pit was a portion of a fossil elephant femur that had been worked to look for all the world like a bat used in the game of cricket. Some suspect this was the hoaxer's attempt to reveal that the whole thing was a prank. But to the English bent on discovering their "roots" the clue was missed and they considered the bone some sort of unknown primitive tool.

Piltdown did have its skeptics. So, the hoaxer supplied the one thing that would convert most of the skeptics: a second discovery at a different location that tended to confirm the original discovery. Enter Piltdown II in early 1915 in another gravel pit a few miles from Piltdown I. No one knows for sure the location of that pit where Dawson discovered Piltdown II. The discovery was not revealed by Woodward until 1917, following Dawson's death the year before. Amazingly, at Piltdown II, the exact pieces came to light that were needed to identify it with Piltdown I and thus confirm Piltdown I.

Evolutionists now like to boast that not everyone accepted Piltdown. Technically they are correct. There were a few, such as Weidenreich and Hrdlicka, who did not accept Piltdown. But the vast majority of paleoanthropologists worldwide did accept Piltdown as legitimate, especially after the confirming discoveries at Piltdown II.

Evolutionists complain that creationists make too much of Piltdown. Frankly, it is not necessary for creationists to do so. It is not the frauds that expose the weaknesses of human evolution. It is the legitimate fossils that clearly falsify human evolution, as the later chapters of this book demonstrate. Furthermore, one does not have to be an evolutionist to recognize that Piltdown really was a dirty trick. Fraud is found in every area of human activity, including religion. We do not hold the entire evolutionist community responsible for the Piltdown fraud, which was committed by only one or very few persons.

What we do hold the evolutionist community responsible for is its continual claim that science is self-correcting when in fact it is not. Piltdown demonstrates that it is not. The Piltdown fossils were discovered between 1908 and 1915. It was not until 1953, thirty-eight to forty-five years later, that Kenneth Oakley, Joseph Weiner, and Wilfred Le Gros Clark discovered that Piltdown Man was a fraud. The British Museum then issued a statement to that effect. That is hardly a case of efficient self-correcting.

There is the possibility that the skull itself was legitimately found in the pit. Radiocarbon dating determined it to be from 520 to 720 years old. Piltdown Common had been used as a mass grave during the great plague of A.D. 1348-9. The skull bones were quite thick, a characteristic of more ancient fossils, and the skull had been treated with potassium bichromate by Dawson to harden and preserve it. This was a common practice for the treatment of fossils at that time.

The other bones and stone tools had undoubtedly been planted in the pit and had been treated to match the dark brown color of the skull. The lower jaw was that of a juvenile female orangutan. The place where the jaw would articulate with the skull had been broken off to hide the fact that it did not fit the skull. The teeth of the mandible were filed down to match the teeth of the upper jaw, and the canine tooth had been filed down to make it look heavily worn.

It was only in 1982 that both the mandible and the canine tooth were determined conclusively, by collagen reactions, to be those of an orangutan.[5] It seems likely that whoever put the canine tooth in the pit knew that the mandible was also that of an orangutan. Orangs are found today only in Borneo and Sumatra.

Some of the mammalian bones probably came from other areas of England, but the mastodon molar is thought to have come from Tunisia and the hippopotamus molar from the Island of Malta. Some of the flints in the pit may also have come from Tunisia. It is obvious that whoever perpetrated the hoax either had traveled extensively or had access to some exotic fossil and archeological collections.

Some researchers feel that the hoax was not sophisticated, and that Dawson was the culprit. Others feel that it was a very professional hoax calling for someone having far more expertise than Dawson. Still others say that the fraud was brilliantly conceived but poorly executed. Sophisticated or not, it cannot be denied that the hoax worked very well. Many feel that Dawson was in some way involved. The British Museum has documented other "discoveries" by Dawson as being fakes. Because of the feeling that Dawson lacked the expertise to pull it off by himself, some investigators think in terms of Dawson and a "Significant Other"—either Dawson and Other in collaboration, or Dawson used by Other as a vehicle for Other's deception.

Nonetheless, there are those who remove Dawson from the list of suspects altogether. Ronald Miller in his book *The Piltdown Men* said that Sir Grafton Elliot Smith of the British Museum was the hoaxer.[6]

Charles Blinderman in *The Piltdown Inquest* fingers Lewis Abbott, a jeweler and amateur geologist.[7] Stephen Jay Gould (Harvard University) believes that the Roman Catholic priest Pierre Teilhard de Chardin was definitely involved (as I do).[8] The most surprising suspect, suggested by John Winslow, is Sir Arthur Conan Doyle, creator of Sherlock Holmes.[9] Doyle, who had been trained as a medical doctor, lived just a few miles from the Piltdown pit.

The amazing thing about the Piltdown hoax is that at least twelve different people have been accused of perpetrating the fraud. All of these suspects had the expertise to do it, all had access to the materials involved, all had opportunity to do it, and all had motives for doing it. It is frustrating to know that at this late date we shall never know with certainty who committed what has been called the most successful scientific hoax of all time.

However, it is not necessary to know who perpetrated Piltdown to know that if science were really self-correcting, Piltdown should have been uncovered long before it was. Like Boule's reconstruction of the Neandertal skeleton from La Chapelle-aux-Saints, there were elements about it that were quite obvious. The file marks on the orangutan teeth of the lower jaw were clearly visible. The molars were misaligned and filed at two different angles. The canine tooth had been filed down so far that the pulp cavity had been exposed and then plugged. If science were really self-correcting, the Piltdown fraud should have been discovered soon after it was committed, rather than thirty-eight to forty-five years later.

Why was the Piltdown hoax so successful? A big-brained ancestor was what evolutionists were expecting to find. Sir Grafton Elliott Smith had predicted that a fossil very similar to Piltdown would be found (it is why he is one of the suspects). If the australopithecines had not come into favor as the preferred evolutionary ancestors of humans, and Piltdown had not become an embarrassment because it no longer fit the scenario, the fraud might still be undiscovered and Piltdown might still be considered a legitimate fossil.

There is a touch of irony in the Piltdown story. When the Piltdown fossils were brought to the British Museum, plaster casts were made of them for display to the public. However, museums seldom indicate that the fossils on display are replicas, and most people thought that they were seeing the real thing. Yet, while people thought that they were seeing actual fossils of their evolutionary ancestors, they were looking at fakes. But more than just looking at fakes, they were looking at *fakes of fakes*.

The literature produced on Piltdown was enormous. It is said that more than five hundred doctoral dissertations were written on Piltdown. The man most deceived was Sir Arthur Keith, one of the greatest anatomists of the twentieth century. Keith wrote more on Piltdown than anyone else. His famous work, *The Antiquity of Man*, centered on Piltdown. Keith put his faith in Piltdown. He was eighty-six years old when Oakley and Weiner called on him at his home to break the news that the fossil he had trusted in for forty years was a fraud.

Keith was a rationalist and a pronounced opponent of the Christian faith. Yet, in his *Autobiography* he tells of attending evangelistic meetings in Edinburgh and Aberdeen, seeing students make a public profession of faith in Jesus Christ, and often feeling "on the verge of conversion."[10] He rejected the gospel, because he felt that the Genesis account of Creation was just a myth and that the Bible was merely a human book. It causes profound sadness to know that this great man rejected Jesus Christ, whose resurrection validated everything he said and did, only to put his faith in what proved to be a phony fossil.[11]

The widespread myth is that science is self-correcting and because of this, it is a superior worldview. In reality, science is not adequately self-correcting, and for very practical reasons cannot be self-correcting in any meaningful way.

5

Looks Isn't Everything

WE HAVE ALL HEARD of that famous fossil Lucy. This three-foot tall australopithecine individual was found by Don Johanson in Ethiopia in 1974. Forty percent of her skeleton was recovered. Since she was believed to be more than three million years old, her completeness was most unusual; at a three-million-year age, a paleoanthropologist expects only a few bits and pieces. Why was she called Lucy? Although she was considered to be a female because of her diminutive size, the main reason was that at the time she was found, the loudspeaker at the base camp was blasting out the Beatles' song *Lucy in the Sky with Diamonds*.

From their evaluation of Lucy and fossils like her, Don Johanson and Tim White (University of California, Berkeley) decided in 1979 that Lucy was our oldest-known direct ancestor. Over a period of three million years, you and I are said to have evolved from something that was very unlike us in size, appearance, and mental ability.

While there is currently much controversy regarding our family tree, many evolutionists still agree with Johanson and White that you and I—modern humans—are genetically related to Lucy. We have Lucy's genes in us. About five million mutational events have occurred in our genes in that three-million-year period to account for the differences, but Lucy's modified genes they are.

In a public forum, anthropologist Vincent Sarich was asked why Lucy isn't here today. He replied, "Why isn't Lucy here? That's simple—because we are. She evolved into us. That's not any problem at all."[1]

Evolution always deals with populations: the collective gene pool. So, when we ask why Lucy isn't here today, we are asking why there are no small, erect, chimplike animals living today that are like Lucy. And the evolutionist's answer is, "Lucy isn't here because we are."

Since contemporary evolutionists are committed to the idea that a population of chimplike animals known collectively as Lucy (*Australopithecus afarensis*) evolved into us, there cannot be any Lucys around. Whether they hold to the slow and gradual view of evolution (phyletic gradualism) or to the newer model of long periods of stability and short bursts of rapid evolutionary change (punctuated equilibria), the result is the same. Why isn't Lucy here? Because we are. Why isn't *Homo erectus* here today? The answer is the same: Because we are. From Lucy, it involved that long trail of three million years and five million mutational events, but here we are.

W. W. Howells (Harvard University) explains how the process takes place in each of the two evolutionary models:

> In 'phyletic gradualism,' change is viewed as gradual and general over the species. . . . In 'punctuated equilibria,' the apparent discontinuity, seen so often in a paleontological succession, is not simply the artifact of a gap in the record but is real. The process of change is not species-wide but results from allopatric speciation [speciation in some other place]. A subspecies, ideally a peripheral isolate of the old species, becomes the new form in some significant respect and replaces populations of the old by migration. Thus the main body of the species does not undergo the gradual change to a new species.[2]

In the gradual model, the entire Lucy population would change into *Homo habilis*, and that entire population would gradually change into *Homo erectus*. The *Homo erectus* population would gradually change into archaic *Homo sapiens*, and they eventually into us. Should there be some isolated groups or individuals along the way who did not inherit those superior mutated genes, they would eventually die out, because they would be less fit to survive in a very competitive environment.

In the newer punctuated equilibria model, a small portion of the Lucy population, probably on the edge of the species range, obtains some favorable mutations. This small "advanced" Lucy population, to use Howells's words, "replaces populations of the old [species] by migration." This replacement of the older population is accomplished by their death, since they are the less fit, and the advanced Lucy population represents the more fit. This process takes place again and again through *Homo habilis*, *Homo erectus*, archaic *Homo sapiens*, and on up to modern humans. (A very famous statement by geneticist Richard Goldschmidt about a reptile laying an egg and a bird hatching out of it has given a false impression of punctuated equilibria. Punctuated equilibria is concerned basically with evolution on the species level, not with the higher categories.)

These two models have their differences. However, in both cases the time element—about three million years—is the same. The total number of mutational events needed to bring about these changes—approximately five million—is the same. And in each case those individuals who did not take part in that evolutionary process must be eliminated through death. In the evolutionary process the less fit must die as the more fit survive. The more fit survive because they are better able to compete for the limited food supply, and they reproduce in greater numbers. In other words, in both models, for species A to evolve into species B, species A must precede species B in time. Furthermore, after species A has evolved into species B, any species A remnants must soon die. It is thus basic to evolution that if species B evolved from species A, that species A and species B cannot coexist for an extended length of time.

The "survival of the fittest" has a flip side. It is the death of the less fit. For evolution to proceed, it is as essential that the less fit die as it is that the more fit survive. If the unfit survived indefinitely, they would continue to "infect" the fit with their less fit genes. The result is that the more fit genes would be diluted and compromised by the less fit genes, and evolution could not take place. The concept of evolution demands death. Death is thus as *natural* to evolution as it is *foreign* to biblical creation. The Bible teaches that death is a "foreigner," a condition superimposed upon humans and nature after creation. Death is an enemy, Christ has conquered it, and he will eventually destroy it. Their respective attitudes toward death reveal how many light years separate the concept of evolution from biblical creationism.

It is possible to determine whether the concept of human evolution is a scientific theory or a philosophy. If it is a scientific theory, it must be capable of being falsified. Since human evolution is alleged to be a historic process, the evidence for it or the falsification of it must come from the fossil record. For instance, if *Homo erectus* people persisted long after they should have died out or changed into *Homo sapiens*, the concept of human evolution would be falsified. If one could show that fossils indistinguishable from modern humans existed long before they were supposed to exist (according to the process of evolution) this also would falsify the concept.

It is the burden of this book that both of these falsifications can be demonstrated. If human evolution is truly a scientific theory, the fossil record shows that it has been falsified. The fact that the evidence is ignored or disguised indicates that the concept of human evolution is a philosophy that is perpetuated in spite of and independent of the facts of the human fossil record.

If we humans evolved from a small chimplike animal like Lucy, it is obvious that we had to pass through a number of stages on this long evolutionary journey. We are classified as *Homo sapiens*. Not only are we said to have come from some form that was not our species, we are said to have come from some form that was not even our genus. This form, Lucy, is called *Australopithecus afarensis* (southern ape from Afar, Ethiopia). However, because of very recent fossil discoveries, some evolutionists are going back to the older view that our immediate non-human ancestor was *Australopithecus africanus*. Evolutionists then propose a sequence going to *Homo habilis* (handy man), *Homo erectus* (erect man), archaic *Homo sapiens* (primitive wise man), and then *Homo sapiens* (wise man).

In theory the progression appears rather tidy. In actuality, it is very untidy. First of all, it may surprise the reader to learn that there is no clear-cut, universally accepted scientific definition for any of these categories, including *Homo sapiens*. While there is some consensus on these categories, there is enough uncertainty to make for some healthy confusion. Anyone who works in this area must be prepared for it.

There is also the question of how long it takes (if one assumes that evolution is true) for a new species to evolve. George Gaylord Simpson estimated that it took a quarter of a million years for a species to evolve within a genus. However, the present sequencing of the hominid fossils would imply that it could take longer. Lucy is dated at about three million years. *Homo habilis* is dated from 2 to 1.5 million years. *Homo erectus* is dated roughly at about 1.6 to 0.4 million years, with *Homo sapiens* in the present. That sequencing implies that it could take up to one million years for a new species to fully develop in human evolution (assuming that there were no species intermediate between those creatures).

Another way evolutionists have attempted to answer the question of how long it takes for a new species to evolve is to estimate the rate of gene flow in a population. Recent estimates of the time one advantageous gene would take to disperse throughout hominid populations in the Pleistocene are 20,000 generations or 400,000 years.[3] Since the evolution of one species from another would require many favorable genetic mutations (the existence of a "favorable" mutation has yet to be conclusively demonstrated), it is obvious that evolution requires vast periods of time even on the species level, and even if several "advantageous" genes were being dispersed throughout the population at the same time. Time thus becomes the key ingredient in the evolutionary process.

We have the right to expect, if evolution were true, that the hominid fossil record would faithfully follow the time and morphology sequence set forth by evolutionists. Since humans are supposed to have evolved from something very different from what they look like today, we have a right to expect that very modern-looking fossils would not show up in Lucy times, or that primitive or archaic fossils would not embarrass the evolutionist by showing up in modern times. We also have the right to expect that if a significant number of fossils are so rude as to show up at the wrong time, the evolutionist would be honest enough to admit that his theory has been falsified. In actuality, many fossils have been that rude. And evolutionists have been less than intellectually honest.

Evolutionists work their own special magic on nonconformist fossils. With the waving of a magic wand, *Homo erectus* fossils can become *sapiens* or Neandertals, and *Homo sapiens* fossils can become australopithecines. To us, this is a serious matter of intellectual integrity. The evolutionist does not see it that way. To him, evolution is true. Hence, fossils must be interpreted accordingly. I will give a few examples of the type of magic wand waving to be found in the scientific literature.

The Dating of the Taung Fossil

In 1924, Professor Raymond Dart, anatomist at the University of Witwatersrand in Johannesburg, South Africa, acquired a skull that had come from a lime works at Taung. Dart immediately recognized that the skull was something unique. After cleaning it and studying it, he announced to the world that he had discovered our evolutionary ancestor. It was the skull and endocranial cast of an extinct primate child which Dart named *Australopithecus africanus*. The skull was so well known in the 1920s that it became the object of a joke. If a man had a blind date, he was asked, "Is she from Taung?"

Although Dart's assessment was initially met with hostility and rejection, the eventual exposure of Piltdown Man as a hoax and the discovery of adult australopithecine fossils at other locations in South Africa turned that rejection into almost universal acceptance. By 1960, it would have been difficult to find any public-school book that touched on human origins that did not have in it a picture of the Taung skull. That popularity has remained. The fossil received much publicity in 1984, the sixtieth anniversary of its discovery. Pictures of Taung are still found in most books dealing with human origins.

Until Lucy was discovered in late 1974, Taung, the type specimen of *Australopithecus africanus*, was considered our oldest direct evolutionary ancestor. Although dating the South African fossils has always been a nasty problem, Taung was generally considered to be between two and three million years old. That age seemed appropriate for it as our evolutionary ancestor.

In 1973, South African geologist T. C. Partridge dropped a bomb. His investigations revealed that the cave from which the Taung skull had come could not have formed prior to 0.87 m.y.a.[4] (m.y.a. means "million years ago"; y.a. means "years ago.") That meant that the Taung skull could be at most only three-quarters of a million years old.

Since it could take up to a million years for the hominids to evolve from one species to another, to go all the way from australopithecines to modern humans in only three-quarters of a million years was out of the question. Further, true humans were already on the scene in Africa at 0.75 m.y.a. Karl W. Butzer (University of Chicago) clearly saw the problem when he wrote:

> If the Taung specimen is indeed no older than the youngest robust australopithecines of the Transvaal, then such a late, local survival of the gracile [a term used to describe the *africanus* fossils] lineage would seem to pose new evolutionary . . . problems.[5]

Anatomist Phillip V. Tobias (then at University of Witwatersrand) pointed to the real problem: ". . . the fact remains that less than one million years is a discrepant age for a supposed gracile australopithecine in the gradually emerging picture of African hominid evolution."[6]

Here was a problem which if allowed to persist could jeopardize the concept of human evolution. It was time for the evolutionist to wave his magic wand. A. J. B. Humphreys wrote:

> One point that needs investigation at the outset, however, is the question of the identification of the Taung skull as *Australopithecus africanus*. The possibility that the Taung skull might represent *Homo habilis* or a more advanced creature than *A. africanus* . . . certainly deserves some consideration in view of this younger date.[7]

Phillip Tobias suggested another way the wand might be waved: ". . . because its brain and its dental characters would exclude it from *H. erectus*, it must seriously be considered whether the Taung child is not a late surviving member of *A. robustus* or *A. cf. robustus*."[8] Tobias then made this amazing confession: "Although nearly 50 yr have elapsed since its discovery, it is true to say that the Taung skull has never yet been fully analyzed and described."[9]

A basic obligation upon the one who discovers a new fossil is to intensively study it and publish descriptions of the study for the benefit of the scientific community. To fail to do this is akin to dereliction of duty. Dart did not do it. Tobias, Dart's successor at Witwatersrand and also his successor as custodian of the Taung skull, did not do it. Yet, for fifty years the public was told that Taung was our evolutionary ancestor. Considering that it has been with us since 1924, we can question if any fossil has been pictured more in the popular press than Taung. Now we are told that the appropriate analysis and description of the fossil have never been done. In spite of that fact, without further study (and only to relieve the embarrassment of the revised date) the suggestion was made that perhaps Taung was closer to humans (*habilis*) or not in the human lineage at all (*robustus*). Looks, obviously, isn't everything.

In 1974, Johanson discovered Lucy. Other material followed. Paleoanthropologists began to focus on the new material from Ethiopia, and the problem of the awkward dating of the Taung skull faded into the background. When Johanson and Tim White revised the hominid family tree in 1979, *afarensis* (including Lucy) replaced *africanus* (in-

cluding Taung) as our direct nonhuman evolutionary ancestor. The
dating of Taung thus became a nonissue. The *africanus* forms were
moved to the australopithecine branch of the family tree, becoming the
link between Lucy and the robust australopithecines.

This comfortable arrangement was severely jolted by the discovery
in 1985 of the famous "black skull," KNM-WT 17000.[10] Australopithecine phylogeny is now in disarray. Dated at 2.5 m.y.a., the "black
skull" seems to have more in common with Lucy and the robust australopithecines. *Africanus* became the odd man out. Many evolutionists are now moving *africanus* (including Taung) back into the human
line, between Lucy and *Homo habilis*.

What happened to Partridge's work showing that Taung could be
only about 0.75 million years old, and hence could not possibly be our
evolutionary ancestor? His work has not even been addressed, let alone
answered. Instead, it is being disparaged. A recent work discussing
Taung states: "An ill-founded attempt at geomorphological dating in
the early 1970s suggested an age of less than 870 k.y. [870,000 years]
for the hominid. . . ."[11]

That "ill-founded" date was supported by a date of about 1 m.y.a.
by thermoluminescence analysis of calcite[12] and uranium-series dates
of 942,000 y.a. and 764,000 y.a. on limestone,[13] but when the evolutionist waves his magic wand, both date and taxon become plastic so
that evolution might be served. Richard G. Klein (University of Chicago) doesn't really face the issue, but he is at least honest when he
writes: ". . . a date for Taung of 2 mya [million years ago] or more may
seem most reasonable, but the argument is obviously circular and the
true age remains uncertain."[14]

Because Taung has been slipped back into the human lineage, Tobias has withdrawn his suggestion that Taung might be a robust australopithecine. However, the full analysis and description of the Taung
skull still has not been published, and the dating problem raised by
Partridge continues to be ignored.

The Kanapoi Elbow Fossil

One of the most flagrant cases of wand waving to deflect evidence
that could be most embarrassing to the idea of human evolution involves a fossil found at Kanapoi, southwest of Lake Rudolf (Turkana)
in northern Kenya. This fossil, known as KP 271, is the lower end of
a left upper arm bone (distal end of the humerus). It was found in

1965 by Bryan Patterson (Harvard University), and is in an excellent state of preservation. The most recent dating of the fossil gives it an age of 4.5 m.y.a.[15] It thus becomes virtually the oldest hominid fossil ever found—older than Lucy and all of the australopithecines. The question is, What is it?

To answer the question of the identity of KP 271, Patterson and W. W. Howells used the method of computer discriminate analysis. They compared KP 271 with the distal ends of the humeri of a modern human, a chimpanzee, and the only other similar fossil they had at the time: *Australopithecus (Paranthropus) robustus*, from Kromdraai, South Africa. Seven different measurements were fed into the computer from each of the four samples. Patterson and Howells published the results of their study in *Science*, 7 April 1967. "In these diagnostic measurements, Kanapoi Hominoid 1 [the original name given to the fossil] is strikingly close to the means of the human sample."[16]

After stating that their computer analysis revealed the Kanapoi humerus to be *strikingly* close to modern humans (they were not comparing it to ancient humans but to modern humans) they made the rather shocking conclusion that KP 271 "may prove to be *Australopithecus*."[17] (They meant the gracile *africanus* form.) Their reasoning was that the upper end of the humerus of *africanus* is quite similar, based on visual assessment, to that of modern humans. Hence, they assumed that the lower end was similar also, even though they did not have the lower humerus portion of *africanus* for comparison with KP 271. The real reason for this strange departure from their data comes out later.

Further computer analysis of many more measurements revealed even more dramatically the similarity of KP 271 to modern humans. Henry M. McHenry (University of California, Davis) wrote: "The results show that the Kanapoi specimen, which is 4 to 4.5 million years old, is indistinguishable from modern *Homo sapiens*. . . ."[18]

Regarding KP271, Pilbeam states:

Multivariate statistical analysis of the humeral fragment aligns it unequivocally with man rather than with the chimpanzee, the hominoid most similar to man in this anatomical region. Professors Bryan Patterson and F. Clark Howell [sic], the describers of this fragment, believe that it represents *A. africanus* rather than *A. robustus*.[19]

McHenry pointed out that multivariate analysis is an excellent diagnostic tool for this portion of primate anatomy:

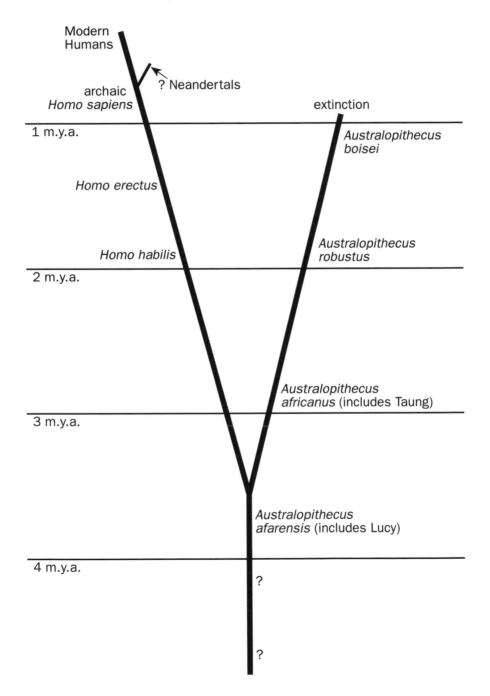

Chart of Human Evolution 1979–1986
(after Johanson and White)

Modern
Humans

? Neandertals

archaic
Homo sapiens

extinction

1 m.y.a.

*Australopithecus
boisei*

Homo erectus

*Australopithecus
robustus*

Homo habilis

2 m.y.a.

*Australopithecus
africanus* (includes Taung)

3 m.y.a.

*Australopithecus
afarensis* (includes Lucy)

4 m.y.a.

?

?

Chart of Human Evolution 1986–

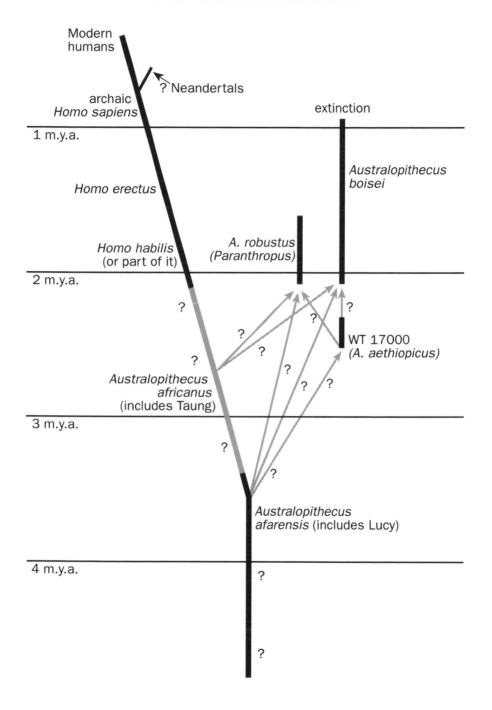

The hominoid distal humerus is ideal for multivariate analysis because there are such subtle shape differences between species, particularly between *Homo* and *Pan*, which are difficult to distinguish in a trait by trait (univariate) analysis.[20]

In her studies on the australopithecines and the living primates, Brigette Senut (Museum of Man, Paris), found reason to challenge some of the earlier diagnostic work on the humeri in primates:

In the field of comparative anatomy of primate postcranial material, it is striking that the forelimb (especially the humerus) has been much less studied than the hindlimb, and only by relatively few authors.

Thus, it appears that these hominids, usually considered as similar to modern man in humeral morphology (Broom et al., 1950) exhibited quite unique features.[21]

Let me review what we have said thus far. We are dealing with the oldest respectable hominid fossil ever found up to the present time.[22] The fossil represents a part of the anatomy where it is relatively easy to discriminate between humans and the other primates, both living and fossil. The appropriate diagnostic tools have been used to evaluate the fossil. The results show unequivocally (to use Pilbeam's term) that the fossil is indistinguishable from modern humans, not just fossil humans. Yet, in their original report Patterson and Howells go against their own empirical evidence and suggest that the fossil represents *Australopithecus africanus*. Why?

Further study strengthened the fact that KP 271 should be ascribed to modern humans. Yet, every textbook or journal article from 1967 to the present time that mentions the Kanapoi fossil calls it *Australopithecus africanus*. Why? We might assume that because so much fossil material has been discovered in East Africa since 1965, the appropriate *africanus* fossil has been discovered which confirmed the original evaluation of Patterson and Howells. As far as I have been able to determine, that is not the case.

Why, then, the universal insistence that this fossil is *africanus* and not *Homo sapiens*? Howells, writing in 1981, fourteen years after the fossil was first ascribed to *africanus*, gives us the reason:

The humeral fragment from Kanapoi, with a date of about 4.4 million, could not be distinguished from *Homo sapiens* morphologically or by multivariate analysis by Patterson and myself in 1967 (or by much more searching analysis by others since then). We suggested that it might represent *Australopithecus* because at that time allocation to *Homo* seemed preposterous, *although it would be the correct one without the time element.*[23]

It is obvious that looks isn't everything. Even though KP 271 is shaped exactly like *Homo sapiens*, the time element is wrong. What determines that? The concept of human evolution. The concept of human evolution decrees that it is impossible for true humans to have lived before the australopithecines—even though the fossil evidence would suggest otherwise—because humans are supposed to have evolved from the australopithecines.

According to the basic principles of the philosophy of science, a theory must be falsifiable if it is a legitimate scientific theory. How could the theory of evolution be falsified? Supposedly if fossils are found that are woefully out of order from what evolution would predict. Many such fossils have been found. KP 271 is just one of them. (More examples will be given later.) However, evolutionists ignore the morphology of fossils that do not fall into the proper evolutionary time period. They wave their magic wand to change the taxon of these fossils. Thus, it is impossible to falsify the concept of human evolution. It is like trying to nail jelly to the wall. That evolutionists resort to this manipulation of the evidence is a "confession" on their part that the fossil evidence does not conform to evolutionary theory. It also reveals that the concept of human evolution is a philosophy, not science.

To the evolutionist there is but one primary fact in the universe: evolution. Everything else is just data. The value of this data does not depend upon its intrinsic quality but upon whether or not it supports evolution and its time scale. Good data is that which supports evolution. Bad data is that which does not fit evolution, and it is to be discarded. It is time to ask the paleontological community, At what point does philosophical bias in the interpretation of the human fossil material become intellectual dishonesty? The interpretation of KP 271 from Kanapoi justifies that question.

Was the original owner of that Kanapoi elbow bone a true *Homo sapiens*? I do not know. I was not there. Neither was Bryan Patterson or William Howells. There is no way at this point that anyone can prove it. That is beyond the scope of science. What we can say with confidence is that *all* of the scientific evidence points in that direction, and there is *no* scientific evidence to the contrary. Science can go only that far. What's wrong with telling a student that? The problem is that there are metaphysical (read *theological*) implications to all of this. Evolution implies a naturalistic, mechanistic origin of things. That the oldest human fossil ever found—skimpy as it is—reveals that man was virtually the same 4.5 million years ago (on the evolutionist time scale) as he is today suggests that humans appeared on the scene suddenly and without evolutionary ancestors. Prior to 4.5 m.y.a. the hominid fossil record is a virtual blank for ten million years. This supports the idea that humans were created by a supreme being.

We are continually told how bad it is to mix religious implications with science. We don't mix them. Those implications are already there. Evolutionists act as if creationists have been trafficking in something very dark and sinister. How many young minds have we led astray by telling them that there is a Creator?

Just for the fun of it, let's grant the evolutionist his point. Let's assume that there is something very bad about telling a student that the oldest human fossil ever discovered supports Special Creation. As bad as that might be, I can think of something even worse. That would be to tell a student that the oldest human fossil ever discovered supports evolution. Because that would be a lie.

6

With a Name Like Neandertal He's Got to Be Good

Praise to the Lord, the Almighty,
The King of Creation!
O my soul, praise Him,
For He is thy health and salvation!
All ye who hear,
Now to His temple draw near;
Join me in glad adoration!

IF YOU ATTEND CHURCH REGULARLY, the probability is high that you have sung this beautiful hymn within the past year. It is found in most hymnals. The fact that there is a relationship between this hymn and the first Neandertal fossil proves that truth is indeed stranger than fiction.

In the late 1600s, an evangelical (Lutheran) theologian and school rector, gifted in poetry and hymn writing, took long walks in the country near Hochdal, Germany. As he strolled, he composed hymns and sang them in praise to God. One of his favorite spots was a beautiful gorge through which the Dussel River flowed, about ten miles east of Dusseldorf. He strolled in this one valley so often that it became iden-

tified with him and was eventually named after him. His name was
Joachem Neander, and the valley became known as the Neanderthal—
the Neander Valley (*tal*, or *thal* in Old German, means "valley," with
the *h* being silent).

Almost two hundred years later, this valley was owned by Herr von
Beckersdorf. As the owner quarried limestone in the valley for the
manufacture of cement, his workmen came across some caves in the
side wall of the gorge. One cave, known as the Feldhofer Grotto, had
human bones in the soil of its floor. Because the prime interest of the
workmen was to quarry the limestone, what probably had been a complete skeleton was largely destroyed. Only the skullcap, some ribs, part
of the pelvis, and some limb bones were saved. The year was 1856.
The first Neandertal had been discovered.[1]

Although the bones were obviously human, they looked different.
Beckersdorf took the bones to a science teacher, J. K. von Fuhlrott,
who was also president of the local natural history society. Recognizing
the antiquity and the extreme ruggedness of the bones, Fuhlrott felt
that they were probably the remains of some poor soul who had been
a victim of Noah's Flood.

Von Fuhlrott, in turn, invited Hermann Schaafhausen, professor of
anatomy at the University of Bonn, to study the bones. He agreed that
the bones represented an ancient race of humans, perhaps barbarians
who lived in Europe before the arrival of the Celtic and Germanic
tribes.

Eventually, the bones came to the attention of Rudolf Virchow, a
professor at the University of Berlin. A brilliant man and a true scientist, Virchow is recognized as the father of pathology. Virchow questioned the antiquity of the bones. He felt that they belonged to a
modern *Homo sapiens* who had suffered from rickets in childhood, arthritis in old age, and had received several severe blows to the head. As
we shall see, Virchow's diagnosis is as valid today as when he first
made it.

William King, professor of anatomy at Queen's College, Galway,
Ireland, however, read an evolutionary history into the bones, and it
was he who eventually gave them their first scientific name: *Homo neanderthalensis*. The name is significant. King believed that the bones
represented a person so primitive that he did not belong to the same
species as modern humans. Hence, Neandertal was placed in the genus *Homo* but in a separate species.

Controversy surrounded the fossils. In 1886, two skeletons were found in a cave near Spy (pronounced "shpee"), Belgium. The Spy fossils made it obvious that the original Neandertal was not some freak, but did represent some form of ancient humans. Found with the human skeletons at Spy were the remains of cave bear, mammoth, and woolly rhinoceros.

In 1908 the first reasonably complete skeleton of a Neandertal type was found. The very faulty reconstruction of it by Marcellin Boule has been described in an earlier chapter. This faulty reconstruction is responsible for some very false views of Neandertal that exist in the popular mind to this day. However, a proper view of Neandertal began to emerge in the 1950s. Although the Neandertals are among the most recent of fossil humans, and although remains of more than three hundred Neandertal-type individuals have been recovered, they are still the most enigmatic humans from the past—especially to the evolutionist.

Because of the richness of the Neandertal fossil record, we do have a general idea of what they looked like. There is a distinct Neandertal morphology: (1) large cranial capacity, the average being larger than the average for modern humans, (2) skull shape low, broad, and elongated, (3) rear of the skull rather pointed, with a bun, (4) large, heavy browridges, (5) low forehead, (6) large, long faces with the center of the face jutting forward, (7) weak, rounded chin, and (8) postcranial skeleton rugged with bones very thick.

The typical Neandertal does differ somewhat from the typical modern human. However, the two also overlap. In fact, there should never have been a question about Neandertal's taxonomic status. When the first Neandertal was discovered, even "Darwin's bulldog," Thomas Huxley, recognized that Neandertal was fully human and not an evolutionary ancestor. Donald Johanson, in his book *Lucy's Child* writes:

> From a collection of modern human skulls Huxley was able to select a series with features leading "by insensible gradations" from an average modern specimen to the Neandertal skull. In other words, it wasn't qualitatively different from present-day *Homo sapiens*.[2]

Although Neandertals are often presented as being inferior to modern humans, Neandertal authority Erik Trinkaus (University of New Mexico) writes:

Detailed comparisons of Neanderthal skeletal remains with those of modern humans have shown that there is nothing in Neanderthal anatomy that conclusively indicates locomotor, manipulative, intellectual, or linguistic abilities inferior to those of modern humans.[3]

The evidence indicates that the Neandertals were people of incredible power and strength—far superior to all but the most avid body builders of today. Trinkaus continues:

One of the most characteristic features of the Neanderthals is the exaggerated massiveness of their trunk and limb bones. All of the preserved bones suggest a strength seldom attained by modern humans. Furthermore, not only is this robustness present among the adult males, as one might expect, but it is also evident in the adult females, adolescents, and even children.[4]

Valerius Geist (University of Calgary) says:

Neanderthal was far more powerful than modern humans. Whereas archeologists can experimentally duplicate the wear pattern on tools such as were used by people from the Upper Paleolithic (the people that followed Neanderthal . . .), the wear patterns on Neanderthal's tools cannot be duplicated. We do not have the strength to do it. Neanderthal's skeleton reflects a supremely powerful musculature.[5]

Conflicting Views of Neandertal

Beyond their incredible physical strength, almost everything else about the Neandertals is the subject of intense debate. It is hard to imagine a greater contrast than the two pictures of the Neandertals presented by Geist and by Jared Diamond (University of California, Los Angeles). Geist pictures the Neandertals as the mightiest of hunters, deliberately provoking and attacking the largest game animals: "Neanderthal's kill patterns, slanted heavily to large-bodied grazers and carnivores and almost devoid of small game, are beyond comparison with any modern hunting culture."[6]

In the absence of any evidence that the Neandertals possessed spears for throwing from a distance, Geist suggests that the Neandertals engaged in close-quarter hunting. While one hunter would divert the animal's attention, one or more other hunters would rush in and attack the sides of the animal where vital internal organs were more accessible to inflicted damage. If the animal had long hair, such as mammoths, woolly rhinos, steppe wisents, and horses, one hunter might actually attach himself to the body of the prey, holding on to the long hair with his powerful grip. With the animal thus distracted, another hunter, using hand-held tools, would rush in and crush the lumbar vertebrae, slash open the chest cavity, or cut the tendon of the hindleg.

Geist explains how this hunting activity is related to the unique shape of the Neandertal skull.

> If great strength, agility, and precision and speed of bodily movements were required for such a hunting technique, those parts of the brain controlling motor functions in the hunter had to be greatly developed. Neanderthal possessed a massive cerebellum and motor cortex compared to modern humans. This pulled the brain case rearward, creating an occiput that reached farther rearward than in modern humans, explaining, in part, the large, long, low brain case and bun-shaped occiput of the Neanderthals.[7]

It is obvious that this kind of precision would be impossible without a quick and fluent language capability. While the Neandertals may not have been as culturally sophisticated as the people who followed, Geist concludes that the Neandertal people were not primitive but the most highly specialized of all the humans of the past.

In a cover story in *Discover*, Jared Diamond presents a contrasting picture of the Neandertals.[8] Diamond maintains that for several million years human evolution went forward at a snail's pace, and that during much of that time humans were not much more than glorified baboons.

> As recently as 35,000 years ago western Europe was still occupied by Neanderthals, primitive beings for whom art and progress scarcely existed. Then there was an abrupt change. Anatomically modern people appeared in Europe, and suddenly so did sculpture, musical instruments, lamps, trade, and innovation. Within a few thousand years the Neanderthals were gone.[9]

Diamond claims that the Neandertals were rather ineffective hunters, their tools and weapons were nothing to write home about, and although their average brain size was about ten percent larger than ours, they obviously were not as smart as we are. Their brains were not "wired" as well as ours. (For many years, evolutionists claimed that the smaller brains of our ancestors meant inferior mental ability, something we now know is not necessarily true. Therefore, it is rather humorous to see evolutionists argue that the larger brains of our Neandertal ancestors also indicate inferior mental ability.) The Neandertals, Diamond claims, lacked art, needles for sewing, boats, long-distance overland trade, and most of all, the precious human quality: innovation. He refers to them as "humans, and yet not really human."[10]

The magnitude of our confusion about the Neandertals can be seen from the fact that while Geist presents them as the mightiest of hunters and Diamond presents them as ineffective hunters, John Shea (Harvard University) presents the Neandertals as basically vegetarians:

> These muscular hominids apparently competed among themselves for access to dense stands of fruits, nuts and other plant foods. The need to defend patches of plant food may have kept them from living in large groups or avidly pursuing seasonal, highly mobile game.[11]

Diamond holds to the view that about 35,000 years ago modern people from Africa invaded Europe (the Out of Africa theory of the origin of modern humans). These people, represented by the Cro-Magnon (pronounced "man yon") fossils found in Europe, possessed innovation. In a few thousand years the Neandertals were gone. The "Great Leap Forward" had taken place and the foundation for all of the culture and technology we know today was laid. While recognizing the physical prowess of the Neandertals, Diamond suggests that in various ways the Cro-Magnon people caused the extinction of the Neandertals, because, in the long run, brains always win over brawn. Geist, however, sees it the other way around. "It was probably only after the Neanderthals' extinction that modern people could colonize the land they once roamed over, for they must have been fighters of stunning abilities, for whom Upper Paleolithic people were no match."[12]

For Diamond, the key that made that Great Leap Forward possible was language; Neandertal did not have language, and Cro-Magnon did. Although Geist does not address the matter of language capability for the Neandertals, his scenario is impossible without it.

However, the fossil evidence now favors Neandertal language capability. At the very time Diamond's article appeared suggesting that the Neandertals lacked speech, *Nature* published a report regarding a Neandertal skeleton discovered at Kebara Cave, Mount Carmel, Israel. The report was of the hyoid bone of the Neandertal individual known as Kebara 2. The hyoid is a small bone lying at the base of the tongue and connected to the larynx by eleven small muscles important in speech. Since the hyoid bone of Kebara 2 is almost identical in size and shape to that of modern populations, the inference is that this part of human anatomy has shown great stability over time. The report continues:

> A related inference would be that the associated larynx beneath the hyoid has scarcely changed in position, form, relationships or size during the past 60,000 years of human evolution. If indeed this inference is warranted, the morphological basis for human speech capability appears to have been fully developed during the Middle Paleolithic, contrary to the views of some researchers.[13]

This conclusion is what creationists would expect, since we believe that Neandertal was a card-carrying member of the human family, a descendant of Adam, and probably a part of the post-Flood population.

The Neandertal Problem

Ever since Darwin, evolutionists have sought to discover the path by which humans arose from their alleged primate ancestors. This, of course, remains the crucial issue in human evolution. In recent years, a second matter has attracted major attention: the path by which our own species, *Homo sapiens*, arose from our alleged more primitive human ancestors. Squarely in the middle of this second issue sets "The Neandertal Problem."

The Neandertal problem is primarily the evolutionists' problem. Simply put, evolutionists don't know where Neandertal came from or where he went. His beginnings, now said to extend back as far as 200,000 y.a.,[14] are as much a mystery as is his alleged rapid disappearance at about 34,000 y.a.

However, the Neandertal problem is only secondarily the question of Neandertal's pedigree out of an assumed *Homo erectus* stock and his apparent disappearance at the beginning of the Upper Paleolithic Age. The major problem is Neandertal's relationship to two contemporary populations, one seemingly more modern in morphology and the other more archaic than he, all living at the same time. The accompanying chart lists some of these fossils on the evolution time scale. (The reader should bear in mind that the author does not subscribe to the evolution time scale). The morphological spectrum represented in the chart is the type of genetic variation in the human family which the creation model would predict. Hence, the creation model seems to explain this data better than does the evolution model.

The older but still somewhat popular evolutionist view regarding Neandertal could be called the Neandertal phase of human evolution. It saw the Neandertals in the mainstream of the evolutionary process, moving from *Homo erectus* to an archaic phase of *Homo sapiens* to Neandertal and on to modern humans. This was the view of one of the early leaders of American anthropologists, Ales Hrdlicka, later held by Franz Weidenreich, and more recently by C. Loring Brace and Milford Wolpoff. A second viewpoint is that the Neandertals were absorbed by more modern *Homo* populations through gene migration and hybridization. Either of these views means that modern humans have Neandertal genes in their make-up.

Two other views consider the Neandertals as an isolated side branch on the family tree. The main branch of human evolution passed by the Neandertals who ended in a European backwater. One view sees environmental factors bringing about their demise without issue. The alternate scenario, perhaps the more popular one today, holds that the Neandertals were exterminated in one way or another by more modern humans who invaded Europe and the Near East from Africa. In either of these latter views, modern humans would not be the genetic descendants of the Neandertals.

This last scenario, the Out of Africa view, also called African Eve or mitochondrial Eve theory, sees modern humans stemming from some "Eve" who lived in Africa about 200,000 years ago. This "mother of us all" founded a modern population which moved out of Africa about 150,000 to 100,000 years ago to eventually replace the Neandertals in Europe and also to replace all other "primitive" humans in the world, including *Homo erectus*.

Several factors have been interpreted as evidence for this replacement of the Neandertals. Anatomically modern human fossils were found at a site known as Qafzeh in Israel. Originally, the date for this site was thought to be about 30,000 y.a., the same as some Neandertal sites in the Near East. This would be about the time that the Neandertals disappeared and anatomically modern humans appeared in Europe. However, recent thermoluminescence dating of the Qafzeh

The Neandertal Problem—Contemporaneousness

	Fossils more modern than Neandertal	Neandertal fossils (partial list)	Fossils more archaic than Neandertal
30,000 y.a.	Lake Mungo, Australia Cro-Magnon, France Velika Pecina, Yugoslavia Springbok Flats, South Africa Darra-i-Kur, Afghanistan Starosel'e U.S.S.R. Fish Hoek, South Africa Ziyang, China Niah, Borneo Brno, Czechoslovakia Combe-Capelle Cave, France	Amud I, Israel Arcy-sur-Cure, France Saint-Césaire, France Shukbah Cave, Israel Le Moustier, France Maurillac, France Vindija Cave, Yugoslavia Ksar Akil, Lebanon Monte Circeo, Italy Shanidar, Iraq	Mugharet el 'Aliya, Morocco Willandra Lakes, Australia Témara, Morocco Eliye Springs, Kenya
	Salawusu, China	Kebara Cave, Israel La Chapelle aux Saints,	Dar es Soltan Cave 2,
65,000 y.a.	Bacho-Kiro, Bulgaria	France	Morocco
	Liujiang, China Ochoz, Czechoslovakia Sala, Czechoslovakia	La Quina, France La Ferrassie, France	Haua Fteah, Libya
100,000 y.a.	Mladec, Czechoslovakia Kulna, Czechoslovakia Jebel-Qafzeh Cave, Israel	Santa Croce Cave, Italy	Narmada, India La Chaise, France KNM-ER 999, Kenya
	Krapina A skull, Yugoslavia Singa, Sudan Wadjak I & II, Indonesia Skhūl 5 & 6, Israel Border Cave, South Africa Klasies River Mouth, S. Africa Mumba Rock Shelter, Tanzania Omo-Kibish I & III, Ethiopia	Krapina, Yugoslavia Skhūl 2, 4, 7, & 9, Israel Tabun, Israel Saccopastore, Italy	Krapina D skull, Yugoslavia Xujiayao, China Florisbad, South Africa Mapa, China Cave of Hearths, South Africa Klasies River Mouth, South Africa Laetoli Hominid 18, Tanzania Omo-Kibish II, Ethiopia Hazorea 2, 4, & 5, Israel Laetoli Hominid 29, Tanzania
150,000 y.a.		Zuttiyeh, Israel	Le Lazaret, France
			Azych, U.S.S.R. Montmaurin, France Hexian, Ding Cun, Changyang, China
	Fontechevade, France		Wadi Dagadle, Djibouti La Chaise, France
		Biache, France Ehringsdorf (Weimar),	Jebel-Irhoud, Morocco
200,000 y.a.		Germany	Rabat & Salé, Morocco

site gave the amazing date of 90,000 y.a.[15] This date not only would imply the very early appearance of modern humans in the Near East, but it also would imply a 60,000 year period when Neandertals and modern humans lived together without any genetic exchange or hybridization.

From the Kebara Cave in Israel came the first complete pelvic bone of a Neandertal. It appears that the structure and orientation of the sockets into which the thigh bones fit were a bit different from those of modern humans. Rak and Arensburg interpret this as differences in "locomotion and posture-related biomechanics"[16] between Neandertal and modern humans. While the true significance of this difference remains unclear, this difference is interpreted by some evolutionists as further evidence that there is no close evolutionary relationship between the Neandertals and anatomically modern humans.

Based on the belief that Neandertals and anatomically modern humans were reproductively isolated from one another for about 60,000 years (though they may have been neighbors geographically), some evolutionists now suggest that the Neandertals were more distinct from modern humans than has been realized. Stephen Jay Gould[17] and Chris Stringer[18] (British Museum—Natural History) even suggest that the Neandertals be removed from our species (*sapiens*) and once again given their earlier designation *Homo neanderthalensis*. However, Stringer does suggest that Neandertals and modern humans ". . . were probably sufficiently closely related to allow hybridization."[19] (This comment reveals why the scientific word *species* and the biblical word *kind* are not the same and should never be used as synonyms.)

The Out of Africa theory on the disappearance of the Neandertals exploded upon the paleontological world in 1987 when three Berkeley biochemists, Rebecca Cann, Mark Stoneking, and Allan Wilson, published a paper in *Nature*.[20] They explored a new way of tracing human origins using tracer DNA from inside the cell called mitochondria (mtDNA). Each of our cells contains many mitochondria, the cell's powerhouses that crank out a cascade of energy-rich molecules. Unlike our regular chromosomes, the chromosomes of mitochondria are passed unchanged from a mother to her offspring. The father's mtDNA ends up "on the cutting room floor." The assumption is that were it not for occasional mutations, everyone in the world would have identical mitochondrial chromosomes. But mutations do occur, and each mutation establishes a new mitochondrial type.

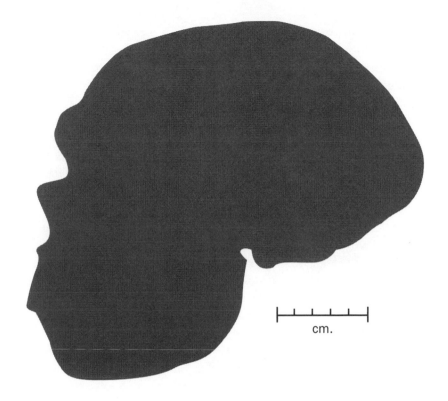

cm.

The Neandertal Skull
from La Chapelle aux Saints, France

The inheritance of mitochondria is very much like the inheritance of surnames, except that mitochondria pass down the female rather than the male line. However, surnames are a good analogy, and I will use them to illustrate the process the Berkeley biochemists used when they sought to determine the evolutionary origin of modern humans.

We start with a small group of people who have the surname Smith. In each generation a random "mutation" occurs in each of the family lines, which changes one letter of the old name. In the first generation, a Smith could become Sbith and in the next generation it could become Qbith. In another line, Smith might become Smjth, and in a third line it could become Smifh. In some lines the name might be lost completely if the males carrying it do not reproduce or if they have only female children. Over many generations, a large population would be traceable back to a single ancestral name—a name that may no longer be in existence but that could be reconstructed from the modified names.

The theory seemed to be rather brilliantly conceived. The Berkeley biochemists made several reasonable although unprovable assumptions. With no mixing from generation to generation, they assumed all changes in the mtDNA were the result of mutations over time. They further assumed that these mutations occurred at a constant rate. On the basis of these assumptions, the researchers believed they had access to a "molecular clock." MtDNA is thought to mutate faster than other DNA, so it was favored because it would lend itself to a more fine-grained index of time.

The original 1987 study used mtDNA from 136 women from many parts of the world having various racial backgrounds. The analysis led back to a single ancestral mtDNA molecule from a woman living in sub-Saharan Africa about 200,000 years ago. A subsequent and more rigorous 1991 study seemed to confirm and secure the theory.

Unfortunately, there was a serpent stalking this "Eve" as well as the first Eve. The researchers used a computer program designed to reveal a maximum parsimony phylogeny. This would be the family tree with the least number of mutational changes, based on the assumption that evolution would have taken the most direct and efficient path—a rather strange assumption considering the presumed random and haphazard nature of evolutionary change. The computer program was, however, far more complicated than the biochemists realized. They did not know that the result of their single computer run was biased by the order in which the data were entered. Others have determined that with thousands of computer runs and with the data entered in different random orders, an African origin for modern humans is not preferred. There is also the suggestion that in the original study the biochemists were influenced in their interpretation of the computer data by their awareness of other evidence that seemed to favor an African origin.

Henry Gee, on the editorial staff of *Nature*, describes the results of the mtDNA study as "garbage." He states that, considering the number of items involved (136 mtDNA sequences), the number of maximally parsimonious trees exceeds one billion.[21] Geneticist Alan Templeton (Washington University) suggests that low-level mixing among early human populations may have scrambled the DNA sequences sufficiently so that the question of the origin of modern humans and a date for Eve can never be settled by mtDNA.[22] In a letter to *Science*, Mark Stoneking (one of the original researchers who is now at Pennsylvania State University) acknowledges that African Eve has been invalidated.[23] There is general recognition that Africans have greater genetic diversity, but the significance of that fact remains unclear.

The African Eve theory represented the second major attempt by biochemists to contribute to the question of human origins. Earlier, Berkeley biochemist Vincent Sarich had estimated that the chimpanzee-human separation took place between five and seven million years ago, based upon molecular studies. Although that date was much later than paleoanthropologists had estimated from fossils, Sarich's date is now almost universally accepted.

In an article written before but published after the recent challenge to African Eve, Wilson (who died in 1991) and Cann (now at the University of Hawaii, Manoa) laud the virtues of molecular biology in addressing human origins. They say that ". . . living genes must have ancestors, whereas dead fossils may not have descendants." The molecular approach, they claim, "concerns itself with a set of characteristics that is complete and objective." In contrast, the fossil record is spotty. "Fossils cannot, in principle, be interpreted objectively. . . ."[24] They conclude that the method of the paleoanthropologists tends toward circular reasoning. They are right. Creationists have asserted that fact for many years.

However, Wilson and Cann were not able to see the logical fallacy in their molecular biology when it addressed phylogeny. This approach, known as molecular taxonomy, molecular genetics, or the newer related field of molecular archaeology, also traffics in circular reasoning. Molecular genetics, hiding behind the respect we all have for the science of genetics and the objectivity of that science, is highly infused with subjective evolutionary assumptions. In this field, the commitment to evolution is so complete that Wilson and Cann understand "objective evidence" as ". . . evidence that has not been defined, at the outset, by any particular *evolutionary* model."[25]

The mtDNA study of African Eve, as well as other aspects of molecular genetics, is based on mutations in the DNA nucleotides. Perhaps we could be forgiven for asking the question, When an evolutionist looks at human DNA nucleotides, how does he know which ones are the result of mutations and which ones have remained unchanged? Obviously, to answer that question he must know what the original or ancient sequences were. Since only God is omniscient, how does the evolutionist get the information about those sequences that he believes existed millions of years ago? He uses as his guide the DNA of the chimpanzee.[26] In other words, the studies that seek to prove that human DNA evolved from chimp DNA start with the assumption that chimp DNA represents the original condition (or close to it) from which human DNA diverged. That is circularity with a vengeance.

It is also necessary for the evolutionist to determine the rate of mutational changes in the DNA if these mutational changes are to be used as a "molecular clock." Since there is nothing in the nuclear DNA or the mtDNA molecules to indicate how often they mutate, we might also ask how the evolutionist calibrates his "molecular clock." Sarich, one of the pioneers of the molecular clock concept, began by calculating the mutation rates of various species ". . . whose divergence [evolution] could be reliably dated from fossils."[27] He then applied that calibration to the chimpanzee-human split, dating that split at from five to seven million years ago. Using Sarich's mutation calibrations, Wilson and Cann applied them to their mtDNA studies, comparing ". . . the ratio of mitochondrial DNA divergence among humans to that between humans and chimpanzees."[28] By this method they arrived at a date of approximately 200,000 years ago for African Eve. Hence, an evolutionary time scale obtained from an evolutionary interpretation of fossils was superimposed upon the DNA molecules. Once again, the circularity is obvious. The alleged evidence for evolution from the DNA molecules is not an independent confirmation of evolution but is instead based upon an evolutionary interpretation of fossils as its starting point.

We humans are enamored with our ability to develop sophisticated experiments and to process massive amounts of data. Our problem is that our ability to process data has outstripped our ability to evaluate the quality of that data. Computers are not able to independently generate "truth," nor can they cleanse and purify data. With the recognition that mtDNA studies are incapable of determining the origin of modern humans, biochemists are now turning to nuclear DNA to help them solve the problem. There are also attempts to recover DNA from Neandertals and other fossil humans. More and more, molecular genetics and sophisticated computer programs are being enlisted in the service of evolution. The results are advertised as independent confirmations of evolution when in reality they are not. Molecular studies are the wave of the future for evolutionary studies. This approach is very convincing, because it appears to be so scientific to those who do not recognize its evolutionary presuppositions.

The Bible is a revelation from God. Genesis is part of that revelation. God's revelation is more than just the passing on of information; it is the imparting of truth which humans *could not know* by any other means. The failure of the African Eve theory is just another illustration of the impossibility of constructing an authentic record of human origins by scientific means. It is because we could not know with certainty our origins any other way that God gave us an authentic revelation of our origins in the Book of Genesis.

Back to Neandertal. Since anatomically modern humans existed in Africa and elsewhere well before the Neandertals, "The Neandertal Problem" is still very much an unresolved problem in contemporary paleoanthropology. Like the Cheshire cat that disappeared while its grin remained, Neandertal has disappeared, but his grin remains to taunt evolutionists.

It is almost universally accepted that the Neandertals became extinct—for whatever reason—between 30,000 and 35,000 y.a. (The Neandertal remains from Saint-Césaire, France, dated at 36,300 y.a., are considered to be the most recent Neandertals.)[29] This *terminus ad quem* is very rigidly maintained even though most of the Neandertal remains are poorly dated. The reasons for distancing modern humans from the Neandertals are philosophical. Since the Neandertal problem is still unsolved, the evolutionist must keep his options open. If he eventually decides that the best solution is to derive modern humans from a Neandertal stock, he must allow enough time for that to happen. Even 30,000 years or so is not enough time for the two evolutionary mechanisms—mutation and natural selection—to work their transforming magic.

However, there is evidence that the Neandertals persisted long after their alleged demise. The Neandertal skull known as Amud I from Upper Galilee, Israel, was found as a burial just below the top of layer BI. If Amud I was buried into layer BI, it follows that he cannot be older than Layer BI but could be younger. The radiocarbon date for Upper BI is 5,710 y.a. Michael Day (British Museum—Natural History) states: "These dates are believed to be too 'young' as the result of contamination by younger carbon."[30] While it is certainly true that younger carbon compromises a radiocarbon date, this is also the standard excuse given whenever a radiocarbon date is too young to fit the system. Day gives no evidence that young carbon was present. It is understood by evolutionists that if a radiocarbon date is too young to fit the evolutionary scenario, that is proof enough that the sample was contaminated, since a "good" date would unquestionably fit the scheme. In dating Amud I, it is bad enough that uranium/ionium growth gives a date of only 27,000 y.a. and uranium fission-track gives a date of only 28,000 y.a., with a margin of error of almost 10,000 years. Anything is better than a date of 5,710 y.a. for a Neandertaler.

Some of the Shanidar, Iraq, Neandertal material from Layers C and D give radiocarbon dates as low as 26,500 y.a.,[31] and the Neandertal Banolas mandible, found near Gerona, Spain, gave a radiocarbon date of 17,600 y.a. After recording this date, obtained by the U.C.L.A. radiocarbon laboratory, *Radiocarbon Journal* made the following comment:

> Comment: for a Neanderthal, present date is too recent. The possibility of more modern travertine contaminating older travertine to yield a more recent composite date, or the relocation of an ancient mandible into travertine is open.[32]

Possibilities are given for the too recent date, but no physical evidence is cited to indicate that these possibilities are valid. The arbitrary assertion that the date is too recent for a Neandertal apparently settles the matter.

If there is any legitimacy to these recent dates for Neandertal, it could mean that Neandertal, like his smaller edition known as *Homo erectus*, persisted until quite recently. That would be additional evidence that the differences between Neandertal and anatomically modern humans had nothing to do with the evolutionary process. For evolutionists, the Neandertal problem remains unsolved.

An Alternative Explanation for Neandertal

Whether the Neandertals were in the main line of human descent, or whether they were a side branch that led to extinction, the evolutionist believes that the somewhat different Neandertal morphology was the result of the evolutionary process. The two evolutionary mechanisms are mutation and natural selection; mutations supply the raw material (new information) upon which natural selection can work. Special Creation and evolution are thus mutually exclusive. If God by

Special Creation supplied the genetic information which accounts for the existence of humans, then evolution is not necessary. If random mutations are able to supply new information upon which natural selection works to produce humans out of a nonhuman stock, then the concept of Special Creation is not necessary.

The evolutionist improperly introduces other mechanisms into the alleged evolutionary process, such as the founder principle, geographic isolation, and genetic recombination. While these are legitimate processes, they are not evolutionary processes. They do not create unique new genetic information. Nor do these processes discriminate between Special Creation and evolution. They would apply in either case. The evolutionist smuggles these nonevolutionary mechanisms into the evolutionary process even though they have nothing to do with evolution. These processes do account for variation, but they cannot produce evolutionary changes that result in increased complexity; that would demand the creation of entirely new genetic information.

It is impossible for the evolutionist to demonstrate that the Neandertal morphology was the result of mutation and natural selection. That is only a dogmatic assertion that is part of his belief system. If evidence could be supplied to show that the Neandertal morphology could be achieved by means other than by mutation and natural selection, the concept of human evolution would be called into question, and the concept of Special Creation would be strengthened. This is what we propose to do.

Over the years the scientific literature has suggested a number of conditions—geographical, environmental, pathological, cultural, and dietary—that could produce a Neandertal-like morphology. Richard Klein writes:

The forward placement of Neanderthal jaws and the large size of the incisors probably reflect habitual use of the anterior dentition as a tool, perhaps mostly as a clamp or vise. Such para- or nonmasticatory use for gripping is implied by the high frequency of enamel chipping and microfractures on Neanderthal incisors, by nondietary microscopic striations on incisor crowns, and by the peculiar, rounded wear seen on the incisors of elderly individuals. Similar, though less extensive damage occurs on the teeth of Eskimos, who also tend to use their anterior jaws extensively as clamps.

Biomechanically, the forces exerted by persistent, habitual, nonmasticatory use of the front teeth *could account in whole or in part* for such well-
known Neanderthal features as the long face, the well-developed supraorbital torus, and even the long, low shape of the cranium. Massive
anterior dental loading could further explain the unique Neanderthal
occipitodmastoid region which perhaps provided the insertions for muscles that stabilized the mandible and head during dental clamping.[33]

In two paragraphs, Klein has given a plausible nonevolutionary explanation for most of the unique features of Neandertal morphology.
Just as the hands of a blacksmith develop calluses as a result of the
unique wear and stress they are subjected to, so the facial and skull
morphology of the Neandertals could be the result of the unique
stresses their jaws and teeth were subjected to when used as tools.
Klein also states: "The long, low shape of the Neanderthal cranium
with its typically large occipital bun probably reflects relatively slow
postnatal brain growth relative to cranial vault growth."[34]

In a statement cited earlier in this chapter, Geist also gave a plausible nonevolutionary explanation for the unique Neandertal skull morphology based on his prowess as a hunter:

> If great strength, agility, and precision and speed of bodily movements
> were required for such a hunting technique, those parts of the brain con-
> trolling motor functions in the hunter had to be greatly developed. Ne-
> anderthal possessed a massive cerebellum and motor cortex compared
> to modern humans. This pulled the brain case rearward, creating an oc-
> ciput that reached farther rearward than in modern humans, explaining,
> in part, the large, long, low brain case and bun-shaped occiput of the
> Neanderthals.[35]

Klein also recognized the effect geographic isolation could have on
the development of the Neandertals when he wrote that "some of the
European mid-Quaternary fossils clearly anticipate the Neanderthals, while like-aged African and Asian ones do not. Clearly, the implication is that the Neanderthals were an indigenous European
development."[36]

Health factors can be reflected in the skeleton, especially a vitamin
D deficiency resulting in rickets. J. Lawrence Angel (Smithsonian Institution) writes: "Pelvis and skull base tend to flatten if protein or Vitamin D in diet is inadequate."[37] This was the diagnosis of Rudolf

Virchow, "the father of pathology," when he examined the rather flat-
tened skullcap of the first Neandertal discovery. He was overruled by
those who favored an evolutionary interpretation. In 1970, Francis
Ivanhoe published in *Nature* an article entitled, "Was Virchow Right
about Neandertal?"[38] He presented a strong case based on diagnostic
evidence that the Neandertals were really modern humans who suffered from rickets.

Still another possible explanation of the Neandertal morphology is
disease, especially syphilis. D. J. M. Wright (Guy's Hospital Medical
School, London) observed that "In societies with poor nutrition, rickets and congenital syphilis frequently occur together. The distinction
between the two is extremely difficult without modern biochemical,
seriological, and radiographic aids."[39]

Based upon his examination of the Neandertal collection at the British Museum, Wright found a number of features in the Neandertal's
morphology compatible with congenital syphilis. These conditions are
seen in both adult and child skulls. Wright specifically mentioned the
original Neandertal skullcap as well as the Gibralter II, Staroselé, and
Pech de l'Aze Neandertal remains.

To suggest that all of the above factors contributed to the Neandertal morphology would be a case of overkill. However, it is well within
reason to suggest that one of them or several in combination are responsible for the group of individuals known collectively as Neandertals. Obviously, to perform controlled experiments on humans today
to determine which factors would produce the skeletal features of the
Neandertals would be both immoral and impossible. But in contrast
to the lack of rigorous scientific evidence that mutation and natural selection could produce these effects, there is a sizable body of scientific
data that suggests one or more of the above-mentioned factors would
constitute a reasonable and nonevolutionary explanation for the Neandertal morphology.

When Joachem Neander walked in his beautiful valley so many
years ago, he could not know that hundreds of years later his name
would become world famous, not for his hymns celebrating creation
but for a concept that he would have totally rejected: human evolution.

7

Evolution's Illegitimate Children:
Archaic *Homo sapiens*

A RELATIVELY NEW CATEGORY has been established by paleoanthro-pologists: archaic *Homo sapiens*. This category includes a minimum of forty-nine fossil individuals who do not fit into either the Neandertal or the *Home erectus* categories. The reasons are that (1) they have a somewhat different skull morphology from the Neandertals, (2) many of this group are dated much earlier than the "classic" Neandertals, although more than half of them are Neandertal contemporaries, and (3) they have a cranial capacity that is too large for them to be classified as *Homo erectus*.

In the past, members of this group have been given various names, such as anti-neandertals, pre-neandertals, Neandertaloids, or African Neandertals. Some people have suggested that the entire category be given the scientific name *Homo sapiens rhodesiensis*, after the best-known fossil in this category, Rhodesian Man. While the original purpose of this category was to distinguish these fossils from the Neandertals (with the differences between this group and the Neandertals being emphasized) the trend now is to suggest that this group has a morphological relationship to the Neandertals and to *Homo erectus*.

Archaic *Homo sapiens* Fossils
(Evolution Time Scale)

+ May be Neandertal	T-Tools/artifacts in association
x May be *Homo erectus*	F-Evidence of use of fire
* Date very uncertain	( ) Min. no. of fossil individuals

5,000 y.a.		Cape Flats skeletal fragments, South Africa[1, 2] (3) T
		Dire Dawa mandible, Ethiopia[3] (1) T
	+	Mugharet el 'Aliya (Tangier) maxilla, tooth, Morocco[4, 5, 6] (2) T
		Eliye Springs skull, Kenya[7] (1)
		Dar-es-Soltan Cave 2 skulls, teeth, Morocco[8, 9] (2) T
	*	Haua Fteah partial mandibles, Libya[10] (2) T
		KNM-ER 999 femur, Kenya[11] (1)
100,000 y.a.	+	Krapina D skull, Yugoslavia[12] (1) T
		Florisbad partial cranium, South Africa[13] (1) T
		La Chaise skeletal fragments, teeth, France[14] (2+) T
		Klasies River Mouth Caves skeletal fragments, South Africa[15, 16] (2+) T F
		Laetoli Hominid 18 skull, Tanzania[17] (1) T
	x	Omo-Kibish 2 skull, Ethiopia[18] (1)
		Le Lazaret skull fragment, teeth, France[19] (2) T
		Azych (Azykhskaya Peshchera) mandible, USSR[20] (1) T
		Montmaurin mandible, teeth, vertebra, France[21] (1) T
200,000 y.a.		Jebel-Irhoud skulls, mandible, Morocco[22] (3) T
		La Chaise skeletal fragments, teeth, lower level, France[14] (2+) T
		Casal dé Pazzi parietal, Italy[49] (1) T
		Eyasi cranial fragments, Tanzania[23] (2) T
	x	Bilzingsleben skull fragments, Germany[24] (1) T F
		Rhodesian (Broken Hill/Kabwe skull, Zambia[25] (1) T
300,000 y.a.		Castel di Guido skeletal fragments, Italy[26] (1) T
		Saldanha (Hopefield/Elandsfontein) cranium, South Africa[27] (1) T
		Atapuerca (Ibeas/Cueva Mayor) skeletal fragments, Spain[28] (2+)
	*	Garba 3 (Melka-Kunturé) cranial fragments, Ethiopia[29] (1)
		Swanscombe skull, England[30] (1) T
400,000 y.a.		Steinheim skull, Germany[31] (1)
	x	Bodo skulls, Ethiopia[32] (2) T
		Arago 21 partial skull, skeletal fragments, France[33] (2+) T
500,000 y.a.	x	Vertesszöllös skull fragment, teeth, Hungary[34] (1) T F
		Ndutu skull, Tanzania[35] (1) T
600,000 y.a.		
		Pertalona skull, Greece[36] (1) T F?
700,000 y.a.		Mauer (Heidelberg) mandible, Germany[37, 38, 39] (1) T

The shift in thinking on the affinities of this group of fossils is documented by David Pilbeam. In 1978, the consensus view in paleoanthropology was that anatomically modern *Homo sapiens* and archaic *Homo sapiens* exhibited "little change from preceding phases," that is, *Homo erectus*. In 1984, the consensus was that anatomically modern *Homo sapiens* was "markedly different from archaic *Homo sapiens* both

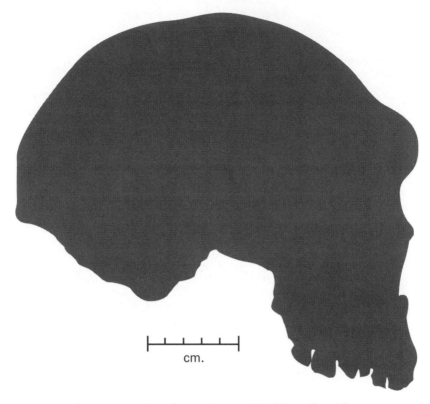

Rhodesian Man Skull, Broken Hill Mine, Zambia
archaic *Homo sapiens*

cranially and subcranially," while archaic *Homo sapiens* is "little different from *Homo erectus*."[40]

In the most general sense, the archaic *Homo sapiens* category can be
described as follows: (1) a low, sloping skull with a cranial capacity of
from approximately 1100 to 1300 cc, (2) very heavy ridges over the
eyes, especially true in Rhodesian Man, (3) the rear of the skull more
rounded and lacking the Neandertal "bun," (4) large, long faces with
jaws that jut forward, and (5) postcranial (body) bones that are essentially either indistinguishable from modern humans or very different
from modern humans, depending upon which paleoanthropologist is
doing the describing.

In reality, the taxon "archaic *Homo sapiens*" is a taxon of conve-
nience. A heterogeneous assortment of fossils was dumped into this
category because the fossils did not fit elsewhere. By making a taxon

of this assortment, evolutionists attempted to give it the appearance of a transitional group between *Homo erectus* and Neandertal or between *Homo erectus* and anatomically modern *Homo sapiens*. However, the dating of these fossils by evolutionists themselves reveals that this is not a true picture. These fossils date from the most recent ones at 5000 y.a. all the way back to the oldest ones at 700,000 y.a. on the evolution time scale (see chart). This chronological spread reveals that these fossil individuals were not part of an evolutionary sequence but were instead contemporaries of *Homo erectus*, Neandertal, and anatomically modern *Homo sapiens*. The dating and morphology of this category help to falsify the concept of human evolution and serve to demonstrate the wide degree of skeletal and cranial diversity that is found in the human family.

Ian Tattersall testifies that these fossils present a problem to the evolutionist who tries to classify them. With delicate understatement he writes:

> The hominid fossil record of the past 300-400 kyr offers a remarkable degree of morphological variety. Yet (late-persisting *Homo erectus* aside), conventional wisdom assigns all these fossils to *Homo sapiens*, albeit of "archaic" varieties.[41]

William Howells, commenting on the lack of formal definition for this fossil category, states:

> Others have agreed that *H. sapiens* should indeed include a number of specimens or populations that were not anatomically modern but that for various reasons (brain size, loss of *erectus* occipital or temporal traits) could hardly be included in *H. erectus*. They are usually given the slightly evasive appellation of "archaic" *H. sapiens*, and in fact seem to be accepted as such simply on the basis of not being *H. erectus*.[42]

Richard Klein doesn't even attempt to give a description of the archaic *Homo sapiens* fossils as a group. He confesses:

> Morphologically, the fossils involved are too variable, even within restricted geographic regions, for summary description.

> Because they are difficult to characterize as a group, they will be treated individually here, according to continent.[43]

As Klein begins to describe each of the individual fossils, he then makes the following statement:

> Readers who are less concerned with the nature of the early *Homo sapiens* sample than with the conclusions that can be drawn from it may wish only to skim the next three sections, attending mostly to the accompanying illustrations.[44]

It's as if Klein wishes to spare the reader the pain of exposure to one of the major anomalies in the concept of human evolution by having him skim over the evidence and proceed to the security of the "party line," where, in the conclusion, the reader is assured that everything is all right.

The Skull Size Argument

In seeking to establish the concept of human evolution, the evolutionist leans heavily on skull morphology and, to a lesser degree in recent years, on skull size. Both are spurious arguments and prove nothing. Typical of the charts and illustrations used by evolutionists is a display at the American Museum of Natural History in New York City. It is titled "Increasing Brain Size" and shows an increase in brain sizes as follows:

Increasing Brain Size
Homo sapiens	1450 cc [cubic centimeters]
Neanderthal	1625 cc
Pithecanthropus [*Homo erectus*]	914 cc
Australopithecinae	650 cc
Gorilla	543 cc
Chimpanzee	400 cc
Gibbon	97 cc

(Bracketed material added for clarity.)

The obvious question is, What is the purpose of this display? or, What does this display say? The obvious answer, since it is a part of the museum's display on "The Evolution of Man," is to show that the hominid brain has enlarged by evolution over time. However, no evolutionist in the world—past or present—believes that it happened in the way the chart implies it did. No evolutionist believes that evolution went from gibbon to chimpanzee to gorilla to the australopithecines to *Homo erectus* to Neandertal and then to modern humans. How many times have I been in the classroom or the lecture hall and heard the speaker attempt to allay the fears of the uninitiated by assuring them that evolutionists do *not* believe that we came from monkeys or apes. They assure us that we came from some transitional form that was the ancestor of both humans and living primates. (The fact that that transitional form—if it ever existed—would readily be called an ape by anyone who saw it was admitted by the famous evolutionist George Gaylord Simpson.) The museum display is an absurd mixing of past and present forms having no relationship to what evolutionists themselves teach. It is a cheap form of propaganda, Madison Avenue style, to convince the uninformed public of the "truth" of evolution.

Although that chart was still in the American Museum as of 1991, that type of illustration is not seen as much in recent years. We now know that relative brain size means very little. The relationship between brain size and body size must be factored in, and the crucial element is not brain size but brain organization. A large gorilla brain is no closer to the human condition than is a small gorilla brain. The human brain varies in size from about 700 cc to about 2200 cc with no differences in ability or intelligence. That variation, more than a factor of three, is an incredible difference in size variation but indicates no difference in quality. Those brain-size charts are meaningless. Yet, the idea of increasing brain size has been injected into the human thought-stream so effectively by evolutionists that most nonspecialists still think of it as significant evidence for evolution. Those charts look so impressive that some evolutionists cannot resist the temptation to continue to use them. The fact that the archaic *Homo sapiens* fossils (those that can be measured) had a brain size intermediate between *Homo erectus* and Neandertal has no significance. The fossil record, as seen by the charts in this book, shows that these individuals lived side by side with fossil individuals in other categories as members of the human family.

The Skull Morphology Argument

Nothing illustrates the futility of basing an evolutionary sequence on skull morphology more than does the skull of Rhodesian Man, probably the best known of all of the archaic *Homo sapiens* fossils.

Rhodesian Man was so named because he was found in 1921 in what was then known as Northern Rhodesia, now Zambia. The fossil is also called Broken Hill Man (after the mine in which he was found), or Kabwe Man (after the city near which he was found). Because the browridges on this fossil skull are more pronounced than those found on any other human fossil, no human fossil appears to be more "primitive," "savage," or "apelike" than does Rhodesian Man. Yet, his brain size of 1280 cc is so large that the fossil demands to be classified as *Homo sapiens*. We need to be constantly reminded that there is nothing in the contours of the skull of an individual that gives clues as to his degree of civilization, culture, or morality.

An indication that Rhodesian Man is not a unique individual is seen by the fact that the Saldanha, South Africa, skull (known also as the Hopefield or Elandsfontein skull) has almost the same "savage" morphology. Although this fossil is considered to be older than Rhodesian Man, there is no question that it also is truly human.

Rhodesian Man had been dated at about 40,000 y.a. Richard Klein gives the newer date as between 200,000 and 400,000 y.a.[45] Yet, there is reason to believe that the fossil is actually quite recent in age. The original 1921 report in *Nature*, telling of its discovery, says: "The skull is in a remarkably fresh state of preservation, the bone having merely lost its animal matter and not having been in the least mineralised."[46]

It is difficult to understand why a fossil buried for 200,000 to 400,000 years (or even 40,000 years) would have no mineralization whatsoever. That fact suggests that the fossil could be quite recent in age, in spite of its "savage" appearance.

Perhaps the most remarkable feature of this fossil is that it was found about sixty feet underground at the far end of a shaft in a lead and zinc mine.[47] The skull was found with the remains of two or possibly three other individuals. The maxilla (upper jaw) of one of those other individuals is considerably more modern in morphology than is Rhodesian Man, indicating a large degree of genetic variation within this small group. The associated postcranial bones are all very modern in appearance.

Found under other circumstances, Rhodesian Man, with his low cranial doming and very heavy browridges, could serve as an excellent illustration of an evolutionary transitional form between apes and humans. The original *Nature* report states: "Its large and heavy face is even more simian [apelike] in appearance than that of Neanderthal man. . . ."[48] Yet, this individual was either mining lead and zinc himself or was in the mine shaft at a time when lead and zinc were being mined by other humans. This smacks of a rather high degree of civilization and technology.

It is amusing that many evolutionists, when reporting on the details of Rhodesian Man, say that he was found in a cave. Technically, I suppose, they are right. A mine shaft is just a cave, of sorts, in the same way that diamonds and emeralds are just pebbles. One wonders if this is a crude attempt to minimize the technical abilities of ancient humans. The Book of Genesis clearly confirms the advanced culture and technology of the ancients, specifically mentioning metallurgy in Genesis 4:22 and music in Genesis 4:21.

In spite of the obvious lesson to be learned from the Rhodesian and Saldanha skulls, evolutionists continue to base much of their evidence for human evolution on the alleged primitive-to-advanced contours of fossil skulls. Creationists maintain that in light of the evidence of the wide genetic diversity in the human family, skull contour is an inadequate basis for determining relationships. Evolution's illegitimate children, the archaic *Homo sapiens* fossils, give eloquent testimony to that fact.

8

Java Man:
The Rest of the Story

THE STORY OF THE GROUP OF FOSSILS called *Homo erectus* by evolutionists begins with Java Man. Java Man! Breathes there a man or woman with soul so dead who has not heard of Java Man? Java Man is like an old friend. We learned about him in grade school. They called him the ape-man and told us that he was our evolutionary ancestor. The artist's drawings of that beetle-browed, jaw-jutting fellow were quite convincing. In fact, the vast majority of people who believe in human evolution were probably first sold on it by this convincing salesman. Not only is he the best-known human fossil, he is one of the only human fossils most people know.

The story has been told many times. Before the turn of the century, a Dutch anatomist, Eugene Dubois, went to the Dutch East Indies (now Indonesia) in search of the "missing link" between apes and humans. In 1891, along the bank of the Solo River in Java, he found a skullcap that seemed to him to have a combination of human and ape features. A year later, about fifty feet away, he found a thigh bone (femur), very human in appearance, that he assumed belonged with the skullcap. Dubois named his "transitional form," which is now dated by evolutionists at about half a million years, *Pithecanthropus erectus*.

However, to the general public he will always be known affectionately as Java Man.

Because Java Man was found so long ago, and because another like him (Peking Man) was not found until 1929, and because he was made to fit the evolutionary picture so beautifully, Java Man is virtually synonymous in the popular mind with human evolution.

Pithecanthropus erectus

Pithecos is the Greek word for "ape." *Anthropos* is the Greek word for "man." *Erectus* means "erect." Hence, the name means erect ape-man. First applied to the original Java Man by Dubois, the term was later given to similar fossils found elsewhere in Java. The term is now obsolete. All these fossils have been placed in the category *Homo erectus* (erect man).

For more than twenty years I have researched Java Man. It has been a frustrating task because basic information is in very short supply. Historians of science have paid little attention to Dubois or his methodology. I have long felt that an authoritative biography of Dubois would help answer some of the basic historical questions, but no such work is available in either Dutch or English. However, the situation has been somewhat corrected. The work by Bert Theunissen, *Eugene Dubois and the Ape-Man from Java*,[1] published in the Netherlands, brings to light information that has hitherto been unavailable to most researchers.

My conclusions on Dubois and Java Man are as follows: (1) Java Man is not our evolutionary ancestor but is a true member of the human family, a post-Flood descendant of Adam, and a smaller version of Neandertal; (2) Dubois seriously misinterpreted the Java Man fossils, and there was abundant evidence available to him at that time that he had misinterpreted them; (3) the evolutionists' dating of Java Man at half a million years is highly suspect; (4) more modern-looking humans—possibly including Wadjak Man—were living as contemporaries of Java Man; and (5) Java Man was eventually accepted as our evolutionary ancestor in spite of the evidence because he could be interpreted to promote evolution. The historical and scientific questions regarding Java Man are as legitimate today as they were when the fossils were first discovered. This chapter will cover the problems concerning the geology and dating of Java Man.

Accurate dating is essential to the proper interpretation of a fossil. Since Dubois claimed that Java Man was *the* missing link between apes and humans, he had to show that it dated at the appropriate time when a certain ape stock was allegedly evolving into humans. If Java Man were rather recent in date, as may well be the case, he could not serve as an evolutionary transitional form because modern humans were already on the scene at that time.

Dubois claimed that the skullcap and the femur came from a rock stratum known as the Trinil layer, named after a nearby village in central Java. He believed that these rocks were below what is known as the Pleistocene-Pliocene (Tertiary) boundary. Dubois was convinced that "real" humans evolved later in the Middle Pleistocene. Hence, his dating of Java Man was quite appropriate for a missing link. However, his interpretation was not exactly straightforward, as the man who later found other Java men, G. H. R. von Koenigswald, tells us:

> When Dubois issued his first description of the fossil Javanese fauna he designated it Pleistocene. But no sooner had he discovered his *Pithecanthropus* than the fauna had suddenly to become Tertiary. He did everything in his power to diminish the Pleistocene character of the fauna, ...
>
> The criterion was no longer to be the fauna as a whole, but only his *Pithecanthropus*. Such a primitive form belonged to the Tertiary!

Pleistocene Epoch

The Pleistocene Epoch is the geological term for the period of the Ice Age or Ages. It is allegedly the time when the bulk of human evolution took place. The Pleistocene is followed by the Holocene Epoch and is preceded by the Pliocene Epoch. Pleistocene dates have been the subject of much debate, especially on the lower end. The date of the Pleistocene/Pliocene boundary was agreed upon by the geological community in about 1980. The generally accepted dates are as follows:

Holocene Epoch	Present to 10,000 y.a.
Upper Pleistocene	10,000 to 100,000 y.a.
Middle Pleistocene	100,000 to 700,000 y.a.
Lower Pleistocene	700,000 to 1.7 m.y.a.
Pliocene Epoch	1.7 to 5.5 m.y.a.

y.a. means "years ago." m.y.a. means "million years ago."

Dubois' view . . . did not go uncontested. But there was no getting at him until he had described his whole collection and laid all his cards on the table. That was why we all had to wait for a study of his finds, and to wait in vain.[2]

The first problem with the geology and dating of Java Man is that Dubois was not qualified to make those determinations. In 1877 Dubois entered the University of Amsterdam as a medical student. He soon lost interest in becoming a practicing physician and majored in anatomy. After graduation he taught anatomy and worked in an anatomical laboratory. In 1884, he also qualified as a medical doctor. Years later he was quoted as saying, "I am always described as a doctor. But I am an anatomist."[3] Actually, he was qualified to be both, but the point is he was not a geologist. He had an interest in geology and paleontology as a child and may have studied them a bit in high school. But there is no record that Dubois had any formal training in geology or paleontology before he went to Java.

It is true that in 1899 the University of Amsterdam appointed him Professor Extraordinary of Geology, that in 1907 it made him a full professor, and that he taught in that area until 1921. However, as far as can be determined, all of his formal training in geology was done after he returned from Java, not before. While in Java Dubois was an amateur geologist at best.

Not only was Dubois not qualified to determine the date and location of the Pliocene and Pleistocene deposits in Java, but the geology and paleontology of Java were virtually unknown. Dubois himself said that the Dutch East Indies were practically virgin territory in this regard.[4] The entire science of geology was still in its infancy, and most of the geological work that had been done was done in Europe. Even there, the date of the Pleistocene/Pliocene boundary had been estimated to be all the way from 100,000 y.a. by Hein to 800,000 y.a. by Lyell. If geology was hardly a precise science in Europe, it certainly was not in Java. Fifty-seven years after Dubois made his discoveries, the famous paleoanthropologist Alan Houghton Brodrick, writing in 1948, said that the stratigraphy of the Trinil beds still was not clear.[5]

If Dubois's lack of expertise in geology, together with a general lack of geological information about Java, wasn't enough of a problem, another factor complicates the picture: the way Dubois collected his fossils. Today it is understood that when an important fossil is discovered,

the utmost care is taken to document the precise rock layer in which the
fossil is located. This is absolutely imperative for both relative and absolute dating of the fossil. The inability to locate a fossil in its geological
context is enough to disqualify it from serious consideration.

Dubois had gone to the Dutch East Indies as an army doctor and
was then given permission to search for fossils. To help him in his
work, Dubois was assigned two corporals from the engineering corps,
who acted as supervisors, and fifty forced laborers. Dubois was usually
at his headquarters, but he made periodic visits on horseback to the
digging sites. He also maintained written contact with the engineers
about the progress of the work. When fossils were found, they were
sent to Dubois for preparation and provisional identification. Dubois
himself did not uncover any of the important fossils ascribed to him,
and he never saw any of them *in situ* (except Wadjak II). He was entirely dependent on his two engineers to determine the position of the
fossils in the deposits—engineers who knew even less about geology
than he did.[6] (Brodrick questions that Java Man came from the Trinil
beds, suggesting that they came from higher up in the rock layers,
which would make them more recent in age.)[7] By today's standards,
these fossils would have been disqualified. Yet, they became for many
years the primary evidence for human evolution.

Perhaps it was Dubois's lack of expertise in geology, together with
the fact that he was not on the scene when the fossils were excavated,
that is responsible for a peculiar characteristic of his reports: a lack of
geological information. His reports always centered on the anatomical
aspects of his fossils, which is not surprising, since he was an anatomist. When the skullcap and the thigh bone were discovered, Dubois
spread the word officially and widely. He gave a rather detailed preliminary description of the fossils themselves, but he gave only the briefest
account of the locality and the geological circumstances surrounding
the discovery. His dating of *Pithecanthropus* was based on the mammalian fossils found with *Pithecanthropus*, but his proof consisted of only
sketchy tidbits of information. It wasn't until late 1895, after he had
returned from Java, that he displayed the first profile drawing and
maps of the excavation site. The scientific community was continually
frustrated because of his lack of geological information. Dubois had
promised that he would publish a comprehensive work on *Pithecanthropus erectus* and the Javanese fossil fauna. It was many years before

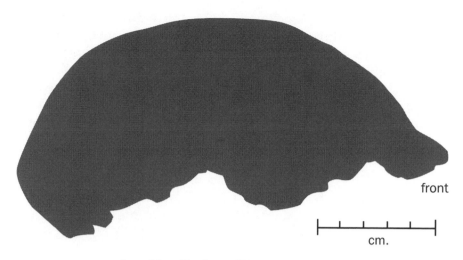

front

cm.

Java Man Skullcap (*Pithecanthropus* I)
Homo erectus

the report on *Pithecanthropus* was published, but he never did produce the promised report on the Javanese mammalian fauna.

Up until 1900, Dubois had been very active in promoting Java Man as the missing link and had allowed full access to the fossils. After 1900, he withdrew completely from the public debate for twenty years, published very little about the fossils, and refused to allow anyone to see them. The reason usually given for this behavior is that Dubois wanted Java Man to be accepted as *the* missing link. Because of the initial controversy over his interpretation, he retaliated by refusing access to the fossils. From what we know of Dubois's personality, this explanation is possible.

Theunissen suggests another explanation.[8] Dubois was very proud, deeply suspicious, and fiercely independent. In 1899, Schwalbe produced a significant publication on the skullcap that far overshadowed anything Dubois had done. Dubois must have sensed that if he allowed continued access to the fossils, others would do the work and get the recognition that rightfully should be his. Hence, the embargo on allowing others to see the fossils. However, Dubois continued to drag his feet. In 1907 and 1908 he published some rather superficial articles on some of the Javanese fossil fauna. His main reason for publishing seems to be that another expedition was working at Trinil at that time, the Selenka Expedition, and Dubois did not want them to steal his thunder. It was not until 1924, thirty-three years after the skullcap had been discovered, that Dubois published a definitive paper on it. Two years later he published a major paper on the thigh bone.

Why Dubois procrastinated on matters so vital to the scientific community cannot be answered with finality. Part of the answer might be found in a comment by one of his students: "Dubois had the habit of just lifting a corner of the veil of a scientific conception, but he was loath to settle down and work it out thoroughly."[9]

9

Java Man:
Keeping the Faith

GEOLOGICAL PROBLEMS were not the only ones swirling around the skullcap of Dubois's Java Man. There was also the problem of identifying that unusual individual. Was it human, an ape, a man-ape, or an ape-man? No one knew. To explain the controversy at the turn of the century, I must first transport you back in time.

To step back into the nineteenth century would cause us culture shock in many ways. One of the shocks would come when we realized what was commonly believed about creation. Before evolution became popular, creation was the accepted scientific model of the universe and of humans. However, it was a type of creationism that few of us would recognize and no biblical creationist did or would endorse.

The concept was known as the Great Chain of Being,[1] patterned not after Moses but after Plato. According to this concept the Almighty had created a great ladder or chain of living things, from single-celled organisms all the way up to humans, each organism being a bit more complex than the one below it. All of nature fit into this ascending organizational scale. Like the keys on a piano, each organism was discrete but a bit higher in the organization and more complex than the one below it. Just as there are no missing keys on a piano, there could be no spaces or gaps in this ladder (no extinction). This Great Chain of Being looked much like an evolutionary progression, but it was static. Each organism was created by the Almighty in its particular slot and did not evolve upward.

Since any imperfections in nature would be a reflection upon the work of the Almighty, it was further believed that nature was perfect. This idea was obviously in conflict with the biblical account of the fall of man into sin (Genesis 3) and its results on nature. There was the further implication that extinction was impossible because that would indicate that the Almighty did not take proper care of his creatures after he had created them. These ideas were in conflict with the Noachic Flood in Genesis 6-9, where the whole purpose of the Flood, as a judgment from God, was to destroy the world as it then was (2 Peter 3:6). It is obvious that in the Great Chain of Being we are dealing not with biblical concepts but with pagan Greek philosophy.

With the discovery of fossils that showed the existence of creatures in the past that were different from those we know today, some people believed that those fossils were either naturally formed objects, fakes, or "sports" (objects placed in the earth by the Almighty to fool us or to test our faith.) Later, when it could no longer be denied that fossils were legitimate remains of past life, some of which was extinct, some people rejected the whole concept of creation. This misunderstanding of the biblical account of creation helped pave the way for the acceptance of evolution as the only viable alternative.

It is not unusual for people to begin with a wrong idea of what the Bible teaches, reject that view, and then reject the entire Bible because "the Bible is unscientific." This is what Darwin did. Most people today believe that Darwin disproved biblical creationism and proved evolution. The Darwinian Revolution, one of the most significant revolutions of all time, is generally thought to be the establishment of the concept of evolution on a solid, empirical base. Not so. In the words of Harvard biologist Ernst Mayr, the Darwinian Revolution was actually a philosophical revolution from a theistic worldview to a worldview in which God was not involved in any way.[2]

Darwin did not reject biblical creation; he knew nothing about it. Even though he studied for the ministry at Cambridge, it is obvious from his writings that he did not have a clue as to what the Bible actually taught regarding Special Creation. Darwin heavily criticized Special Creation in *The Origin of Species*. He claimed that the imperfections of nature demonstrated that a wise and all-powerful God could not have done such a sloppy job. He seemed oblivious to the fact that those imperfections of nature were the result of the Fall, and that the world is not now the way God originally made it.

Darwin's abysmal ignorance of what Genesis teaches is seen in that it was not until he was fifty-two years old, two years after he published the *Origin*, that he realized that the much ridiculed date of 4,004 B.C. for creation was not a part of the text of Genesis but was instead the work of Archbishop James Ussher, who lived from 1581 to 1656.[3] (Creationists no longer accept Ussher's date for creation.) Darwin first rejected an unbiblical, philosophical creationism and then rejected biblical creationism as well. It is a classic case of throwing out the baby with the bath water. The philosophy of nature he then developed had as its cardinal principle that it is "unscientific" to believe in supernatural causation. Darwin's purpose was to "ungod" the universe.

The concept of the Great Chain of Being by its very nature became a "setup" for evolution. All one had to do was to change that static chain to a dynamic one, with the forms gradually evolving upward from one into another, and one had the basic evolutionary scenario. That philosophical preparation explains why evolution was accepted so rapidly after the publication of Darwin's *Origin*.

The Great Chain of Being was responsible for even more mischief. It allowed for the endorsement of slavery. When the nations of Africa and the East were opened up and world trade routes developed, western Europe learned about the many "savage" tribes that inhabited large portions of the earth. The differences in culture and language of these "savages" was proof to the chauvinistic western Europeans that these strange peoples were inferior races. The "savages" were fitted into the Great Chain of Being above the apes and below the Europeans. There was no evolutionary significance in that placement. Europeans believed that the Almighty had created the "savages" as true humans but as inferior races. Hence, since the Almighty had created them as inferior races, it was proper for the superior races of western Europe and the United States to keep them in their place; that had been ordained by the Almighty. Some even went so far as to claim that the Almighty created these inferior beings without souls, to be used by the superior races much as they would use domestic animals.

To justify this outrageous idea, some even appealed to the Bible. They claimed that there was a curse upon Ham, one of the sons of

Noah, and that Ham was the father of the blacks (Gen. 9:25-27). However, even a superficial reading of those verses shows that God's judgment was not on Ham but on Canaan. Ham had four sons (Gen. 10:6): Cush, Mizraim, Put, and Canaan. Cush (Ethiopia) is the biblical term for the black race. Mizraim is the common Hebrew word for Egypt. Put refers to Libya. Canaan was the father of the Canaanites who settled in Palestine and in North Africa around ancient Carthage. God's judgment was upon the Canaanites (who were white) and was subsequently carried out. God has the right, as God, to judge nations as well as individuals. He has done so in the past and he will do so in the future, but it is done for deeds, not race. There is absolutely no basis in the Bible for racism. The Bible clearly declares that all humans are made in the image of God (Gen. 1:26, 27; 9:6), and Paul declares that "From one man he made every nation of men" (Acts 17:26).

It was against this background of the Great Chain of Being and its pagan idea of inferior races that the first Neandertal discoveries and Java Man were interpreted at the turn of the century. Many, if not most, in the anthropological community saw the Java Man skullcap as truly human.[4] Those anthropologists did not know then of the tremendous genetic variability that exists in humans as we know it today. Yet, they did not consider Java Man to be a transitional form. They considered him to be truly human and able to reproduce with any other member of the human family. The only qualification was that they considered the owner of that skullcap as racially inferior, in the same way that they characterized Africans, Southeast Asians, and native Australians at that time.

Many anthropologists at the turn of the century also noted the remarkable similarity in shape between the Java Man skullcap and the Neandertal skulls, which would indicate that Java Man and the Neandertals, although differing in size, represented the same race of humans. Up until his return from Java, Dubois had ignored the Neandertal remains because he felt they were pathological. With the discovery of more Neandertal material, Dubois was forced to reconsider them, and even he admitted the close resemblance between his own Java Man and the Neandertals.

One of Dubois's earliest papers on Java Man was presented in 1895 in Dublin shortly after his return from Java. It was entitled, "On *Pithecanthropus erectus*: a Transitional Form Between Man and the Apes." In it Dubois emphasized what was to be his theme for the rest of his life: Java Man was not a human being, nor was he an ape; he was a true intermediate or transitional form possessing features of both humans and apes.

Sir Arthur Keith, the famed Cambridge University anatomist, was asked to comment on Dubois's paper. He replied that the chief question to be settled was whether or not the skullcap was human. In answering that question, one had to determine the criterion of a human skull versus an ape skull. To his mind there were two basic differences: first, the very large cranial capacity of human skulls as compared to ape skulls, and second, the large muscular ridges and processes, connected with the chewing apparatus, which ape skulls have compared to human skulls. On both points Keith declared that the Java Man skullcap was distinctly human.[5] The cranial capacity of the anthropoid apes never exceeds 600 cc and averages about 500 cc. On the other hand, the cranial capacity of Dubois's Java Man was estimated at 1000 cc, which is well within the range of humans living today.

Sir William Turner, also responding to Dubois's paper, said that since the Java Man thigh bone had a pathological growth on it, possibly the skullcap was pathological also. He pointed out that in the Edinburgh University Museum is the cast of a microcephalic woman with a frontal flattening very much like that of Java Man.[6]

The main question about the skullcap of Java Man was, Is it human? However, there was no question about the nature of the thigh bone, found a year after the skullcap and fifty feet from it. From the time of its discovery, virtually every authority except Dubois felt that it was indistinguishable from the modern human femur. The great question on the femur was, Did it belong with the skullcap? It seemed far too modern in morphology to be associated with the rather archaic shape of the skullcap.

It was here that Dubois's weakness in geology and the shortcomings of his methodology put him on the defensive. He did none of the actual digging, nor was mapping done at the time of excavation. All quadrant maps and diagrams were made after the fact. In Dubois's first report of his find to Dutch authorities in August 1892, he stated that the femur was located ten meters from the place where the skullcap had been found a year earlier. In early September he changed this figure to twelve meters. His official report later that same month, and his official publication on *Pithecanthropus*, mentions fifteen meters. In a much later 1930 publication, he again mentions twelve meters.[7]

One of the most amazing facets of the Java Man saga is this: In all the years of the twentieth century, the skullcap and the femur together have been presented to the public as Java Man. Yet, the association of the skullcap with the femur has *always* been questioned by the most respected anatomists from the time of their discovery until today.

In 1938, Franz Weidenreich described several femoral fragments of Peking Man. (Both Peking Man and Java Man are now called *Homo erectus*.) Whereas the skulls of Peking Man and Java Man were quite similar, the Peking Man femora differed from the Java Man femur in the very places where the Java Man femur was similar to modern humans. Since the association of the Peking Man skulls and femora was undisputed, Weidenreich concluded that the Java Man femur was not a true *Homo erectus* femur but was instead a modern one.[8]

The most recent assessment of the Java Man femur comes to the same conclusion. Michael Day and T. I. Molleson compared the Java Man femur, the Peking femora, and the femur known as Olduvai Hominid 28 (OH-28) found by Louis Leakey in Olduvai Gorge, Tanzania, in unquestioned association with other *Homo erectus* material. They state that OH-28 and the Peking Man femora, although truly human, are much more similar to each other than either is to the Java femur. Their conclusion is that OH-28 and Peking Man represent a *Homo erectus* anatomy, whereas the Java femur is more modern.[9]

Here, then, is the problem faced by evolutionist paleoanthropology. If the Java skullcap and femur actually belong together, then it is difficult to maintain a species difference between *Homo erectus* and *Homo sapiens*. The distinction would be an artificial one, and it would compromise these fossils as evidence for human evolution. If, on the other hand, the skullcap belongs to *Homo erectus*, and the femur belongs to *Homo sapiens*, it shows that these two forms likely lived together as contemporaries. It likewise removes these fossils as evidence for human evolution, because fluorine analysis indicates that the fossils are both the same age.[10]

Recent suggestions are that the femur came from a much younger stratigraphic horizon than did the skullcap.[11] But this naive attempt at time separation—one hundred years after the fact—reveals the awkwardness evolutionists feel regarding these fossils. This suggestion comes not from the physical evidence at the scene but from a transparent attempt to salvage Java Man as evidence for human evolution.

The Java Man skullcap and femur are evidence that the distinction between *Homo erectus* and *Homo sapiens* is an artificial one, that these two forms are both truly human, and that they lived as contemporaries. The differences attributed to evolution are instead evidence of the wide genetic variation found in the human family.

For the rest of his life Dubois maintained that the skullcap and the femur belonged together and that *Pithecanthropus* was unique. When fossils similar to his Java Man were found, Dubois rejected the evidence out of hand. He labored hard to find tiny areas where his fossils differed from anything else that had been found so as to defend the uniqueness of his discovery.

G. H. R. von Koenigswald wrote of Dubois: ". . . on this point he was as unaccountable as a jealous lover. Anyone who disagreed with his interpretation of *Pithecanthropus* was his personal enemy."[12] Von Koenigswald should know. He made the unfortunate mistake of finding a number of fossils very much like Dubois's original Java Man in—of all places—Java.

Both evolutionists and creationists report that towards the end of his life Dubois renounced *Pithecanthropus* as the transitional form and declared that it was just a giant gibbon. Although this report is not true, Dubois himself is responsible for the misunderstanding. To distance *Pithecanthropus* from similar fossils that were later discovered in both Java and China, Dubois so emphasized what he believed to be gibbonlike characteristics of *Pithecanthropus* that others thought he was saying that *Pithecanthropus* was just a giant gibbon. However, to the end of his life Dubois "kept the faith," believing that his beloved *Pithecanthropus* was uniquely *the* missing link.[13]

10

Wadjak Man:
Not All Fossils Are Created Equal

CHANCES ARE you have not heard of Wadjak Man (now spelled Wajak). Whether or not that is a significant omission in your life, you alone must decide. If you decide that it is, you can blame Eugene Dubois for your ignorance of Wadjak.

They called Ronald Reagan the "Teflon" president because the adversities that would have tarnished the popularity of a lesser man did not seem to stick to him. If Reagan was the Teflon president, Dubois was certainly the Teflon paleoanthropologist. Neither his sins of commission nor his sins of omission seemed to stick to him. One of his sins of commission was his handling of Wadjak Man. It was so serious that it should have jeopardized his credibility as a scientist. Instead, his fellow scientists excused it.

When Dubois went to the Dutch East Indies, he first went to Sumatra, another large Indonesian island northwest of Java. He found animal fossils there but no hominids. He was planning to search in Java when he heard of a remarkable discovery that confirmed the wisdom of those plans. B. D. van Rietschoten, a Dutch mining engineer who was looking for marble, had discovered a fossil human skull near the village of Wadjak, on the south coast of east Java, in 1888. Van

Rietschoten sent the skull to the curator of the Natural Science Museum in Batavia (now Djakarta), and he in turn sent it on to Dubois. Dubois was delighted. The fossil is now known as Wadjak I. Upon his arrival in Java, Dubois found a second skull at the same site. Known as Wadjak II, it is not quite as complete as Wadjak I. Later he found fragments of a skeleton there.

The same lack of precision plagues the geological details of this site as it did that of *Pithecanthropus*. Theunissen calls the site a rock shelter and later calls it a cave.[1] Von Koenigswald wrote that the fossils were found in a cave opened up in the Wadjak marble quarry.[2] But Sir Arthur Keith wrote that the fossils were found in one of the terracelike deposits on the limestone bluffs on the shoreline of an ancient freshwater lake.[3] Marble is a metamorphosed limestone, so it is not unusual for limestone and marble to be found together. But the accounts do seem to be contradictory. If they can be reconciled, the solution is not readily apparent. Carleton Coon has stated: "As the site has since been destroyed by quarrymen, the exact date may never be known."[4] And it is the date that is of vital importance.

The Wadjak skulls, found before Dubois discovered *Pithecanthropus*, are truly those of modern man, morphologically speaking. Whereas *Pithecanthropus* had a cranial capacity of 1000 cc, the Wadjak skulls have cranial capacities of 1550 cc and 1650 cc respectively. Although Dubois called them *Homo wadjakensis*, there has never been any doubt that they belong to the same species as modern humans and should be classed as *Homo sapiens*.

The Wadjak skulls were found in 1888 and 1890. Dubois brought them back from Java in 1895 and kept them sequestered in his home in Haarlem, Holland. He made no public announcement about these fossils until May 1920, thirty years after they were found. The motivation then for Dubois's revelation of his Wadjak discoveries was that Stuart A. Smith published a monograph on Talgai Man, claiming that he had discovered the first "proto-Australian." Dubois's massive ego could not let that claim go unchallenged. He unveiled the Wadjak skulls and said that he had discovered the first "proto-Australian" years before Smith did.[5]

Why did Dubois keep his discoveries at Wadjak a secret for thirty years? Most evolutionist writers who mention Wadjak do think that Dubois's actions were a bit strange. Theunissen is the most casual about it.[6] He says that Dubois went to the Dutch East Indies to find the missing link. Since the big-brained Wadjak was obviously not the missing link, Dubois was not that much interested in the fossils. He did not even remove the hard sediment that still partly covered the fossils until 1910. By that time, Dubois's view of Neandertal had changed and Wadjak became worth studying.

Wadjak presented another "window of opportunity" for Dubois, says Theunissen. Marcellin Boule was busy moving the Neandertals out of man's direct ancestry, because they were too "primitive" for their advanced date. Dubois had already claimed to have discovered the missing link. By revealing Wadjak and emphasizing their advanced features over Neandertal, Dubois could claim credit for having discovered a second direct human ancestor, although a more modern one. The time was thus ripe to tell the world about Wadjak.

Theunissen has done well in explaining why Dubois revealed Wadjak, but he has not explained why Dubois hid Wadjak in the first place. Sir Arthur Keith let the cat out of the bag. After he described Dubois's strange behavior, he wrote:

> . . . we cannot question his honesty; the Wadjak fossil bones were discovered under the circumstances told by him. There can be no doubt that if, on his return in 1894, he had placed before the anthropologists of the time the ape-like skull from Trinil side by side with the great-brained skulls of Wadjak, both fossilised, both from the same region of Java, he would have given them a meal beyond the powers of their mental digestion. Since then our digestions have grown stronger.[7]

Keith had a way with words. He was saying in a gracious way that if Dubois had revealed the Wadjak fossils at the time he revealed *Pithecanthropus*, his beloved *Pithecanthropus* would never have been accepted as the missing link. Dubois was well aware of that fact. There is evidence that Wadjak was approximately the same age as *Pithecanthropus*, so to sell *Pithecanthropus*, Dubois had to hide Wadjak.

Wadjak was not the only occasion on which Dubois manipulated fossils to protect his claim for *Pithecanthropus*. Earlier he had found a jawbone fragment at Kedoeng Broeboes. Jawbone fragments are sometimes difficult to diagnose, and Dubois did have problems with it. He first ascribed it to the genus *Homo*. But when he wrote his initial description of *Pithecanthropus erectus* in 1894, he recognized that he had a problem. Theunissen describes how he handled it:

He did make a very glancing reference to the jaw fragment from Ke-doeng Broeboes which he had earlier attributed to the genus *Homo,* now suggesting that this fossil might also have belonged to Pithecanthropus. It is fairly easy to fathom the reason for this change of opinion. It "must have belonged" to Pithecanthropus because Dubois believed that the Kedoeng Broeboes sediment layers were as old as those at Trinil. The existence of a *Homo* fossil in layers of the same age as those in which the ape-man had been found, would have considerably weakened his argumentation that *Pithecanthropus erectus* was a transitional form between ape and Man. It was obvious, however, that his doubts about the jaw fragment had not dissipated, since he omitted any description or illustration; we can imagine that he was anxious not to present his critics with an easy target.[8]

To preserve the uniqueness of *Pithecanthropus* as the missing link, Dubois had to make sure that no fossils of more modern morphology could be assigned to the same stratigraphic level or given the same date. He did this by (1) changing the assignment of the Kedoeng Broeboes jawbone from *Homo* to *Pithecanthropus* while at the same time withholding any description or illustration of it so that no one could challenge his assignment, and (2) hiding Wadjak for thirty years so that no one would know he had found such fossils until *Pithecanthropus* had been thoroughly established in the human thought stream as our evolutionary ancestor.

In 1982 creationist Duane Gish debated C. Loring Brace on the campus of the University of Michigan. Gish had stated that Dubois's secrecy regarding Wadjak "can only be labeled as an act of dishonesty." Brace responded:

> Dubois did publish preliminary accounts of his Wadjak material in 1889 and 1890 before his Trinil discoveries were even made, and he recapitulated these in print in 1892 before becoming involved in what he correctly realized was the far more significant *Pithecanthropus* issue. If there is a question of honesty involved, it has nothing to do with Dubois.[9]

Yes, there is a question of honesty involved. Dubois was deceptive. Technically, however, Brace told the truth. But he also gave a very false impression.

It is possible to lie by telling the truth. It is done often. Suppose a man owes you one hundred dollars. Because you need the money, you call him to find out when he can pay you. His wife answers the phone and tells you that he is out. You take that to mean that he is unavailable. You don't know that he is standing just outside the front door of his house so that his wife can "honestly" say that he is "out." She justifies herself in that she technically told the truth. But she really lied, because she intended that you would think that "out" meant "unavailable." She lied by telling the truth.

Brace said that Dubois published accounts of the discovery of Wadjak in 1889, 1890, and 1892. Brace knew full well that the two thousand persons attending that debate would take "publish" to mean that Dubois had made an announcement to the world, or at least to the anthropological community, about Wadjak. When someone publishes about a fossil discovery, that is normally what is meant. However, this is not what happened. This publishing was nothing more than Dubois's quarterly and annual reports to the Director of Education, Religion, and Industry of the Dutch East Indies government, the department that had granted him permission to work in Java. These were bureaucratic reports and were not intended for the public or for the scientific community. They may not have been read even by the bureaucrats. Did Brace know that? Of course he did. Trivia buffs will recognize that the dates of Dubois's government reports are probably some of the least significant facts in the history of the universe. Anyone who knows those dates certainly knows what kind of reports they were. If they had been public reports, why would Keith and other anthropologists have made excuses for Dubois's keeping Wadjak a secret?

Dating the Wadjak Skulls

The mystery, however, is still not solved. What was there about Wadjak that caused Dubois to keep it a secret for thirty years? Allow me to take you on a behind-the-scenes excursion. We are going to undertake a project: the dating of Wadjak Man. We are going to do it, moreover, by pretending that we are evolutionists. I have long felt that if I were going to critique evolution, I owed it to my readers, as well as to my own integrity, to learn to think like an evolutionist, to honestly try to understand where he is coming from. I have tried very hard to get inside the brain of the evolutionist and understand his system. I do not claim to have succeeded. I claim only to have honestly tried.

As "evolutionists" we recognize an awesomeness, a pagan elegance in unaided nature's capacity to produce the incredible variety of living things we see about us. There is a logic to evolution that cannot be denied. It is not difficult to imagine that the more complex forms we see about us came from simpler forms by gradual change over time. We are familiar with some of the thousands of experiments that tend to support evolution. In just looking at nature, it seems obvious that evolution is true because things can be arranged in an evolutionary pattern. The fact that some doubt evolution is frustrating. It is hard to deal with people who deny the obvious and will not open their minds.

Thanks to Darwin, we know that it is unscientific to appeal to supernatural forces to explain the universe. Living in a scientific age, we must oppose with all our might those who would turn back the clock and corrupt science by trying to inject religion into it under the term *scientific creationism*. The very idea of scientific creationism is a contradiction in terms. If creationists only understood the scientific method they would see the light. It is obvious that we evolutionists have done a very poor job of teaching science. We must do a better job in the future, for science itself is at stake.

Our immediate task as evolutionists is to date the Wadjak fossils. To start, we consult Michael Day. He writes of Wadjak: "In view of the degree of mineralization of the skulls and the modern fauna, the earliest possible date is probably late Pleistocene."[10]

We notice several aspects of Day's remarks. By "late Pleistocene" he probably means the Upper Pleistocene, although his term is a bit vague. When he refers to "the earliest possible date," he means that the fossil could be younger than late Pleistocene but certainly not older. He mentions two points upon which his evaluation is based: the mineralization of the skulls and the modern fauna. By modern fauna he means that most, if not all, of the fossil animals, especially mammals, found in association with the skulls are living today. If many of the fossil animals found in association with human fossils are extinct, this seems to indicate a long intervening time span by evolutionary standards. This is then referred to as an older fauna.

Dubois recognized that the Wadjak skulls were highly mineralized, or fossilized,[11] which would imply that the fossils did have a bit of age, although fossilization rates are dependent upon a number of conditions. Thus a date of middle Upper Pleistocene, about 50,000 y.a., would certainly be a reasonable assessment, and a few evolutionists have given Wadjak that date.

Day further states that the fossil fauna associated with the skulls ". . . does not differ significantly from that found in modern Java."[12]

This could argue for a date more recent than 50,000 y.a. In the *Catalogue of Fossil Hominids*, the date is given as "late Upper Pleistocene or early Holocene."[13] Since the Holocene is the more recent epoch after the Pleistocene, that would indicate that Wadjak could be dated at about 10,000 y.a. This is the generally accepted date of Wadjak. Thus Wadjak is considered to be too recent to have played any part in human evolution, and it is for this reason that Wadjak is virtually unknown. It simply is not an important fossil as far as evolutionists are concerned.

Michael Day, and the editors of the *Catalogue*—Oakley, Campbell, and Molleson—are extremely competent scholars. We can safely place our faith in their date for Wadjak. The editors of the *Catalogue* list their resources for their dating of Wadjak: Dubois and von Koenigswald. We place our faith in the evaluation of Oakley, Campbell, and Molleson. They place their faith in the accuracy of Dubois and von Koenigswald.

Since there is a bit of fuzziness about the date of the Wadjak fossils, we could wish that they had been found after the advent of the radiometric dating methods. If these methods had been used on the Wadjak fossils, there would be no doubt about the date. We could trust the results. That is why they are called absolute dating methods. We "evolutionists" have heard that creationists challenge the very foundations of the radiometric dating techniques, but creationist writings are such a confused mixture of science and religion that they are not to be trusted. Besides, creationists have this "infallibility of the Bible" thing that locks them into a preconceived philosophical framework. They are not free to accept data that conflicts with the Bible. We evolutionists are not constrained by some rigid philosophical framework. We are free to go with the facts wherever they may lead.

Back to Wadjak. It seems that we can safely conclude that the date for the Wadjak skulls is about 10,000 y.a., and that our project is complete. We have consulted many other authoritative scholars, and there is almost universal agreement on that date for Wadjak.

The *Catalogue* does mention that found among the fossil fauna at Wadjak was a species of tapir, *Tapirus indicus*. A tapir is a hoofed mammal that looks like a pig but is considered to be more closely related to the rhinoceros. It is a bit curious that of all the fossil fauna found at Wadjak, this is the only one mentioned in the *Catalogue*. Yet, nothing is said about its being significant. If it is not significant, why is it mentioned? If it is significant, why wouldn't they give at least a hint as to the reason?

Wadjak seems like such an open-and-shut case that it is very easy to let the tapir item pass by. However, we are pretending to be evolutionists, and evolutionists, by definition, are scholars. So, almost for the fun of it, we decide to check it out. After much research and a feeling that we are wasting our time, we come across a 1951 article by American Museum of Natural History anthropologist Dirk Albert Hooijer, in which he writes: "*Tapirus indicus*, supposedly extinct in Java since the Middle Pleistocene, proved to be represented in the Dubois collection from the Wadjak site, central Java, which is late—if not post—Pleistocene in age."[14]

Hooijer clearly states that Wadjak is late or post-Pleistocene—that is, about 10,000 y.a. Yet, a species of tapir that became extinct in the Middle Pleistocene was found by Dubois among the fossil fauna. Now we understand why Dubois hid the Wadjak fossils rather than reveal them and try to explain away *Tapirus indicus*. Fossil forms that are known to have become extinct, or believed to be extinct, within a specific time frame can be important indicators in dating fossil assemblages.

A question comes to mind. If *Tapirus indicus* was found in the Wadjak faunal assemblage, and if this species of tapir is believed to have become extinct in the Middle Pleistocene, why couldn't Wadjak be Middle Pleistocene? In spite of Dubois's original claim that *Pithecanthropus* was of Pliocene age, it is now universally placed in the Middle Pleistocene. That would mean that the big-brained Wadjak skulls and smaller-brained *Pithecanthropus* skullcap could be approximately the same age. Perhaps we have just answered our own question. Wadjak and *Pithecanthropus*, both in the Middle Pleistocene, would not fit the evolutionary scenario well at all; in fact, it would be evidence against it. We certainly understand now why Dubois hid the Wadjak skulls for thirty years.

Since you and I still endeavor to think like evolutionists, we recognize that we have a problem. Is there a solution? We think of the vast array of evidence and experiments that support evolution, and that virtually the entire scientific community endorses it as being true. It seems incredible that one tiny fact should threaten such a vast array of evidence. Certainly if there were serious challenges to the fact of evolution, these would be published in the accepted scientific literature. But we do not read of any. Of the two possibilities—that evolution is wrong, or that this one fact is wrong—it seems more logical that this one little fact is wrong. Let's start there.

We can explore a number of possibilities for error in the matter of *Tapirus indicus*. (1) Is it possible that this species of tapir did not become extinct in the Middle Pleistocene, but actually lived into the Upper Pleistocene or even into the Holocene before becoming extinct, and that fossils of this tapir from these time periods have not been discovered yet? The answer obviously is yes. An argument based on extinction is really an argument from silence, and that type of argument always carries with it a bit of risk. (2) Although Dubois was an anatomist, he would not have been as familiar with tropical animals, and the tapir is a tropical animal. Is it possible that the fossils in question belonged to another species of tapir, or to another type of animal altogether, and that Dubois misinterpreted them? Once again, the answer is yes. (3) Is it possible that the fossils of this particular tapir came from another locality, and that they were put in the Wadjak collection by mistake, by Dubois or someone else? Yes. There may even be other possibilities for a legitimate mistake to have been made regarding these fossils. Since there are so many possibilities for mistakes, and since evolution is so well established, it seems best to go with evolution rather than with the negative implications of this one little questionable fact.

After mentioning the discrepancy in the Wadjak fauna, Hooijer comments on the faunal dating of the fossils in Java:

> It would seem to be only logical to assume that, upon further accurate monographic studies of the various elements to the Pleistocene fauna of Java, the faunal differences between the Lower, Middle and Upper divisions will dwindle more and more, and the picture of the evolution of the fauna will become less and less cataclysmic.[15]

Later he writes:

> Thus, it seems that in Java this method of Pleistocene chronology fails. The method has certainly part of the truth, but in Java it is completely overshadowed by the epeirogenic movements [earth uplifts and downwarps] of much greater magnitude.[16]

Hooijer is saying that the differences in the fossil fauna of the Upper, Middle, and Lower Pleistocene of Java are not that great. This, together with the fact that most of the fossils upon which the faunal

dating in Java is based were collected by hired nationals who had absolutely no knowledge of geology, causes Hooijer to say:

> The finds of Early Man in Java, therefore, cannot be exactly dated, and
> the only thing we can honestly say is that they are Pleistocene in age, and
> most probably neither lower-most nor late Pleistocene.[17]

It is shocking to learn that our problem is far greater than just the
dating of Wadjak. Between 1936 and 1941 von Koenigswald found a
number of fossils similar to Java Man, now all classified as *Homo erectus*. Some were found in the Trinil beds at Sangiran, some in the lower
Djetis beds, and some in the higher Ngandong beds. All of these were
found before the advent of radiometric dating (although some of the
beds have been dated radiometrically well after the fact). All of these
Javanese human fossils were dated according to the alleged evolution
of the fossil fauna. Based upon this, much has been made of human
evolution within the taxon *Homo erectus* in Java. It is a favorite subject
for displays in museums, such as San Diego's Museum of Man. Yet,
all of us evolutionists know that the Javanese fossils are very poorly
dated. When we are faced with such uncertainties, perhaps the best
thing to do is to go with the one thing we know for sure, the fact of
human evolution—and interpret the fossils accordingly.

The problem is that you and I do not have the time, the money, or
the expertise to go to Java to check these things out to our own satisfaction. Even if we did go to Java, it would do no good. The Wadjak
site has long since been removed for marble. We have no choice but to
put our faith in the skill, accuracy, and evaluation of the experts. Since
we know that evolution is true, it seems logical to take the position on
the Wadjak fossils that is most consistent with it. We thus accept the
date of 10,000 y.a. for Wadjak.

We have a passing thought. We found a little detail that seemed to
militate against Wadjak being late Upper Pleistocene or Holocene. We
ruled against it on philosophical grounds. Is it possible that in much of
the evidence that we lean on in support of evolution, there were also
anomalies that researchers missed or ignored? No, we feel sure that
such discordant items would have been published. However, in our report on Wadjak, we will ignore the tapir incident. Our report is more
for the public, and the public likes direct answers without a lot of un-

certainties and conditions. Too much of that makes it look as if we evolutionists don't know what we are doing.

We return now from our imaginary experience as evolutionists. I have attempted to describe the process an evolutionist goes through in dating Wadjak. As I show in the chapter on dating KNM-ER 1470 (appendix), the newer radiometric dating methods do not clarify the picture. Because scientists refer to radiometric dating as absolute dating, they grant an assumption of precision and accuracy to those methods which they do not deserve. They are usually considered as independent verifications of the age of something. Since the age of *Pithecanthropus I*, the original Java Man, is usually given in the literature as about 500,000 y.a., it might be of interest to know how that figure was obtained. Kenneth Oakley, referring to the category of radiometric dating known as A-3 (Absolute 3 category), wrote:

A3 Dating: The age of a specimen in years inferred by correlation of the source bed with a deposit whose actual age has been determined. Thus the original *Pithecanthropus I* remains from river gravel at Trinil, Java, can be dated as c 500,000 years old on the basis of the K/Ar age of leucite in volcanic rock found elsewhere in Java but containing Trinil fauna (von Koenigswald, 1968).[18]

Thus, the radiometric age of Java Man is not an independent confirmation of the faunal date, but is directly related to the estimates regarding the Trinil fauna. In light of Hooijer's conclusions about the unreliability of dating by the Trinil fauna, the radiometric date for Java Man—although almost universally accepted—is worthless. This heightens the possibility that Wadjak and *Pithecanthropus* could be approximately the same age.

I'm sure that an evolutionist would cry "foul" on several of my points in the dating of Wadjak. First is the point that in many situations he goes on faith rather than by facts. In spite of his protests, it is true. We personally check out and examine only a tiny portion of the things we accept as fact. Everything else we accept on faith, which really means we trust in the credibility and accuracy of someone else.

The second point is my contention that he evaluates facts according to his theory or his philosophy. He believes it is the other way around, that his facts have developed his theory. However, he is mistaken. But I dare not criticize him for doing it, lest I condemn everyone. The old idea that theory always comes from facts is universally believed, even though it is totally inaccurate.

As early as 1935 the Austrian philosopher of science (later at the University of London), Sir Karl Popper, demonstrated in his monumental work *The Logic of Scientific Discovery*[19] that scientists do not work according to the so-called scientific method, and that they could not work that way even if they wanted to. To say you can start with observations but without a theory is absurd. Scientists simply do not go around collecting observations and data indiscriminately and then try to fit them into theories. They must start with some theory or concept. This then gives them direction in the collecting of data.

If this idea still sounds strange to you, let me assure you that it is indeed the way science works. Perhaps another illustration will help. Writing about a conversation with Albert Einstein, German physicist Werner Heisenberg, who had been emphasizing the importance of observations in the formulation of scientific theory, recalls his surprise at Einstein's response:

> "But you don't seriously believe," Einstein protested, "that none but observable magnitudes must go into a physical theory?"

> "Isn't that precisely what you have done with relativity?" I asked in some surprise. "After all, you did stress the fact that it is impermissible to speak of absolute time, simply because absolute time cannot be observed; that only clock readings, be it in the moving reference system or the system at rest, are relevant to the determination of time."

> "Possibly I did use this kind of reasoning," Einstein admitted, "but it is nonsense all the same. Perhaps I could put it more diplomatically by saying that it may be heuristically useful to keep in mind what one has actually observed. But on principle, it is quite wrong to try founding a theory on observable magnitudes alone. In reality the very opposite happens. *It is the theory which decides what we can observe.*"[20]

Obviously the context of that quotation was not a discussion of human fossils. However, it makes no difference. From physics all the way to anthropology, theory strongly influences facts. That is the way science works because that is the way people work—evolutionists, creationists, and everyone in between. But no one ever says it. Why? Because it sounds so wrong. It sounds so wrong that we don't even like to think it. But it is not wrong. It is the way we operate. It is the only way we can operate. The only deception is self-deception when we try to tell ourselves that we don't operate that way—that we, above all others, are objective and without bias. In theory, facts determine theory; but in fact, theory determines facts. Ultimately, *everything* is philosophy—or theology.

Meanwhile, it does not seem fair that Dubois's Java Man enjoys worldwide fame while Wadjak Man lives in virtual obscurity. But it does serve to prove that not all fossils are created equal.

11

The Selenka Expedition:
A Second Opinion

ALTHOUGH DUBOIS WAS CERTAIN that he had found what he set out to find, the missing link, his fossils were the source of heated debate in the scientific community at the end of the nineteenth century.

First was the problem of whether the very archaic looking skullcap and the very modern looking femur belonged in the same category, let alone to the same individual. Did their being found fifty feet apart constitute "being together" as Dubois claimed? The bones could just as well have been the result of chance association in flood deposits on the bank of the Solo River.

Second, the nature of *Pithecanthropus* was also a mystery. Was he human, an apish man, a mannish ape, or an extinct primate with no relationship to humans at all? The limited information did not allow a definitive answer. The Wadjak skulls would have helped, but no one knew about them at the time.

Third was the question of where *Pithecanthropus* fit into the alleged human evolution sequence. Dubois felt that he was in the direct line of human evolution. Some scientists considered him an evolutionary dead end with no direct bearing on human origins; most felt that he was already fully human but belonged to an inferior race.

Last was the matter of his age. The geology of Java was virtually unknown, and Dubois's information was frustratingly skimpy. Dubois
assigned the fossil-bearing stratum at Trinil to a much earlier geological period than later investigations determined it to be.

A Munich zoologist, Professor Emil Selenka, saw clearly that the solution to the problem of *Pithecanthropus* was not continued discussion
but further exploration. Selenka was one of the leading scientists of his
day. Sir Arthur Keith said that he "did more than any man of his time
to advance our knowledge of the higher primates."[1] Selenka's dream
was to have an expedition go to the place in Java where Dubois had
found *Pithecanthropus* and search for more fossils. He was in the process of organizing such an expedition when he died. His wife successfully completed the project, which became known as the Selenka-
Trinil Expedition of 1907-08.

Frau M. Lenore Selenka was a professor and academician in her
own right. As she undertook the expedition, she enlisted Professor
Max Blanckenhorn of Berlin as her associate. Sponsorship came from
the Berlin Academy of Science and from at least one academic institution in Munich. However, the bulk of the expenses for the expedition
came from Frau Selenka's private purse.

Besides Selenka and Blanckenhorn, seventeen other specialists were
involved in the study of the forty-three large crates of fossils that were
returned to Germany. These specialists produced an excellent scientific report of 342 pages, edited by Selenka and Blanckenhorn, published in 1911, and entitled *Die Pithecanthropus-Schichen auf Java*.
Few, unfortunately, have ever heard of it. The report has suffered the
fate decreed for all evidence that is contrary to evolution: consignment
to the lower reaches of oblivion.

British creationist A. G. Tilney, who died in 1976, first made me
aware of the Selenka-Trinil Expedition report. Tilney served as secretary of the Evolution Protest Movement (now the Creation Science
Movement) in England and was a modern languages scholar. He
searched over sixty libraries on the continent of Europe before he finally located a copy of the report in the Volkschulla Library in Aachen,
(West) Germany.

The report has never been translated into English. However, being
a modern languages professor, Tilney was able to study it. Unfortunately, he did not make a full translation of it before his death. My information on the expedition report comes from Tilney's pamphlet on

the subject[2] and from a review of the report by Sir Arthur Keith, published in *Nature*.[3]

The field of paleoanthropology has had more than its share of scandals and sloppy scholarship. In contrast, the manner in which Frau Selenka organized and executed her expedition and published its results, Keith said, "commands our unstinted praise." The thoroughness and scientific integrity of the expedition are exceeded only by the obscurity into which it has fallen. Although the purpose of the expedition was to confirm Dubois's findings, its results actually contradicted his claims.

According to Tilney, the seventeen specialists who contributed to the expedition's report were also on the scene of the excavations in Java. If that was the case, it was truly a sizable project. Keith mentioned only that "scientific investigators were sent out." If Tilney was right, then Frau Selenka was a pioneer in the multidisciplinary approach to paleoanthropology. Not until the 1970s did F. Clark Howell (University of California, Berkeley) institute the practice of bringing a number of specialists, such as paleobotanists, paleoecologists, paleozoologists, geologists, as well as paleoanthropologists to a fossil site. This practice is considered a quantum leap in the study of the past. Richard Leakey, known for his organizational ability, has popularized this multidisciplinary approach in his work in East Africa. Yet, Frau Selenka may have pioneered this concept in 1907, more than sixty years earlier.

In 1907, the Selenka-Trinil Expedition journeyed to the banks of the Solo River, an area known as "the hell of Java," and found the stone that Dubois had used to mark the site of his original discoveries. Seventy-five national workers were hired, and barracks were built for them. A Dutch sergeant who had worked for Dubois was also employed by Frau Selenka.

Extensive mining and digging operations were required, as the fossil-bearing layer of the Trinil formation was under thirty-five feet of volcanic sediment. More than 10,000 cubic meters of material were removed in the search for more remains of *Pithecanthropus*. But, *Pithecanthropus* was not to be found.

What they did find was more significant than what they did not find. Three of the specialists, Dr. E. Carthaus, Frau H. Martin-Icke, and Dr. J. Schuster, concluded that Dubois had seriously overestimated the age of the stratum in which *Pithecanthropus* was found. Paleontologist Martin-Icke reported that 87 percent of the gastropods found in it were of modern forms, and botanist Schuster testified that the flora was not too dissimilar from that of today. This would indicate a rather recent age for *Pithecanthropus*, and that alone would eliminate him as the missing link.

In the very same stratum in which *Pithecanthropus* was found, Dr. Carthaus discovered splinters of bones and tusks, foundations of hearths, and pieces of wood charcoal. As a geologist, he felt that the *Pithecanthropus* stratum was rather recent, and that modern humans and *Pithecanthropus* (Java Man) had lived at the same time. This was another blow to the missing link status of *Pithecanthropus*.

The most striking discovery was by Dr. Walkhoff, an anthropologist. In the dry bed of a tributary of the Solo River, about two miles from Dubois's famous discovery, he discovered the crown of a human molar. The dentine within the enamel cap had been replaced by a fossilized organic matrix. Although the tooth belonged to a relatively modern human, Walkhoff concluded that its condition indicated an even greater age than the age assigned to *Pithecanthropus*. The tooth is known as the Sondé fossil.

All of the members of the Selenka-Trinil Expedition were evolutionists. The purpose of the expedition was to confirm Dubois's findings of fossil evidence for human evolution. However, Frau Selenka, the leader of this exemplary expedition, concluded that modern humans and *Pithecanthropus* both had lived at the same time and that *Pithecanthropus* played no part in human evolution. This is the same conclusion that would have been reached had Dubois revealed Wadjak at the time he paraded *Pithecanthropus* before the public. As it was, the Wadjak skulls were still sequestered beneath the floorboards of Dubois's home. It would be another ten years after the release of the Selenka-Trinil Expedition report before the Wadjak skulls would see the light of day.

Perhaps the most remarkable part of the report was its description of the violent volcanic eruptions from nearby Mount Lawu-Kukusan and subsequent flooding that took place in that part of Java every thirty years or so. The geologic activity was so intense that the report concluded that the volcanic Trinil sediments which contained *Pithecanthropus* were far too young to yield any information on human origins. Native traditions tell of the Solo River actually having changed its course in the thirteenth or fourteenth centuries, which could mean that the Trinil beds were only about 500 years old, not the 500,000 years old believed today. Because volcanic matter is heavily mineralized, the report states that the degree of fossilization of *Pithecanthropus* was the result of the chemical nature of the volcanic material, not the result of vast age.

The summary chapter of the report was written by Dr. Max Blanckenhorn. In it he apologized to the reader because what they had hoped would be a corroboration of Dubois's findings seemed more like a debunking of Dubois's work. He used the German word for "fruitless" to describe their failure to substantiate Dubois's claim that the famous Java Man he had discovered was our evolutionary ancestor.

It did not occur to Blanckenhorn that their work was not fruitless but had produced very positive results. It showed that humans have wide morphological variation, something that anthropologists have only recently come to appreciate. It is possible that the variation in early humans could have exceeded what it is today. Although the Selenka Expedition is universally considered to have failed in its primary purpose, in true scientific terms it was a resounding success. Had the wide variation in the human family been appreciated at that time, many of the later mistakes in anthropology could have been avoided.

In light of the expedition's findings, it is interesting to see how evolutionists have handled the Selenka report. With one exception, the newer works on paleoanthropology ignore the Selenka report completely. About half of the books written between 1945 and 1975 mention the expedition or the report, but do so in such a way that it would be impossible for the English reading researcher to discover what the Selenka report actually said. It is an amazing conspiracy of silence.

The Selenka report is most often mentioned for its excellent description of the fossil fauna of the Trinil deposits (a description that Dubois was expected to give but never did). All evolutionist works that refer to the geological findings of the Selenka report misrepresent those findings, which were that the Trinil beds appeared to be recent Pleistocene deposits (this, I assume, would mean Upper Pleistocene). Some writers actually claim that the expedition supported Dubois's assessment that the Trinil beds were Pliocene. Other authors state that the expedition reduced the age of the beds to Lower or Middle Pleistocene age from Dubois's original claim of Pliocene age.

Some authors report that no *Pithecanthropus* fossils were found, but they do not mention that other human fossils and artifacts were found.

Other authors state flatly that no artifacts were found at all, when in truth artifacts were found.

Alan Houghton Brodrick mentions that a *Homo sapiens* tooth was found,[4] but since he gives no details, the reader would not know that the tooth was believed by the Selenka Expedition to be older than *Pithecanthropus*. On the other hand, Carleton Coon mentions the Sondé tooth as probably coming from the Trinil beds,[5] but he is very fuzzy on the category to which it belongs (*Homo sapiens*) and does not associate it with the Selenka Expedition. Once again, the reader would miss the true impact of the discovery.

Only two books reveal the true significance of the Sondé tooth discovery. The first one is a 1924 work by MacCurdy:

> The Selenka Expedition of 1907-08 . . . secured a tooth which is said by Walkhoff to be definitely human. It is a third lower molar from a neighboring stream bed and from deposits older (Pliocene) than those in which *Pithecanthropus erectus* was found. Should this tooth prove to be human, *Pithecanthropus* could no longer be regarded as a precursor of man.[6]

The other book is the recent work by Theunissen. Notice that his account does not entirely agree with MacCurdy's.

> The only discovery of (subfossil) hominid remains was that of a fully human tooth, uncovered at Sondé, some distance from Trinil. However, the age of the specimen was uncertain. Dubois joined in the discussion arising from this find and—naturally—refused to accept that the tooth could possibly have come from the same layer as the *Pithecanthropus* remains. *He even suggested that there was fraud somewhere.*[7]

The man who could have, and should have, given us a full account of the Selenka Expedition was G. H. R. von Koenigswald. He spoke German, had access to the report, was a noted paleoanthropologist, and worked many years in Java. Yet, he disappoints us. However, he has added a vital bit of information: "The spot at which the Selenka expedition pitched its camp is still easy to find. It is strewn with innumerable broken beer bottles, testifying to the expedition's thirst."[8] This was a German expedition. Did he expect them to drink Tetley Tea?

The members of the Selenka Expedition went to Java to confirm Dubois's findings. They thought they had failed. Actually, they succeeded in revealing the true nature of the human fossil record—that the human family had wide morphologic diversity—and that Java Man was not our evolutionary ancestor.

I have written at length on the background and details of Dubois and his Java Man. I have done so to show that one of the conclusions of this book—that *Homo erectus* was not our evolutionary ancestor but lived as a contemporary of more modern humans—is not original with me. The evidence has existed for one hundred years. Dubois himself had three warnings of it: (1) The Trinil femur and skullcap evidenced wide morphological diversity in a single contemporaneous population. Dubois rejected that evidence. (2) Wadjak and *Pithecanthropus* demonstrated the possibility that humans having wide morphological diversity had lived together. Dubois retaliated by hiding Wadjak for thirty years. (3) The Selenka Expedition confirmed that humans with wide morphological diversity had lived together in recent times. Dubois responded by claiming that the Selenka findings were fraudulent.

Dubois went to the Dutch East Indies to prove human evolution by discovering fossil evidence. He then undertook a campaign to sell his fossils to the world. He succeeded. The result was that he set the study of paleoanthropology back one hundred years.

Was Dubois a bad man? Perhaps yes, perhaps no. Theunissen puts it this way: "If Dubois deceived others, he deceived himself just as much."[9]

12

Homo erectus:
A Man for All Seasons

JAVA MAN *(PITHECANTHROPUS I)* was the first of at least 222 similar fossil individuals that have been discovered to date. It would be impossible to exaggerate the importance of this group of fossils known collectively as *Homo erectus*. For the evolutionist, *Homo erectus* is the major category bridging the gap between the australopithecines (which everyone recognizes as nonhuman) and the archaic *Homo sapiens* and Neandertal fossils (which everyone recognizes as truly human). Thus, *Homo erectus* is indispensable to the evolutionist as *the* transitional taxon. (*Homo habilis* is a flawed taxon that will be given a respectful burial in a later chapter.)

Surprisingly, *Homo erectus* furnishes us with powerful evidence that falsifies the concept of human evolution. Three questions are crucial. First, is *Homo erectus* morphologically distinct enough to warrant its being classified as a species separate from *Homo sapiens*? The evidence clearly says no. By every legitimate standard applicable, the fossil and cultural evidence indicate that it should be included in the *Homo sapiens* taxon.

Second, are the *Homo erectus* fossils found in the relevant time frame so as to serve as a legitimate transitional form? The clear answer from the fossil record is again no. The *Homo erectus* charts in this chapter are among the most complete listing of fossils of *Homo erectus* morphology to be found in the scientific literature. When compared with the other charts in this book, they show that *Homo erectus* individuals have lived side by side with other categories of true humans for the past two million years (according to evolutionist chronology). This fact eliminates the possibility that *Homo erectus* evolved into *Homo sapiens*. That this two-million-year contemporaneousness has been largely camouflaged is a tribute to the skill of evolutionist writers.

The third question is whether or not there are adequate nonevolutionary explanations for *Homo erectus*, archaic *Homo sapiens*, and Neandertal morphology. The answer is Yes. There is a touch of humor in the fact that the nonevolutionary factors that adequately explain this morphology are some of the same factors that evolutionists themselves use to explain *Homo erectus*-like fossils when these fossils mischievously

Homo erectus Fossils—Holocene—Upper Pleistocene
(Evolution Time Scale)

* Date uncertain
+ Information limited

T- Tools/artifacts in association
F- Evidence of use of fire
( ) Min. no. of fossil individuals

6,000 y.a.		Mossgiel cranium, Australia[1] (1) T F
		Cossack skull, mandible, limb fragments, Australia [2] (1)
8,000 y.a.	Holocene	+ * Lake Nitchie people, Australia [3] (2+)
10,000 y.a.		* Java Solo (Ngandong) people, Indonesia[4, 5, 6] (15) T
		Kow Swamp people, Australia[7] (40) T F
		Cohuna cranium, Australia [8] (1)
		+ * Coobool Crossing people, Australia [9] (2+)
		Talgai cranium, Australia [10, 11] (1)
25,000 y.a.		
		Willandra Lakes WHL 50 cranium, Australia [12, 13, 14] (1)
		Témara (Smugglers' Cave) cranial fragment, mandible, Morocco[15, 16](2) T
50,000 y.a.		
75,000 y.a.		
100,000 y.a.		Narmada cranium, India[17] (1) T

Homo erectus Fossils—Middle Pleistocene
(Evolution Time Scale)

x May be archaic *Homo sapiens*
* Date very uncertain

T- Tools/artifacts in association
F- Evidence of use of fire
( ) Min. no. of fossil individuals

100,000 y.a.
    Xujiayao (Hsu-chia-yao) skull fragments, China[18, 19] (2) T
    Mapa (Maba) skull, China[20, 21] (1) T
    Cave of Hearths mandible, arm fragment, South Africa[22] (1) T F
* Hazorea 2, 4, 5 skull fragments, Israel[23] (2) T
    Laetoli Hominid 29 mandible, Tanzania[24] (1) T?
    Hexian skull fragments, China[25] (2)
    Ding Cun (Ting-Tsun) skull fragment, teeth, China[26, 27] (1) T
    Changyang maxilla, tooth, China[28, 29] (1) T

200,000 y.a.
    Wadi Dagadle maxilla, Djibouti [30] (1)

x Rabat (Kebibat) maxilla, mandible, skull frag., Morocco[31] (1)
    Salé cranium, maxilla, Morocco[32, 33] (1)
    Sidi Abderrahman (Casablanca) mandible, Morocco[34, 35] (1) T
    Dali (Tali) cranium, China[36, 37, 38] (1) T
    Zhoukoudian (Peking Man) layers 1, 2, China[39] (20?) T F
* Kapthurin (Baringo) BK-67, BK-8518 mandibles, limb frag., Kenya[40] (2) T
x Jinniu Shan (Yinkou/Gold Ox Hill) partial skeleton, China[41, 42] (1) T F?

250,000 y.a.

    Thomas Quarries 1 & 3 skull, mandible, Morocco[43,44] (2) T

400,000 y.a.
x Yunxian (EV 9001, 9002) crania, China[159, 160] (2) T
* Luc Yen Teeth, (North) Vietnam[45] (1)

    Arago (Tautavel) 13 & 44, mandible, hip bone, France[46, 47] (1) T
    Zhoukoudian (Peking Man) layers 9, 10, China[48] (20?) T F
    Lang Trang teeth, (North) Vietnam[49] (3)

500,000 y.a.
    Olduvai Hominid 23 mandible fragment, Tanzania[50, 51] (1) T
    Java Man (*Pithecanthropus* I) skullcap, Indonesia[52] (1) T

* Lainyamok maxilla, teeth, femur, Kenya[53] (1) T

600,000 y.a.
    Ternifine (Tighennif/Palikao) skull, mandibles, Algeria[54, 55] (3) T

    Lantian (Chenchiawo) mandible, China[56] (1) T
    Yunshien teeth, China[57] (1)
    Java (Sangiran/Trinil) people, Indonesia[58, 59] (40+)
    Olduvai Hominid 12 cranial fragments, Tanzania[60] (1) T

700,000 y.a.
    Olduvai Hominid 28 femur, hip bone, Tanzania[61] (1) T

Homo erectus Fossils—Lower Pleistocene—Pliocene
(Evolution Time Scale)

x May not be *H.e.*
* Date uncertain
+ Information
  very limited

T- Tools/artifacts in association
F- Evidence of use of fire
( ) Min. no. of fossil individuals

700,000 y.a.

Yayo (Koro Toro) skull fragment, Chad[62, 63] (1)
Olduvai Hominid 22 mandible, Tanzania[64, 65] (1)
Yuanmou teeth, China[66, 67] (1) T F
Olduvai Hominid 2 skull fragments, Tanzania[68, 69] (1) T
Lantian (Gongwangling) skull, tooth, China[70] (1) T F

1 m.y.a.   *+   Dmanisi mandible with teeth, Georgia, USSR[158] (1) T

Olduvai Hominid 29 teeth, phalanx, Tanzania[71, 72] (1) T
Olduvai Hominid 51, mandible, Tanzania[73, 74] (1)
Gomboré II (Melka-Kunturé) cranial fragment, Ethiopia[75] (1) T

Olduvai Hominid 9 (Chellean Man) cranium, Tanzania[76] (1) T
Olduvai Hominid 36 lower arm bone, Tanzania[77, 78] (1)

1.3 m.y.a.

*   Sambungmachan (Java) skullcap, Indonesia[79, 80] (1) T

Omo L-996-17 cranial fragments, Ethiopia[81] (1)

1.5 m.y.a.

KNM-ER 992 mandible, Kenya[82, 83] (1) T
KNM-ER 803 skeletal fragments, Kenya[84, 85] (1)
Olduvai Hominid 15 teeth, Tanzania[86, 87] (1)
KNM-ER 3883 cranium, Kenya[88, 89] (1) T
KNM-ER 737 femur, Kenya[90, 91] (1)
KNM-ER 820 mandible, Kenya[92, 93] (1)
KNM-WT 15000 skeleton, Kenya[94, 95] (1)
KNM-ER 1466 cranial fragments, Kenya[96, 97] (1)

KNM-ER 807 maxilla fragments, teeth, Kenya[96, 97] (1)
1.7 m.y.a.   KNM-ER 1821 cranial fragments, Kenya[96, 97] (1)

KNM-ER 1808 skeleton, Kenya[96, 98] (1)
KNM-ER 730 skull fragments, Kenya[99, 100] (1) T
KNM-ER 3733 cranium, Kenya[101, 102] (1)
KNM-ER 1507 mandible, Kenya[103, 104] (1)
Swartkrans SK-15, 18a, & 18b skel. frag., S. Africa[105, 106, 107] (1) T F
x   Swartkrans SK-84, SKX 5020 thumbs, South Africa[108] (2) T F
*   Swartkrans SK-45, 847 (Telanthropus) skull, S. Af.[109, 110, 111] (1) T F
Java (Djetis) crania, teeth, Indonesia[112, 113] (2)
KNM-ER 1809 femoral shaft, Kenya[114, 115] (1)
KNM-ER 2598 cranial fragment, Kenya[116, 117] (1)
KNM-ER 3228 hip bone, Kenya[114, 118] (1)

2 m.y.a.   +   Damiao mandible, teeth, China[119] (1)

Pliocene

show up in the wrong time frame. In this chapter we will discuss the time frame of *Homo erectus*.

The fossil charts in this book are central to the falsification of the theory of human evolution. Both the dates of these fossils and their assignment to specific categories are well documented from recent and reliable evolutionist sources. Many fossils have a degree of uncertainty on their dates. Only the most notorious cases are marked "date uncertain." When a fossil has a range of dates, its placement on the chart represents an average (a fossil dated between 400,000 and 600,000 y.a. would be placed at 500,000 y.a.). In some cases there is strong evidence for going against the evolutionist date, and that evidence is documented. It is important to emphasize that the uncertainties in the dating of the individual fossils are not serious enough to affect the conclusions reached in this book, because the fossil sampling is so large and the margin of uncertainty in the dating of each of the fossils is so small relative to the human evolution time scale.

The evolutionary category to which a fossil belongs can also be a problem. For example, since the category *Homo erectus* includes a suite of morphological characteristics, it is obvious that there must be enough of the fossil individual recovered to make a proper diagnosis. Usually, the less material recovered, the more precarious the diagnosis. It is also generally true that the older a fossil is, the less of its material is preserved. While there are striking exceptions, this means that the older fossils are often the more difficult to diagnose. In the fossil charts, the assignment usually represents the most recent evolutionist scholarship. However, because evolutionists have done much fudging when a fossil is not found in its "proper" time slot, I have not hesitated to go against the grain when the evidence justifies it. In such cases, I have given ample documentation.

When we address the matter of the dates of the *Homo erectus* fossils, we are confronted with an interesting situation (perhaps the kindest way to put it). The general dates recently assigned to *Homo erectus* by evolutionists are:

Ann Gibbons	300,000 to 1.7 m.y.a.[120]
Milford Wolpoff	400,000 to 1.4 m.y.a.[121]
William Howells	400,000 to 1.5 m.y.a.[122]
Richard G. Klein	400,000 to 1.8 m.y.a.[123]
C. Loring Brace	500,000 to 1.5 m.y.a.[124]
Jared Diamond	500,000 to 1.7 m.y.a.[125]

These dates for *Homo erectus* represent a "comfort zone" for evolutionists. It is not difficult to see that these dates position *Homo erectus* in the relevant time period to serve as that much needed transitional taxon that progresses toward modern humans.

However, even these accepted dates for the span of *Homo erectus* present problems. One is the length of the time span. The minimum date (300,000 y.a.) to the maximum (1.8 m.y.a.) represents a span of 1.5 million years, an incredible span of stability for a species which is supposed to be part of an evolutionary continuum. None of the individual writers quoted makes it that large. The largest span quoted is 1.4 m.y. (Gibbons and Klein) and the shortest is 1 m.y. (Wolpoff and Brace). Even a span of one million years makes evolutionists nervous, as seen in the following quotation by Susman, Stern, and Rose regarding two *Homo erectus* innominate (hip) bones from East Africa.

> The morphological similarity of O.H. 28 and KNM-ER 3228 supports their assignment to the same taxon *but for the large discrepancy in their ages.* The age of KNM-ER 3228 is roughly 1.5 m.y. (or greater) while O.H. 28 is dated at around 0.5 m.y. The possibility thus exists that 3228 represents the taxon *H. habilis* while O.H. 28 represents *H. erectus* and that locomotor anatomy grades subtly from one taxon to the other. KNM-ER 3228, a male, suggests considerable sexual dimorphism within *H. habilis* and that dimorphism was subsequently reduced in *H. erectus.*[126]

The fossil hip bones, KNM-ER 3228 and O.H. 28, are very similar and are normally assigned to *Homo erectus.* The problem is that they are dated one million or more years apart. The absence of any evolutionary change over a million-year span is the difficulty. The solution is to wave the magic wand and turn KNM-ER 3228 into *Homo habilis* without a shred of supporting physical evidence. However, *Homo habilis* is now known to be considerably smaller than *Homo erectus* (about half the size). The solution is to call 3228 a male (without evidence), and to ascribe the large size difference in *Homo habilis* to sexual dimorphism. The implication then is that all of the other *Homo habilis* fossils (with the exception of 1470, 1481, and 1590, which we will consider later) are female because of their small size. That assumption is also without physical evidence. A further undocumented assumption is that in evolving from *Homo habilis* to *Homo erectus,* sexual dimorphism changed a great deal, while the locomotor anatomy of the two taxa changed hardly at all. I will allow my readers to pass judgment on the validity of those assumptions. The most recent dating separates the O.H. 28 and KNM-ER 3228 hip bones by 1.2 million years.

A different type of problem is the oldest age limit for *Homo erectus*, suggested by Klein (1.8 m.y.a.), which causes this category to trespass dangerously into the private time domain of *Homo habilis*. Nor are evolutionists comfortable in having *Homo erectus* come all the way up to 300,000 y.a. Most evolutionists believe that we have *Homo erectus* genes in us, whether we evolved from the entire *erectus* population (neo-Darwinism) or from a small segment of that population (punctuated equilibria). With the new view that modern humans originated in Africa at about 200,000 y.a., to have humans evolve from *erectus* (at 300,000 y.a.) to *sapiens* (at 200,000 y.a.) in just 100,000 years is out of the question.

However, there is a more massive problem that will not go away. A minimum of 222 fossil individuals having a *Homo erectus* morphology have been discovered to date. If we accept the widest range of dates suggested by evolutionists for *Homo erectus* (300,000 to 1.8 m.y.a.), only 108 out of a total of 222 *Homo erectus* fossil individuals fall within those dates. Six *Homo erectus* fossil individuals have probable dates *beyond* 1.8 m.y.a., and a minimum of 108 fossil individuals are definitely dated more recently than 300,000 y.a. Put another way, at least 51 percent of the fossils having a *Homo erectus* morphology fall outside the widest range of dates given by evolutionists for that category.

It is likely that the 51 percent figure should be larger. In all aspects of my work, I have tried to be as conservative as possible so as not to be guilty of overstating my case. At the Upper Pleistocene site at Coobool Crossing, Australia, a number of fossils are said to exhibit the robust morphology similar to other *Homo erectus* discoveries in Australia.[127] The literature also indicates that about 130 fossil individuals are found at that site, and that the fossil population there is a mixture of robust and gracile individuals.[128] Since no published material to date indicates how many individuals exhibit this robust morphology, I have entered it on the chart as (2+) and counted it as 2, indicating that more than one *erectus*-like individual is at that site. Since it is likely that many more than two *erectus*-like individuals are at that location, the 51 percent figure is actually a very conservative minimum percentage.

Furthermore, the 51 percent figure is based on dates from 300,000 to 1.8 m.y.a. for *Homo erectus*. However, no evolutionist writer gives that wide a time distribution for *erectus*. Ann Gibbons, who dates *erectus* at 300,000 y.a., takes it back only to 1.7 m.y.a. Richard Klein, who takes *erectus* back to 1.8 m.y.a., starts it at 400,000 y.a. Hence, an even higher percentage of *erectus* fossils falls outside the dates given by any particular evolutionist.

When people become aware of the massive misrepresentation of the dates for the *Homo erectus* fossil material, they act perplexed. But the factual evidence is so clear that it cannot successfully be challenged. The perplexity usually gives way to the question, Why do evolutionists do this? The answer is obvious. If the date range of all the fossils having a *Homo erectus* morphology were commonly published on a chart as they are in this book, it would be clear that human evolution has not taken place. However, it is possible that evolutionists are not being intentionally deceptive. The reason may be deeper and more complex. Because of evolutionists' faith in and commitment to evolution, I believe we are seeing a psychological phenomenon. Evolutionists give us the dates they want *Homo erectus* to have, the dates they wish *Homo erectus* would have. I suspect it is more a case of self-deception on the part of evolutionists than it is an attempt to deceive others. It indicates how deeply their faith has colored their facts.

On the far end of the *Homo erectus* time continuum, *Homo erectus* is contemporary with *Homo habilis* for 500,000 years. In fact, *Homo erectus* overlaps the entire *Homo habilis* population, as the *Homo habilis* chart in this chapter shows. Thus the almost universally accepted view that *Homo habilis* evolved into *Homo erectus* becomes impossible. Later, we will demonstrate that *Homo habilis* is a flawed taxon, or category, because it is a mixture of fossils that can legitimately be called human (KNM-ER 1470, 1481, and 1590) and other fossils that are definitely not human. However, even if *Homo habilis* were a legitimate taxon, and 1470, 1481, and 1590 were proper members of that taxon, *Homo habilis* could not be the evolutionary ancestor of *Homo erectus* because the two groups lived at the same time as contemporaries.

In chapter 5 we discussed the process by which one species allegedly evolves into another species (regardless of whether the evolutionist believes in phyletic gradualism or punctuated equilibria). For *habilis* to evolve into *erectus*, *habilis* must precede *erectus* in time. Furthermore, after *habilis* has evolved into *erectus*, *habilis* must be eliminated by death, because *erectus* is supposedly the better fit of the two in the intense competition for limited resources. Yet, the fossil record shows that (according to evolutionist dating) *Homo habilis* and *Homo erectus* existed side by side as contemporaries for half a million years. The fos-

Homo erectus/*Homo habilis*—Contemporaneousness
(Evolution Time Scale)

	Major *Homo habilis* Fossils Each entry is one individual * Date uncertain (x considered not to be *Homo habilis* by the author)	*Homo erectus* (See *H.e.* chart for details)
1.5 m.y.a.	Olduvai Hominid 13 cranial fragments, Tanzania [129]	KNM-ER 992
		KNM-ER 803 O.H. 15
1.6 m.y.a.		KNM-ER 3883 KNM-ER 737
	Olduvai Hominid 16 skull fragments, Tanzania [130, 131]	KNM-ER 820 KNM-WT 15000 KNM-ER 1466
1.7m.y.a.		KNM-ER 1808
		KNM-ER 730 KNM-ER 807 KNM-ER 1821
1.8 m.y.a.	* Sterkfontein Stw 53 skull, South Africa [132] Olduvai Hominid 6 skull fragments, Tanzania [133, 134, 135] Olduvai Hominid 7 skull, hand, Tanzania [133, 134, 135] Olduvai Hominid 8 foot bones, Tanzania [133, 134, 135] Olduvai Hominid 35 tibia, fibula, Tanzania [133, 134, 135] Olduvai Hominid 49 radius, Tanzania [133, 134, 135] Olduvai Hominid 62 skeletal fragments, Tanzania [135, 136]	KNM-ER 3733 KNM-ER 1507 SK-15, 18a, 18b SK-84 SKX-5020
1.9 m.y.a.	Omo L 894-1 cranial fragments, Ethiopia [137] KNM-ER 1805 cranium, mandible, Kenya [82, 138] KNM-ER 1813 cranium, teeth, Kenya [82, 138] Olduvai Hominid 4 teeth, Tanzania [139, 140] Olduvai Hominid 24 cranium, Tanzania [139, 140] Olduvai Hominid 52 cranial fragment, Tanzania [139, 140] KNM-ER 1802 mandible, tooth, Kenya [141, 142] Swartkrans SK-68 tooth, South Africa [105, 143]	SK-45,847 KNM-ER 1809 Java (Djetis) KNM-ER 2598 KNM-ER 3228
2.0 m.y.a.	x KNM-ER 1590 partial cranium, Kenya x KNM-ER 1470, 1481 cranium, leg bones, Kenya KNM-WT 15001 cranial fragment, Kenya [144, 145]	Damiao

sil record also shows that *Homo erectus* lived alongside the group known as archaic *Homo sapiens* for the entire 700,000 years of archaic *Homo sapiens* history and that *Homo erectus* lived alongside a more modern form of *Homo sapiens* for two million years (according to evolutionist chronology).

When a creationist emphasizes that according to evolution, descendents can't be living as contemporaries with their ancestors, the evolutionist declares in a rather surprised tone, "Why, that's like saying that a parent has to die just because a child is born!" Many times I have seen audiences apparently satisfied with that analogy. But it is a very false one. In evolution, one species (or a portion of it) allegedly turns into a second, better-adapted species through mutation and natural selection. However, in the context of human reproduction, I do not turn into my children; I continue on as a totally independent entity.

Furthermore, in evolution, a certain portion of a species turns into a more advanced species because that portion of the species allegedly possesses certain favorable mutations which the rest of the species does not possess. Thus the newer, more advanced group comes into direct competition with the older unchanged group and eventually eliminates it through death. The older group is not able to compete successfully for the limited resources available. This competition and eventual death of the less fit is indispensable to the evolutionary process. However, in the human reproductive process, I do not compete with my child. I devote all of my resources to the *survival* of my child—not to his death. The analogy used by evolutionists is without logic, and the problem of contemporaneousness remains.

Terms like *Homo erectus* and *Homo habilis* are convenient terms to use in reference to groups of fossil material. But it is obvious that when evolutionists give dates for *Homo erectus* that do not fit the fossil material, or when they say that *Homo habilis* evolved into *Homo erectus,* contrary to what the fossil material shows, they are using those terms in a manipulative manner without regard for the fossil material in those categories. The most recently dated *Homo habilis* fossil is Olduvai Hominid 13, dated at 1.5 m.y.a. A number of *habilis* fossils cluster around the 1.8 m.y.a. date, and another group clusters around the 1.9 m.y.a. date. KNM-ER 1470, 1481, and 1590 are dated just under 2 m.y.a. All of the above fossils were dated by potassium-argon (K-Ar). The Sterkfontein STw 53 skull is of uncertain date, but considered to be in the 1.5–2.0 m.y.a. range based on the fauna. The Swartkrans

SK-68 tooth is dated between 1.8 and 2 m.y.a., likewise based on the associated fauna. The most recently discovered *Homo habilis* fossil, the KNM-WT 15001 cranial fragment, is dated between 1.85 and 2.35 m.y.a. (K-Ar). Thus, 2.35 m.y.a. is the oldest suggested date for any *Homo habilis* fossil, and even that particular fossil may be as much as a half million years younger, with a midrange of 2.1 m.y.a.

In 1991, Bed I at Olduvai Gorge was redated using the ^{40}Ar-^{39}Ar laser-fusion technique. At least nine *Homo habilis* individuals have come from this bed. The reasons given for redating Bed I are that "precise age estimates have been elusive," and "its detailed chronology is largely unknown."[146] Since Bed I had already been redated several times, and the Leakeys and others have worked at Olduvai Gorge for over thirty years, the need to redate Bed I does not inspire confidence in the dating methods. Although the reasons given for redating are certainly valid ones, I suspect that there was also the hope that redating could put a bit more age on *Homo habilis* to improve his credibility as a transitional taxon. Unfortunately, there was no evidence for an older date for *habilis*, and the need was expressed for even further dating work on Lower Bed I.

All of the *Homo habilis* material thus falls between 1.5 and about 2 m.y.a. In the *Homo erectus* fossils is a clustering of fossils at about 1.8 m.y.a. (K-Ar). The SK-45 mandible and the SK-847 partial cranium (both possibly belonging to the same individual) are dated between 1.8 and 1.9 m.y.a. (fauna). The Java (Djetis) fossils are radiometrically dated at 1.9 m.y.a. KNM-ER 2598 is dated from 1.85 to 1.95 m.y.a., and KNM-ER 3228 is dated between 1.9 and 2 m.y.a., both radiometrically. The Damiao, China, mandible is tentatively dated at 2 m.y.a. Thus, the oldest *Homo habilis* fossil has a midrange date virtually identical to the age of the oldest *Homo erectus* fossil. In all, twenty-four *Homo erectus* individuals are in the 1.5–2.0 m.y.a. range that are contemporaneous with the entire *Homo habilis* fossil material. *This does not constitute an evolutionary sequence.* (We have followed Klein's more recent date for Yuanmou Man on our fossil charts, but Chinese authorities date Yuanmou Man at 1.7 m.y.a. The recently discovered Dmanisi mandible with teeth from Georgia, U.S.S.R., is said to be either 0.9 m.y. or 1.6 m.y. old. We have placed it on the chart at 0.9 m.y. old. If the earlier dates for these two fossils are legitimate, it would place twenty-six *Homo erectus* individuals contemporary with the entire *Homo habilis* material.)

Anyone who is concerned about truth in packaging ought to be concerned over the way the relationship of *Homo habilis* to *Homo erectus* is presented in the evolutionist literature. Klein's *The Human Career,* as an example, is both recent and representative. In contrast to his detailed discussion of the individual *Homo erectus* and archaic *Homo sapiens* fossils, Klein gives virtually no attention to the individual *Homo habilis* fossils and gives the dates of virtually none of them. He is thus able to manipulate the taxon as a group by disregarding the dates of the individual fossils which comprise the group. On a bar graph Klein shows *Homo habilis* extending out of *Homo erectus* and ancestral to it, with the bar graph going back farther in time than any of the *Homo habilis* fossils in the category would warrant.[147] In a profile chart showing skulls of the individual taxa, the *Homo habilis* skull is below the *Homo erectus* skull, implying that *Homo habilis* is ancestral to *Homo erectus.* However, the individual fossil evidence shows this to be *absolutely false.* Evolutionists demonstrate a marvelous faculty for snatching fantasy from the jaws of truth.

If the fossil evidence at the far end of the *Homo erectus* spectrum shows no evidence of its having evolved from *Homo habilis,* the evidence at the near end of that spectrum shows no evidence of its having evolved into *Homo sapiens. Homo erectus* is in reality a form of *Homo sapiens* and should be so classified, as the next chapter will demonstrate.

Although the most recent date usually given for the disappearance of *Homo erectus* is about 300,000 y.a., at least 106 fossil individuals having *Homo erectus* morphology are dated by evolutionists themselves as being *more recent* than 300,000 y.a. Of those 106 fossil individuals at least sixty-two are dated *more recently than 12,000 y.a.* This incontrovertible fact of the fossil record effectively falsifies the concept that *Homo erectus* evolved into *Homo sapiens* and that *Homo erectus* is our evolutionary ancestor. In reality, it falsifies the entire concept of human evolution.

The most dramatic illustration of this condition of the fossil record is the Kow Swamp, Australia, fossil material initially discovered in 1967 and published in 1972 in *Nature.*[148] (The Cohuna cranium, having the same morphology, was discovered in Kow Swamp in 1925.) Others that are included in the group known as robust, *erectus*-like Australian fossils are the Talgai cranium discovered in 1886, the Mossgiel cranium discovered in 1960, and the Cossack skull discovered in 1972. Newer Australian sites yielding robust fossils include

Willandra Lakes and Coobool Crossing. The Java Solo fossils from the Ngandong Beds exhibit this same robust, *erectus*-like morphology. Their date is highly controversial, but a date of less than 10,000 y.a. based on the associated artifacts is justified.[149] These discoveries, however, are just the tip of the iceberg. At least four other locations near Kow Swamp are said to contain material of similar robust (*Homo erectus*) morphology but have not yet been explored in detail. Two of these sites are at Gunbower and Bourkes Bridge.[150] A third site is near the Murray River, and a fourth is at Lake Boga.[151]

The evolutionists' response has been interesting and predictable. Although these robust Australian fossils are said in the literature to have *Homo erectus* features (Kow Swamp and Cohuna,[152, 153] Mossgiel,[154] Talgai,[155] and WHL-50 from Willandra Lakes[156]), the evolutionist has waved his magic wand and called them *Homo sapiens* because of the very late date. Since *Homo erectus* long ago was supposed to have evolved into *Homo sapiens*, it is simply unthinkable that any *Homo erectus* fossils could still be around so recently. Thus, any thinking person would know that these fossils are *Homo sapiens*, no matter what they look like. The most common explanation now given by evolutionists for this *erectus*-like morphology is cranial deformation (discussed in chapter 14). We are asked to believe that evolution produced the *Homo erectus* morphology in the Lower and Middle Pleistocene, but that cranial deformation was responsible for this very same *Homo erectus* morphology after these individuals had allegedly evolved into *Homo sapiens* in the Upper Pleistocene.

While evolutionists have not yet developed a formal definition for *Homo erectus*, a suite of characteristics is generally accepted:

1. Skull low, broad, and elongated
2. Cranial capacity 750—1250 cc
3. Median sagittal ridge
4. Supraorbital ridges
5. Postorbital constriction
6. Receding frontal contour
7. Occipital bun or torus
8. Nuchal area extended for muscle attachment
9. Cranial wall unusually thick overall
10. Brain case narrower than the zygomatic arch
11. Heavy facial architecture

12. Alveolar (maxilla) prognathism
13. Large jaw, wide ramus
14. No chin (mentum)
15. Teeth generally large
16. Post-cranial bones heavy and thick

Where there is material for comparison, the Kow Swamp fossils, as well as the other robust Australian fossils, fit the above description well—allowing for reasonable genetic variation. They qualify as *Homo erectus*, as the evolutionist uses the term.

The evolutionists' attempt to explain the Kow Swamp fossils (and others) by calling them the result of an isolated population that was removed from the evolutionary mainstream also fails. While most of these robust fossils were found in southeast Australia, the Cossack skull was found on the west coast of Australia, two thousand miles away, and the Java Solo people were found three thousand miles away. We are dealing with a continent-wide phenomenon. Furthermore, the Cossack skull has a maximum age of 6500 y.a., but a minimum age of just a few hundred years.[157] Thus, it is possible that *Homo erectus*, whom the evolutionist claims is our evolutionary ancestor, walked the earth just a few hundred years ago. He is truly a man for all seasons.

13

Homo erectus: All in the Family

Is *HOMO ERECTUS* MORPHOLOGICALLY DISTINCT enough to warrant its being classified as a separate species? There have always been evolutionists who have asked that question. Now their tribe is increasing. Michael Day, reviewing G. Philip Rightmire's 1990 book, *The Evolution of Homo erectus*, writes:

> Of the three stages we know of the evolution of man (the australopithecine ape-men, *Homo erectus* the first true men, and early *Homo sapiens* our own species) *Homo erectus* of the Middle Pleistocene would have seemed the most clearly understood and the most taxonomically stable of them all a relatively few years ago—not any more. Important new finds as well as new ways of thinking about hominid taxonomy have thrown this 'species' into the same turmoil as all of the others.[1]

Day then enumerates the many questions that now embroil *Homo erectus*. One is, "Does *Homo erectus* exist as a true species or should it be sunk into *Homo sapiens*?"[2]

A number of evolutionists have expressed the fact that *Homo erectus*, while a bit different, is not so different from modern humans as to war-

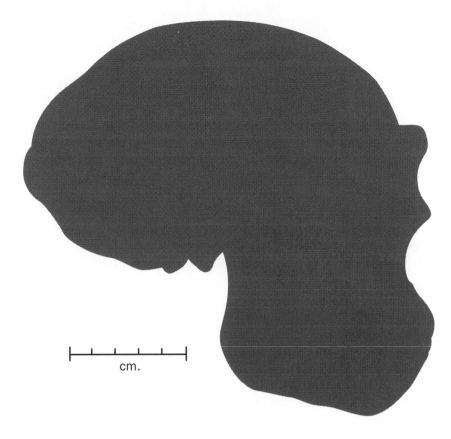

Peking Man, Zhoukoudian, China
(Weidenreich reconstruction)
Homo erectus

rant a separate species designation. Gabriel Ward Lasker (Wayne State University) has written:

> *Homo erectus* is distinct from modern man (*Homo sapiens*), but there is a tendency to exaggerate the differences. Even if one ignores transitional or otherwise hard to classify specimens and limits consideration to the Java and Peking populations, the range of variation of many features of *Homo erectus* falls within that of modern man.[3]

William S. Laughlin (University of Connecticut), in studying the Eskimos and the Aleuts, noted many similarities between these peoples and the Asian *Homo erectus* people, specifically *Sinanthropus* (Peking Man). He concludes his study with a very logical statement:

> . . . when we find that significant differences have developed, over a short time span, between closely related and contiguous peoples, as in Alaska and Greenland, and when we consider the vast differences that exist between remote groups such as Eskimos and Bushmen, who are known to belong within the single species of *Homo sapiens*, it seems justifiable to conclude that *Sinanthropus* belongs within this same diverse species.[4]

Milford Wolpoff has been one of the most vocal evolutionists calling for the "sinking" of the taxon *Homo erectus* into *Homo sapiens*. He writes in conjunction with Wu Xin Zhi (Institute of Paleoanthropology, Beijing) and Alan G. Thorne (Australian National University): "In our view, there are two alternatives. We should either admit that the *Homo erectus*/*Homo sapiens* boundary is arbitrary and use nonmorphological (i.e., temporal) criteria for determining it, or *Homo erectus* should be sunk [into *Homo sapiens*]."[5]

They then quote Franz Weidenreich. Weidenreich wrote the original descriptions of *Sinanthropus pekinensis* and made the very fine plaster casts of those fossils before the originals were lost. Writing in 1943, Weidenreich recognized that this classic *Homo erectus* material was not that different from *Homo sapiens*: ". . . it would not be correct to call our fossil '*Homo pekinensis*' or '*Homo erectus pekinensis*'; it would be best to call it '*Homo sapiens erectus pekinensis.*' Otherwise it would appear as a proper "species" different from '*Homo sapiens*' which remains doubtful, to say the least."[6]

The authors of the Time-Life book on *Homo erectus*, entitled *The First Men* comment on the postcranial anatomy of *Homo erectus*:

> His bones were heavier and thicker than a modern man's, and bigger bones required thicker muscles to move them. These skeletal differences, however, were not particularly noticeable. "Below the neck," one expert has noted, "the differences between Homo erectus and today's man could only be detected by an experienced anatomist."[7]

A remarkable discovery occurred in Kenya, in 1984, of an almost complete *Homo erectus* skeleton dated at about 1.6 m.y.a. It was the first time that a *Homo erectus* postcranial skeleton was uncontested as to its association with a *Homo erectus* cranium. This skeleton brought some surprises. Designated KNM-WT 15000, it is thought to be that of a twelve-to-thirteen-year old boy. Yet, he was 5'4" to 5'6" tall. Had he grown to maturity, it is estimated he could have been six feet tall. This was a new insight into the stature of *Homo erectus*, who was always assumed to have been smaller than modern humans.

Susman, Stern, and Rose give a more recent assessment of *Homo erectus* anatomy: "changes in locomotor anatomy from *H. erectus* to modern man were relatively minor and by earliest *H. erectus* times body size was essentially modern. . . ."[8]

The only true test of whether or not *Homo erectus* and *Homo sapiens* are members of the same species is the test of interfertility, which for obvious reasons is impossible to apply. Most Bible scholars would agree that if Adam and Eve were somehow restored to life, we would be able to produce offspring with them. The intervening years would not have affected the interfertility of the human race from its beginning. The biblical word *kind* is based upon this interfertility. We thus speak of mankind and humankind. Unfortunately, fossils do not reproduce. However, Donald Johanson expresses this opinion:

> It would be interesting to know if a modern man and a million-year-old *Homo erectus* woman could together produce a fertile child. The strong hunch is that they could; such evolution as has taken place is probably not of the kind that would prevent a successful mating. But that does not flaw the validity of the species definition given above, because the two cannot mate. They are reproductively isolated by time.[9]

While Johanson is just expressing an opinion, note that the major obstacle he sees to a *sapiens-erectus* mating is the time element. We have already stressed that a species distinction based solely on the time element is an evolutionary concept. It is valid only if evolution is valid. The fact that Johanson believes that *erectus* and *sapiens* could mate if they were living at the same time is actually a confession that the differences between them are not great. Furthermore, the fossil record shows that *sapiens* and *erectus* were living at the same time.

Johanson's statement reveals the semantic confusion rampant today. If one million years would not produce significant genetic change to inhibit conception, then the differences between *Homo erectus* and *Homo sapiens* are not the result of evolution but instead represent genetic variation within one species. Furthermore, although I am genetically isolated from my great grandmother because of time, this does not mean that she and I are in different species. A species distinction based primarily on time is an evolutionary necessity but absurd nonetheless.

When we compare the crania of *Homo erectus* with those of archaic *Homo sapiens* and Neandertal, the similarities are striking. My own conclusion is that *Homo erectus* and Neandertal are actually the same: *Homo erectus* is on the lower end, sizewise, of a continuum that includes *Homo erectus*, archaic *Homo sapiens*, and Neandertal. The range of cranial capacities for fossil humans is then in line with the range of cranial capacities for modern humans. Modern humans have a cranial capacity range from about 700 cubic centimeters all the way up to about 2200 cc.[10] This range—a factor of three—is an amazing spread and is most unusual in the biological world. It is recognized that this spread has virtually nothing to do with intelligence, because human intelligence is more dependent on how the brain is organized than on sheer brain size alone.

The cranial capacity of *Homo erectus* goes from about 700 cc (Java Modjokerto infant) to about 1200 cc (the largest Peking Man skull). The Neandertal cranial capacity begins at about 1250 cc (Saccopastore I, Italy) and goes to about 1740 cc (Amud I, Israel), with a few Neandertals possibly going a bit higher. If the archaic *Homo sapiens* fossils are factored in, they would fit in at the transition, with cranial capacities of about 1100 cc to about 1300 cc. Even when these three categories are considered as a single unit, they still do not reflect the cranial capacity range of modern humans.

Homo erectus and the Neandertals are very similar in cranial morphology. In the question period following a lecture by Neandertal authority Erik Trinkaus, I asked him, "Other than brain size, what are the differences in cranial morphology between *Homo erectus* and Neandertal?" His reply was, "Virtually none."[11]

In an article on *Homo erectus* C. Loring Brace writes: ". . . some scholars even treat Neandertal as late *erectus*."[12] And Harry L. Shapiro (American Museum of Natural History) observes: ". . . when one examines a classic Neanderthal skull, of which there are now a large number, one cannot escape the conviction that its fundamental anatomical formation is an enlarged and developed version of the *Homo erectus* skull."[13]

Jared Diamond, telling of the work of Alan Walker (Johns Hopkins University), says: "Walker's analysis of skull shape shows that Neanderthals were much more similar in this respect to *H. erectus* than to *H. sapiens* [modern *Homo sapiens*]."[14]

It is because of the similarities of the various categories of fossil humans that evolutionists have had great difficulty in assigning certain fossils to a specific category. The African archaic *Homo sapiens* fossils have been referred to in the past as "African Neandertals." Many of the Asian *Homo erectus* fossils have been termed "Asian Neandertals." A running battle is currently going on over the precise assignment of many fossils. The problem of deciding which fossils properly belong in the *Homo erectus* taxon is so vexing that Jerome Cybulski (National Museum of Man, Ottawa) laments:

> Indeed, one may well wonder whether agreement will ever be reached as to which fossils do belong to or represent the taxon, and on what morphological-cum-phylogenetic grounds fossil hominids are or are not to be regarded as *Homo erectus.*[15]

An evolutionist would claim that in an evolutionary continuum we would expect to have a number of fossils "on the line," since the transition points of the various species are arbitrary because of the nature of the evolutionary process. We agree. If evolution were true, that would be the case. However, there is another, more satisfying way in which to interpret the data: a morphological continuum that includes just one species of humans, called *Homo sapiens*. In a court of law, a case must be proven beyond all reasonable doubt. The fact that the data can be explained in at least two ways shows that the evolutionist is far from proving his case. That he *never* considers the other arrangement shows that he is far from impartial.

There is a way, however, by which we can discriminate between the two possible explanations of the data and thereby determine which is the more likely to be correct. That is to place *all* of the relevant fossil material on a time chart according to the probable dates for each of the fossil individuals and to evaluate the results as to whether the evidence favors an evolutionary or a morphological continuum. When this is done, as it is done in this book, the evidence is strongly in favor of a morphological continuum, both horizontally across species and vertically over time. The horizontal continuum shows that anatomically modern *Homo sapiens*, Neandertal, archaic *Homo sapiens*, and *Homo erectus* all lived as contemporaries over extended periods of time. The vertical continuum shows that as far back as the human fossil record goes the human body has remained substantially the same and has not evolved from something else.

This condition is what the creation model would predict. It is what we would expect if creation were true. The evidence, in fact, is so strong for the creation model of human origins that it is extremely unlikely that any future fossil discoveries would weaken it. This is because no future fossil discoveries in the 1– 4.5 m.y.a. time period could cancel out the solid body of factual evidence that has already been accumulated. Up to now, new fossil discoveries have only strengthened the creationist position. It is understandable why evolutionist books no longer carry this type of human fossil chart. Charts of bits and pieces of the human fossil record abound in evolution books, but one will look in vain in an evolutionist work for a time chart that places *all* of the relevant human fossil material on a time chart according to the morphological description of the individual fossils.

The Archeological Evidence

Archeological evidence also demonstrates that the distinction between *Homo erectus* and *Homo sapiens* is an artificial one. Although there are limitations in the archeological record, all of the evidences that one could reasonably expect to be discovered to demonstrate the full humanity of *Homo erectus* have already been found in association with him.

Of the seventy-seven localities where *Homo erectus* fossils have been found, more than half (forty-two) of these sites have also yielded stone tools. At eleven of the *Homo erectus* sites there is evidence of the controlled use of fire. Most significant is that at one of the oldest sites, Swartkrans, South Africa, thought to date between 1.5 and 2 m.y.a., *Homo erectus* fossils have been found in association with both tools and fire. While it is technically impossible to prove that the stone tools and the fire at a given site were made by *Homo erectus* individuals, the sheer number of associations makes it unreasonable to believe otherwise. Three Upper Pleistocene *Homo erectus* sites show evidence of burial, one of a cremation, one of the use of red ochre, and another of the use of bone chopping tools. We hardly dare ask the archeological record for more evidence of the true and full humanity of the individuals having a *Homo erectus* morphology.

One of the most popular myths of human evolution is that stone tools testify to the increasing mental and conceptual abilities of humans as they evolved. The most "primitive" stone tools, Oldowan, once were identified by evolutionists with *Homo habilis,* the most "primitive" of humans. Acheulean tools (named for the French site at which they were discovered) were associated with the more evolved *Homo erectus.* Neandertal was said to be associated with the even more advanced Mousterian tool kits. The most sophisticated and artistic stone tool were identified with Cro-Magnon and other relatively modern peoples. Thus, stone tools were once considered an almost independent confirmation of the evolutionary development of the human mind.

Things are different now. Almost every basic style of tool has been found with almost every category of human fossil material. Stringer and Grun write: "The simplistic equation of hominids and technologies in Europe has thus been abandoned."[16] The fallacy of the evolutionary archeologist was to equate simple with primitive. Louis and Mary Leakey were among the first to identify the "primitive" Oldowan tools with the primitive *Homo habilis,* both thought to be about two million years old. However, Mary Leakey tells of discovering Oldowan-type tools in Kanapoi Valley, northern Kenya, associated with potsherds and hut circles that gave evidence of being rather recent. She writes:

> . . . the occurrence of an industry restricted to heavy duty tools of Lower Palaeolithic facies associated with pottery and hut circles, is an anomaly hard to explain. It may be noted, however, that a crude form of stone chopper is used in the present time by the more remote Turkana tribesmen in order to break open the nuts of the doum palm.[17]

On these "primitive" Oldowan tools Lawrence Robbins (Michigan State University) commented: "It is interesting that these oldest of technological items were among the most successful inventions for they continued to be manufactured throughout the entire Stone Age."[18]

If these Oldowan tools were so successful and efficient that they were used throughout the entire Stone Age (Paleolithic, Mesolithic, and Neolithic), if they were the best tools for certain jobs, and if they are still the best tools for certain jobs in some parts of the world today, is it intellectually honest for evolutionists to refer to them as primitive and use them as evidence for the evolution of the human brain?

One of the most common Acheulean tools is the hand ax, which in the past has been almost exclusively identified with *Homo erectus*. So complete was this identification with *Homo erectus* that if hand axes were found at a habitation site, it was called a *Homo erectus* site even though no *Homo erectus* fossils were found there to so identify it. Now it is known that Acheulean tools, including hand axes, have also been found with *Homo sapiens* fossils. Furthermore, many Asian *Homo erectus* fossils are found with tools considered to be more primitive than those of the Acheulean Culture. (In Asia, it is now believed, bamboo tools were used more extensively than stone tools, which could account for stone tools being less frequent and more "primitive" there.)

The Acheulean hand ax, however, was truly used worldwide. It is found from northern Europe to southern Africa, and from the Mediterranean to India and Indonesia. It is also mystifying. Although it is called a hand ax, no one knows for sure what its use was. In shape it resembles a giant almond, pointed at one end and round on the other. The pointed end is thinner, the rounded end thicker, but overall it is rather flat like the almond. Because the rounded end is thicker, it has an eccentric center of gravity; in lengthwise cross section, it looks like a very tall and skinny teardrop. The length ranges from a few inches to well over a foot. I have seen some that were rather crudely made and others that were works of art. It has a cutting edge all around its perimeter, and as far as we know, it was never hafted (used with a handle). The assumption is that it was some type of chopper; hence its name. The problem is that since it is sharp all around, it could do as much chopping on the hand using it as it did on the object being chopped.

Eileen M. O'Brien (University of Georgia) had a better idea.[19] Her experiments led her to conclude that the hand ax was actually a flying projectile weapon, thrown discus style and used in the hunting of large game. To test this idea, she had a fiberglass replica made of one of the largest hand axes in the collection at the National Museums of Kenya, Nairobi. It was about a foot long and weighed a bit over four pounds. She then had several discus throwers practice with it. When thrown, the hand ax spun horizontally as it rose, like a discus. However, when it reached its maximum altitude it flipped onto its edge and descended that way. It landed on its knife-sharp edge 93 percent of the time, and

70 percent of the time it landed point first. The average throw was over one hundred feet, and it was usually accurate to within two yards right or left of the line of trajectory. The Olympic record for the discus, which weighs about the same as O'Brien's hand ax replica, is well over two hundred feet. O'Brien believes that ancient humans could have attacked large animals over two hundred feet away with great accuracy.

One of the puzzling aspects of hand axes is that they are found in great numbers in places that used to be streams, rivers, or lakes. It would be logical for bands of ancient humans to attack animals when they came to water. Axes that landed in the dirt could be retrieved. However, axes that were overthrown and landed in the water usually would not be recovered, which could explain why we find them in those places today.

Also significant is that the Acheulean hand ax first appears in the archeological record at about the same time as evidences of large animal kills—hippopotamus, elephant, and *Dinotherium* (an extinct elephant-like animal with large tusks in the lower jaw). Bands of ancient humans, throwing four-pound hand axes two hundred feet, could inflict heavy damage on even the largest and toughest game.

Although it was round, the discus of the ancient Greeks was unhafted, edged all around, and made of stone. To some people, the discus throw seems a strange and unlikely sport. O'Brien suggests the possibility that the Olympic discus throw is a carry-over from the hand-ax hunting technique of ancient humans. She asks: ". . . is it possible that the ancient Greeks preserved as a sport a tradition handed down from that distant yesterday?"[20]

If that is the case, that "distant yesterday" may not have been so long ago.

14

Back to the Future

ARE THERE ADEQUATE NONEVOLUTION EXPLANATIONS for the distinctive morphology of the fossil humans *Homo erectus*, archaic *Homo sapiens*, and Neandertal? Although we previously addressed some possibilities in the chapter on the Neandertals, the most likely explanation for the morphology of fossil humans is related to the biblical record of earth history.

Many people believe there is a serious conflict between science and the Bible. A classic work on this subject is Andrew White's *A History of the Warfare of Science with Theology*. However, the famous scientist and educator James B. Conant made this interesting observation on White's book:

> . . . the warfare White describes has been for the most part *a series of battles in regard to the interpretation of the past.* . . . As long as one sticks to physics, chemistry, and experimental biology, a cautious approach to science such as I have employed will offend but few.[1]

The overwhelming majority of people working in science and technology deal with the present, not the past. The overwhelming majority of books and journal articles of a scientific nature also deal with the present, not the past. In truth, there is simply no conflict between the

144

Bible and scientific discoveries and observations in the present. The only conflict between science and the Bible involves the scientific community's interpretation of the past.

While science thrives on observation and experimentation in the present frame of reference, it has no mechanism to observe the past with the same authority it has to observe the present. The scientific method (or methods) applies to the past only indirectly, if at all. In the absence of historical records, all data regarding the past involves interpretations which may or may not be correct. Not without reason did Oxford scholar R. G. Collingwood say that the study of the past is really a study of the human mind,[2] indicating that there is a high degree of subjectivity in all scientific reconstructions of the past. It is unfortunate that many scientists think that because they are able to make authoritative statements about present processes of nature, this allows them to speak dogmatically about the past history of our planet, ignoring historical documents such as Genesis.

The failure of the scientific community to recognize the high degree of subjectivity in its interpretations of past events is the major cause of the "warfare" between the Bible and this area of science. While Christians may not always have handled this warfare well, the *cause* of the conflict cannot be laid upon our shoulders. Believing that we have a reliable historical record of two crucial past events, creation and the Flood, the Christian accepts the contributions of science to the present with gratitude and thanksgiving while he challenges the authority of science to make dogmatic statements about the past.

Typical of the confusion that comes from not understanding that the nature of the scientific method and the nature of biblical revelation are two distinct and separate things is the statement by Eugenie C. Scott (National Center for Science Education, an organization founded to combat scientific creationism): "The scientific method is vastly superior to revelation (or other epistemologies) as a means to discover the workings of the natural world."[3]

Scott is absolutely right in what she says, but she doesn't have a clue as to why she is right. She is right, but she is comparing apples with oranges. The primary purpose of revelation is not to tell us about the "workings of the natural world." Why should God give us a revelation of things we can discover for ourselves when we utilize (among other ways) the scientific method? God's purpose in biblical revelation is to give us information on things we could not know by any other means.

For this reason God has given us a revelation of two momentous historic events: creation and the Flood. Knowledge of these two singular, unrepeatable events is beyond the scope of the scientific method. Scott confuses the issue. She has lumped information of the past through God's revelation with information of the present obtained by the scientific method. She then indicates that she prefers the information provided by the scientific method above the information provided by the Word of God, whether the scientific method is adequate for the task of dealing with past singularities or not.

The Ice Age

One of the remarkable features of the biblical revelation, when it is interpreted literally, is its internal consistency in recording past events that help to explain the present world. Nowhere is this consistency seen more clearly than in the ability of the worldwide Genesis Flood to provide the only adequate explanation for a great geological mystery: the Ice Age. Few realize that the Genesis Flood and the Ice Age are intimately connected in terms of cause and effect. It was the severe disruption of the global climate by the Genesis Flood that caused the Ice Age to develop immediately afterward. (The scientific community speaks of four ice ages during the Pleistocene, but the three earlier ones are open to challenge.)

It would be natural to assume that all that is needed to produce an ice age is a series of very cold winters. Not so. In many areas of the world the winters are very cold; yet, these areas may have little snow. In contrast to our present climate, the two basic requirements for an ice age are (1) much cooler summers, and (2) much more snowfall. Probable contributing mechanisms would be warmer winters and warmer oceans. This combination is an unlikely scenario, and the scientific community, limiting itself to present processes, has hard sledding in trying to account for the cause of the Ice Age. Over sixty theories have been proposed to explain the Ice Age, all with serious defects. However, from computer simulations and from what we know of atmospheric science, the Genesis Flood was capable of producing the Ice Age.

Relatively warmer winters were probably one of the mechanisms that supplied the abundant snowfall necessary for the Ice Age. First, the northern oceans had to remain ice free to provide the moisture source for that abundant snowfall. Today, the Arctic Ocean is normally frozen, and colder winters would cause major portions of the North Atlantic to freeze as well. Thus, cold winters cut off the moisture sources for an ice age. Furthermore, cold air carries very little moisture. Air temperatures in winter had to be warm enough to carry the massive amounts of water to be precipitated as snow. Those who live in northern areas know that very cold weather seldom brings large amounts of snow, whereas some of the heaviest and wettest snowstorms occur in spring.

No matter how heavy the snowfall might have been in winters, if the summers had been warm enough to melt all of the snow, no Ice Age would have occurred. An ice age cannot develop in one year. There must be an accumulation of snow over many years without significant summer melts. Hence, very cool summers were a basic necessity for the Ice Age.

I was raised in Fargo, North Dakota. Fargo is in the midst of some of the richest farmland on earth—the Red River Valley, which is an ancient lake bed formed by glacier melt water. East, in Minnesota, there is also much evidence of ancient glaciers. Today, however, in winter the temperature sometimes drops to forty degrees below zero. In summer, one hundred-degree weather is not uncommon. That is not an ice-age producing climate. The mystery of the Ice Age is this: in the very geographic areas where there is incontrovertible evidence of past continental glaciers, the climate today is such that it is impossible for an ice age to be produced there. Further, it stretches the imagination to create a scenario that would have allowed glaciers to be produced there in the past. Certainly, as far as the Ice Age is concerned, the present is not the key to the past. A logical inference is that since the Ice Age was caused by conditions which do not exist today, only historical records such as the Book of Genesis can give us clues as to what those past conditions were.

Oxygen isotope and other data give scientific evidence that the deep ocean was warmer for extended periods of the past. While the scientific community cannot explain this fact, warm oceans are exactly what the Genesis model would predict for the early earth. The waters above the expanse (atmosphere), created on day two of creation week (Gen. 1:6-8), would cause a relatively uniform climate worldwide, including warm oceans and warm polar regions. The oceans would have become even warmer at the time of the Flood when hot water, which was a part of the "fountains of the deep," mixed with them. These warm oceans served as a mechanism to supply the abundant moisture necessary for the heavy snowfall of the Ice Age.

After the Flood, the oceans, while gradually cooling, would remain warm for hundreds of years. However, the land surfaces would cool very quickly for the following reasons: (1) The waters above the atmosphere, having contributed to the rains for the Flood, would now be gone and their insulating effect would be removed; (2) the intense volcanism of the Flood would continue for a time after the Flood and would put high amounts of particulate matter and aerosols into the atmosphere which would reflect the sun's rays back into space; (3) immediately after the Flood the earth would be barren and denuded, and barren land is highly reflective of the sun's rays; (4) the heavy cloud cover produced by the warmer oceans would also be highly reflective of the sun's rays; and (5) the rapidly building snow cover would itself have a cooling effect on the land. The result is that the land surfaces would cool very quickly after the Flood.

This formula of very cool summers and very heavy snowfall, together with the probable mechanisms of warmer winters, warmer oceans, and cold land surfaces becomes the secret for the Ice Age. It explains why the Ice Age started immediately after the Flood, why there had not been an ice age before that time (evolutionists' claims of pre-Pleistocene ice ages notwithstanding), and why there cannot be an ice age again. It also explains why the scientific community, which rejects the Genesis Flood, has never been able to adequately explain the cause of the Ice Age. Evangelicals who believe in a local Noachic Flood are also hard pressed to explain the cause of the Ice Age and seldom address the issue.

Creationist Michael J. Oard (National Weather Service) has meticulously detailed the cause and nature of the Ice Age in his monograph *An Ice Age Caused by the Genesis Flood*.[4] While those rejecting the Genesis account of early earth history are at a loss to explain the cause of the Ice Age, they are in some agreement with creationist atmospheric scientists such as Michael Oard and Larry Vardiman (Institute for Creation Research) as to what the Ice Age climate was like. And that is what is relevant to our subject.

The Ice Age Climate

During the Ice Age, continental ice sheets covered large sections of North America, northern Europe, and northwest Asia. The ice occurred even on the high mountains of the tropics. The climate over much of the earth was cold, damp, and rainy. Thick cloud cover caused by the warm oceans together with heavy post-Flood volcanism robbed the earth of much of its sunshine. This condition would have lasted until the slowly cooling oceans gradually approached their present temperature range. It is significant that the Book of Job, with its setting after the Flood (Job 22:16) and probably before Abraham, has more references in it to snow, ice, and violent weather than any other book of the Bible.

Michael Oard estimates that the time from the end of the Flood to the time when the largest volume of ice and snow covered the land (glacial maximum) was about five hundred years, and the time for deglaciation about two hundred years. He writes: "Thus, the total length of time for a post-Flood ice age from beginning to end, is about 700 years."[5] However, Oard cautions that these are difficult figures to come by, ". . . because there are too many variables that are poorly known."[6]

The human responses to the harsh climate of the Ice Age would have been (1) to seek out natural shelters such as caves, (2) to construct shelters out of whatever material was available, and (3) to wear heavy clothing, probably animal skins, to cover much or all of the body. The lack of access to sunshine because of the heavy cloud cover, their need for shelter, and the wearing of heavy clothing would have predictable results: rickets.

Rickets is a deforming bone disease caused by a lack of vitamin D. In humans, vitamin D is produced in the deep layers of the skin through irradiation by the ultraviolet component of sunlight, the very component of sunlight most effectively filtered out by heavy, thick clouds.

The only significant dietary sources of vitamin D are fatty fish and egg yolk. The archeological record gives no evidence that *Homo erectus*, archaic *Homo sapiens*, or the Neandertal peoples consumed these foods except sporadically. On the other hand, the Cro-Magnon people, with their very modern morphology, give evidence that fish contributed substantially and routinely to their diet. Further, by Cro-Magnon times, the Ice Age would have been in its final stages with relatively cool oceans, less cloud cover and volcanism, a dryer climate, and more sunshine.

Rudolf Virchow

When the first fossil human was discovered (the original Neandertal) several competent medical authorities stated that the peculiar apish shape of the bones was caused by rickets. In 1872, Rudolf Virchow published a carefully argued and factual diagnosis that the original Neandertal individual had been a normal human who suffered from rickets in childhood and arthritis in adulthood. Virchow's diagnosis has never been refuted. It was ignored in his day because by coincidence the Neandertal morphology was what evolutionists believed a transition between primates and humans would look like. Today, too, Virchow's report is ignored. We assume that his medical knowledge was unsophisticated. Furthermore, rickets is relatively unknown, and Virchow is hardly a household name.

Rudolf Virchow (1821-1902) was professor of pathology at the University of Berlin. He was the first to recognize the cell as the basic unit for alteration in disease. He founded the science of cell pathology; he is known universally as the father of the science of pathology. His discoveries, among others, include embolism and leukemia. He redefined sarcoma and melanoma. His microscopic studies added tremendously to our knowledge of connective tissue, inflammation, and tumors. He cleared up the life cycle of the trichinae. His discoveries were important foundations for surgery and drug therapy.

A political activist, Virchow was a member of the Prussian National Assembly and the German Reichstag. Through his political influence he was a pioneer in public health, successfully fighting for better hospitals, schools, meat inspection, and sanitation procedures. "During the last third of the 19th Century, Virchow probably was the best-known medical man in the world."[7] Nevertheless, despite all of this, Virchow was guilty of an unpardonable sin: He was quite skeptical of Darwin's theory of natural selection, feeling that it lacked sufficient demonstration.

Virchow was also an anthropologist. He was responsible for the emphasis in his day on laboratory research in anthropology, and was involved in anthropological field work in Germany and Greece. He was one of the founders of the German Anthropological Society, and of the Berlin Society for Anthropology, Ethnology, and Prehistory, the latter of which, from 1869 until his death, he served as president and editor of its journal. He was personally familiar with the original Neandertal fossils and expertly acquainted with the disease of rickets. *This was the man who diagnosed the first Neandertal as a case of rickets.*

Virchow was well acquainted with rickets because rickets was particularly common in the industrial parts of Europe in the eighteenth and nineteenth centuries. The same industrial pollution that darkened the barks of the trees in England which in turn caused the ratios in the peppered moth population to change from light to dark (falsely claimed by evolutionists as an illustration of evolution in action)[8] also obscured the sunlight in these industrial areas. The result was that many children, especially those having inferior diets, suffered from rickets.[9]

The relationship between a sunless climate and a high incidence of rickets was well known by medical authorities in Virchow's time. However, a vitamin D deficiency as the cause of rickets was not identified until after World War I. Because rickets is clinically most active in humans between the ages of six and twenty-four months, vitamin D is now added to milk in most western countries. The result is that rickets is virtually unknown today in the United States. However, a friend of mine, born in a poor section of Boston in 1913, recalls rickets as being a rather common disease during his childhood years.

A more recent identification of fossil humans and rickets was made by Francis Ivanhoe in a paper in *Nature*. Ivanhoe said that ". . . every Neandertal child skull studied so far shows signs compatible with *severe* rickets."[10] These include the child remains from Engis (Belgium), La Ferrassie (France), Gibraltar, Pech de l'Aze (France), La Quina (France), Starosel'e (U.S.S.R.), and Subalyuk (Hungary). Less extreme cases are seen in the child remains from Teshik-Tash (U.S.S.R.), Shanidar (Iraq), and Egbert (Lebanon). The rickets skull morphology, seen in these children, has carried over into the adult Neandertals and other fossil humans. The gross bowing of the long bones of the body, so typical of rickets, is seen in both Neandertal children and adults.

If a number of human fossil remains are compatible with the disease of rickets, it is fair to ask why this has not been generally recognized, especially in light of our advanced medical knowledge. First, it is *because* of our medical progress that the disease has been overlooked in fossil humans. Medically, we no longer think in terms of rickets. In the very countries where evolution is taught most extensively, rickets has been practically eliminated. In all of my life, I have seen only one person who, I suspect, has had rickets. Even a medical doctor, dealing with flesh-and-blood people, does not see the bones of his patients directly useless he takes x-rays. Hence, when we see pictures of the fossil bones of Neandertal or *Homo erectus*, we do not make the identification.

A second reason why the rickets diagnosis has been ignored is easy to understand in retrospect. It is one thing to recognize a single specimen as being pathological, such as the original Neandertaler. However, when more Neandertals were discovered in Europe, and the Java and Peking remains were found in Asia, it would have been hard to think in terms of a worldwide pathology unless one also thought in terms of a worldwide *cause* for that pathology. It is more than coincidence that at the very time of these fossil discoveries, the Book of Genesis was coming under intensive attack, and the credibility of the creation and Flood accounts was being severely challenged. Hence, the one event that could explain a worldwide pathology—the Genesis Flood—was being discredited in intellectual circles. In light of that background it is understandable why the pathological condition of post-Flood humans came to be interpreted as a "normal" phase of human evolution.

The third reason why the rickets diagnosis was ignored has to do with the evolutionists' expectations of that day. For many years the Piltdown fraud was not recognized as such, because Piltdown Man was the type of fossil that evolutionists expected to find. In fact, the fraud was committed to advance a certain philosophical concept of human evolution. It was only when that particular philosophical concept became obsolete that the Piltdown fossils were studied more objectively and the fraud discovered. The Neandertals, Java Man, and Peking Man were also the kind of fossils evolutionists were looking for. By a unique coincidence, rickets gives to the skull and to the long bones of the body the superficial apish cast the evolutionists were expecting to find in the "missing link."

Because the human fossils are used today as evidence for human evolution, it is natural for us to assume that the concept of human evolution grew out of the fossil evidence. (That is how the scientific method says it should happen.) But such was not the case. When Charles Darwin published his famous *Origin* in 1859, he had been working on his theory for thirty years. (His grandfather, Erasmus Darwin, was also a well-known evolutionist.) Although Darwin did not directly address human evolution in the *Origin*, he clearly had humans in mind, as the last sentence of the *Origin* reveals. The original Neandertal fossils were found in 1856. While Darwin was undoubtedly aware of them, there is no evidence that he had ever seen the fossils when he published the *Origin*, or even when he published *The Descent of Man* in 1871.

In other words, the allegedly scientific concept of human evolution was well established before the relevant fossil evidence was discovered. Eugene Dubois was also thoroughly committed to human evolution before he went to Java to look for evidence. The human fossils, when discovered, were then used to try to prove human evolution, which had already been well developed on philosophical grounds. Thus, the human fossils were never evaluated in a neutral atmosphere to determine what they really were. It is a classic case of using ambiguous evidence to prove a preconceived idea.

Because the Neandertal and *Homo erectus* fossils were the type of thing that evolutionists were expecting to find, it has been relatively easy for them to explain the morphology of the Lower and Middle Pleistocene fossils as phases of human evolution. It has been harder for them to explain that same morphology in Upper Pleistocene fossils, because humans at that point were supposed to have a more modern appearance. This is the reason for the "Neandertal Problem." It is also the reason for a lesser known but more serious problem for evolutionists—the Upper Pleistocene *Homo erectus*-like fossils.

Recent fossil discoveries in Australia reveal a condition that defies evolution as an explanation. Two populations were living side by side in very recent times. One population had a very modern morphology, and the other had a *Homo erectus*-like morphology. The *erectus*-like fossils include the Mossgiel individual (discovered in 1960 and dated at about 6000 y.a.), the forty Kow Swamp individuals (first discovered in 1967 and dated at about 10,000 y.a.), and the Cossack skeletal remains (discovered in 1972 and dated from just a few hundred years ago to about 6500 y.a.).

We can sense the evolutionists' bewilderment as they write about these fossils. Jeffrey Laitman (Mt. Sinai School of Medicine) mentions fossil authorities who speak of the "extreme disparities" found between these two groups.[11] And Richard G. Klein declares that "the range of variation [between these two groups] is extraordinary. . . ."[12]

The range of explanations given for these two contemporaneous groups exceeds the range of variation in the fossils themselves. At the

time the Kow Swamp report was published, an editorial in *Nature* suggested many possibilities for the *erectus*-like morphology: (1) These
fossils could represent a small inbred community; (2) the thick bones
of these fossils could be the result of differential survival—thicker
bones would survive intact longer than thinner bones; (3) the thick
cranial bones could be the result of a nutritional problem; (4) low-
grade anemia; (5) genetic factors; (6) endocrinal factors; or (7) a
pathological condition.[13] The writer certainly covered all the bases.
However, in doing so he gave away the store. The admission that one
or more of these factors could produce a *Homo erectus*-like morphology
is also an admission that the concept of human evolution is not needed
to explain that morphology—which is what creationists have claimed
all along. All of the explanations suggested in that editorial are nonevolutionary.

The authors of the Kow Swamp report, A. G. Thorne (Australian
National University), and P. G. Macumber (Geological Survey of Vic-
toria), give their nonevolutionary explanation. In a sure and confident
manner they say that ". . . the Kow Swamp series represents an isolated and remnant population."[14] But at the very time (1972) they were
writing those words, the Cossack skeletal material, having that very
same morphology, was being discovered on the west coast of Australia
two thousand miles away. Hence, we are not dealing with an "isolated
and remnant population." The authors of the Cossack article write
that the Cossack discovery ". . . indicates that this morphology was not
a regional variant but continental in distribution."[15]

Many evolutionists have suggested that these diverse Australian
populations are the result of two or even three migrations into Australia from elsewhere.[16] However, this explanation does not solve the
problem; it just pushes the problem back to the Asian land mass.
Evolutionists date the first humans in Australia at a bit before 40,000
y.a. Hence, even if these successive waves of migration were separated by as much as 20,000 years, the differences in morphology cannot
be ascribed to evolutionary processes. On the evolution time scale
20,000 years is nothing. It just means that these two morphologically
diverse groups, which may have been separated geographically, had
been living as contemporaries on the Asian mainland. Hence, the
dual migration hypothesis doesn't even address the real problem, let
alone solve it.

The most common explanation (also a nonevolutionary one) for these Upper Pleistocene *Homo erectus*-like fossils is the idea of cranial deformation. The artificial deformation of bones is well known in human history. A classic example is the distortion of the feet of Chinese women caused by foot binding in childhood. In the Americas a common cause of skull deformation was the strapping of an infant to a cradle board. In South America Inca infants of noble birth would have their heads bound so as to give the heads a pointed shape. The purpose was to distinguish the nobility from the commoners.

Richard Klein suggests cranial deformation as a likely explanation for the Kow Swamp fossils.[17] The *Catalogue of Fossil Hominids* also suggests this as a possibility for the skull of the Kow Swamp 2 individual.[18] As to the particular method of cranial deformation, Chris Stringer says that the practice of headbinding ". . . was certainly responsible for some of the peculiarities in cranial shape amongst the Kow Swamp people."[19] Phillip J. Habgood (University of Sidney) suggests that the method used was one of repeated pressure to the front and the back of the infant's cranium. He feels that this method, unlike binding, would allow for the degree of variable deformation that one sees between the Kow Swamp 5 and the Kow Swamp 2 individuals.[20] Thorne, while acknowledging the possibility of cranial deformation, likewise sees no evidence of long-term binding in these fossils.[21]

Freedman and Lofgren, describing the Cossack cranium, say that it is very similar to the Kow Swamp crania but very different from other recent Western Australian skulls. They further state that the differences in the fossils do not appear to be due to artificial cranial deformation.[22]

To use cranial deformation as the explanation for the large number of Upper Pleistocene *Homo erectus*-type fossils seems contrived. That explanation is never given for fossils of similar morphology in the Lower and Middle Pleistocene. Why is it valid for one geologic period and not for the others? Nor is that explanation given for the Neandertal and archaic *Homo sapiens* fossils possessing a similar morphology. If cranial deformation can produce that morphology, then evolution is not needed. It is time to inform the evolutionist that he can't have it both ways. It is only the pressure of evolution that spawns ideas which are without one shred of factual evidence.

Textbook illustrations of valid cases of artificial cranial deformation are quite different from the typical *Homo erectus* morphology.[23] I have seen many museum specimens of artificial cranial deformation, and they, too, are very different from a *Homo erectus* morphology. In fact, it is hard to imagine a method of artificial cranial deformation that would result in an *erectus*-like skull. No type of cranial deformation could produce the thick cranial walls that are so typical of *erectus*-like individuals.

It is surprising how unscientific evolutionists can be when their theory is under stress. There is no excuse for ignoring the large body of evidence of rickets as the more probable explanation for the morphology of many fossil humans. Rickets could also explain why fossil populations existed simultaneously who had different morphologies because of different diets and climate conditions. It is time to go back to the future—back to Virchow.

15

Homo habilis: The Little Man Who Isn't There

HOMO HABILIS HAS ALWAYS BEEN A PROBLEM. It has been a problem to many evolutionists, and it has certainly been a problem to me. Fortunately, thanks to a recent fossil discovery, we now know why.

Louis and Mary Leakey had worked at Olduvai Gorge, Tanzania, for many years. The gorge, part of the East African Rift System, had produced many stone tools and animal fossils, but no hominids. Yet, Louis felt that hominid fossils had to be there. One day in 1959, because Louis was ill, Mary went out alone. At a certain spot she saw teeth sticking out of the ground. Excavation revealed a large cranium having some resemblance to the South African robust australopithecines.

The stone tools found in association with the fossil led Louis to believe that this individual was a tool maker. And to Louis, tool making meant just one thing: man! Believing that they had found the first tool maker, Louis named the fossil, *Zinjanthropus*, "East Africa Man." The ridiculously large molars indicated that the individual probably lived on nuts and berries, and so it became affectionately known as "Nutcracker Man."

Some of us suspect that Louis knew all along that "Zinj" was just a variant of a robust australopithecine. But the financial support Louis desperately needed to continue his work does not come from the discovery of fossil primates. It comes from finding human ancestors. The long financial association the Leakeys had with the National Geographic Society began at this time. Telling of the discovery of "Zinj" in *National Geographic*, Louis began his report: "The teeth were projecting from the rock face, smooth and shining, and quite *obviously human*."[1] Included in the article was an artist's painting of what "Zinj" might have looked like. It was quite a piece of work. Whereas our eyes are about midway between our chin and the top of our head, there was hardly any head at all showing above "Zinj's" eyes. He had virtually no brain. But the skill of the *National Geographic* artists was such that the portrait was almost believable.

The potassium-argon (K-Ar) dating method was being perfected by Garniss Curtis (University of California, Berkeley) at about the time "Zinj" was discovered in Bed I at Olduvai. The date—1.8 m.y.a.—was a shocker. No one at that time believed that humans went back that far.

It was not long before the Leakeys began to find the remains of another type of fossil individual. This type was a far better candidate for human ancestry. Louis began to realize that "Zinj" really was just a super-robust australopithecine, and it is now known as *Australopithecus boisei*. What Louis claimed was "obviously human" turned out to be obviously nonhuman.

Louis gave these newer fossil individuals some colorful names. There was Cinderella (Olduvai Hominid 13), George (Olduvai Hominid 16), Twiggy (Olduvai Hominid 24), named after a flat-chested English actress, and Johnny's Child (Olduvai Hominid 7), so named because it was found by Louis's son Jonathan. These newer fossils consisted of cranial fragments, hand bones, and foot bones. The foot bones seemed to indicate bipedality. The hand bones seemed to indicate manual dexterity. The associated stone tools, formerly attributed to "Zinj," were now ascribed to these newer individuals.

In 1964, Louis Leakey, Phillip Tobias (University of Witwatersrand) and John Napier (University of London) announced in *Nature* a new human ancestor: *Homo habilis*.[2] Since some of those fossils were found in Bed I, they were also dated at 1.8 m.y.a. From the start, those fossils were the subject of intense controversy. Some felt they were just a mixture of australopithecine and *Homo erectus* fossils, and, hence, did not constitute a new taxon. Even those who were sympathetic to the new category recognized that the fossils were a mixture of juvenile and adult material, and juvenile material is difficult to evaluate.

However, a philosophical problem was also at the center of the controversy. At that time, the accepted scenario for human evolution went from *Australopithecus africanus* (including Taung) to *Homo erectus* and then on to *Homo sapiens.* Many evolutionists felt that there was not "room" between *africanus* and *erectus* for another species, nor was there need for one. But Louis was marching to the tune of a different drummer. Louis believed in "old *Homo.*" Louis did not believe that humans had evolved from the australopithecines, at least not from the ones that had been discovered thus far. He believed that the transition from primates to humans took place much farther back in time. In Louis's evolutionary scheme, there was not only room for a new taxon, there was a desperate need for one. In fact, Louis felt that he had discovered the true ancestor of modern humans.

Louis Leakey was at least consistent. He recognized that for evolution to go from *africanus* to *erectus* to *sapiens* presented a problem. The cranium of *africanus,* although very small, is thin, high domed, and gracile. The *erectus* cranium is thick, low domed, and robust. The *sapiens* cranium is thin, high domed, and gracile. Thus, to go from *africanus* to *erectus* to *sapiens* represents a reversal in morphology. And a reversal is an evolutionary "no-no." It was for this reason that Louis believed that neither *Homo erectus* nor the Neandertals were in the mainstream of human evolution. Both these robust groups, he felt, were evolutionary cul-de-sacs that led to extinction. The *Homo habilis* cranium, on the other hand, was thin, high domed, and gracile. By going from *habilis* directly to *sapiens,* Louis avoided the reversal problem. Although most evolutionists have accepted *habilis* into the hominid family, they have also retained *erectus.* Hence, they still have a reversal problem in going from *habilis* to *erectus* to *sapiens.*

The concept of reversals in the fossil record is intriguing. A rule in evolution, Dollo's Law, says that reversals are not supposed to happen. Evolutionary lineages are believed to go from a generalized to a specialized condition. The basic idea in evolution is that a species is successful when it is able to adapt to its environment. The better it adapts, the more specialized it becomes to that environment. Specializations usually involve feeding and defense mechanisms. The species not able to adapt dies out. However, adaptation has a flip side. Major extinctions in evolutionary history follow catastrophes or extensive environmental changes. Those species that have become well adapted (specialized) to the former environments cannot change rapidly enough to meet the new situation, nor can they reverse themselves and go back to their former generalized condition. Hence, mass extinction.

Evolutionists have a logical reason for holding to this "you-can't-go-back" idea. Mutations are, they say, the raw material for evolutionary change. When mutations occur in an organism, those mutations represent permanent changes in the genetic structure of the organism. Whether the mutational events are for better (there is no observational or experimental support for "good" mutations) or for worse, the genes that had programed the former condition are gone. Through mutations, those genes are permanently changed and have become different genes which program for something a bit different. To believe that chance mutations could occur that would exactly restore the former genes would be like believing in the tooth fairy. Hence, reversals have not been considered a part of evolutionary theory.

This lack of reversals gives to evolution a one-way directionality which is basic to the system. Phylogenetic trees and evolutionary relationships have been developed on this generalized-to-specialized concept. It is also basic to fossil and rock correlation on a worldwide scale. If a rather generalized fossil organism is found in a certain rock layer, and what appears to be a more specialized form of that same organism appears in an unconnected rock layer some distance away, it is assumed that the specialized form evolved from the generalized form, never the other way around. On the basis of that first assumption— that the specialized form evolved from the generalized form—it is further assumed that the rock containing the specialized form is more recent in age than the rock containing the generalized form. The importance of these concepts in worldwide stratigraphy cannot be overemphasized.

There is a personal side to this reversals story. Some years ago, I did a major research project on the possibility of reversals in the fossil record. In the paleontological literature I discovered a number of well-documented cases of reversals in the fossil record of insects, worms, ammonites, fishes, mammals, and humans. These were cases where organisms had seemingly gone from a specialized to a more generalized condition. Because the paleontological literature is so vast, I suspect that the results of my research are just the tip of the iceberg, that further research would reveal many more documented cases of reversals.

In the paper I wrote on the subject, I pointed out the tremendous implications reversals pose for worldwide rock correlation (where ammonite fossils are used extensively) and for evolutionary relationships. If reversals should prove to be extensive, it could call into question evolutionary relationships as they are presently understood.

I mailed my paper to Richard H. Bube (Stanford University), editor of the *Journal of the American Scientific Affiliation,* published by an organization of which I am a member. The ASA is composed of Christians who work in the sciences. Its orientation is largely Theistic Evolution. My paper was rejected. Although it was well documented, Bube said that I must be mistaken. He apparently recognized the serious implications reversals have for evolution and preferred to disparage my work. The paper was later published in the *Creation Research Society Quarterly.*[3]

That was fifteen years ago. Today reversals are openly talked about in the evolutionary literature. Richard Klein, citing McHenry, states that to go from *habilis* to *erectus* to *sapiens* means reversals in both cranial thickness and in brow ridge development.[4] However, in spite of the many references in today's literature to reversals, seldom is there any hint that evolutionists understand the serious implications these reversals have for their theory. So much for reversals.

Early on, some evolutionists claimed that *Homo habilis* had been launched mainly on the power of Louis Leakey's personality. Those who have heard Louis lecture will readily admit that he had the charisma to do it. However, the turning point toward acceptance of *Homo habilis* into the hominid family came in 1972 when Louis's son Richard, working at a site on Lake Rudolf (now Lake Turkana) in northern Kenya, found the famous fossil skull and leg bones known as KNM-ER 1470 and 1481. (KNM stands for Kenya National Museum, where the fossils are kept; ER stands for East Rudolf, where the fossils were found; the numbers are the museum acquisition or catalog numbers.)

Before skull 1470 was discovered, the volcanic strata above where the fossil was found, the KBS Tuff, had already been dated at 2.6 m.y.a. Since skull 1470 was found below this tuff, Richard Leakey estimated its age at about 2.9 million years. Shocking about the fossil was its large cranial size (about 800 cc) and its very modern morphology, which includes high doming and thin cranial walls. The skull was so different from what evolution theory would predict that Richard Leakey said: "Either we toss out this skull or we toss out our theories of early man. It simply fits no previous models of human beginnings."[5] In reporting the new fossil discovery, *Science News* wrote:

> Leakey further describes the whole shape of the brain case as remarkably reminiscent of modern man, lacking the heavy and protruding eyebrow ridges and thick bone characteristics of *Homo erectus*.[6]

In his *Nature* report, Richard Leakey stated: "The 1470 cranium is quite distinctive from *H. erectus*. . . ."[7] In fact, comparisons show that skull 1470 is more modern than any of the *Homo erectus* fossils—even the Kow Swamp material, which is only about 10,000 years old.

On the other hand, skull 1470 obviously is not an australopithecine, although some, such as Alan Walker, have suggested that it might be. Its cranial capacity is far beyond even the largest known australopithecine. What is disconcerting to evolutionists is that 1470's cranial capacity is well within the range of modern humans. That 1470 is not a unique specimen is shown by the fact that other fossils of the same age have been found with close affinities to 1470. KNM-ER 1590 consists of dental and cranial fragments. Although this cranium is from an immature individual, it is as large as 1470. Hence, in adulthood it would have been even larger. KNM-ER 1802, a mandible, may also belong to this group.

Not only does skull 1470 qualify for true human status based on cranial shape, size, and cranial wall thickness, there is also evidence on the inside of the skull of a Broca's area, the part of the brain that controls the muscles for producing articulate speech in humans.

> The two foremost American experts on human brain evolution—Dean Falk of the State University of New York at Albany and Ralph Holloway of Columbia University—usually disagree, but even they agree that Broca's area is present in a skull from East Turkana known as 1470. Philip [sic] Tobias, . . .renowned brain expert from South Africa, concurs. . . .So, if having the brains to speak is the issue, apparently *Homo* has had it from the beginning.[8]

Soon after casts were available, I purchased one of skull 1470 from the National Museums of Kenya, Nairobi. As I studied it, I sensed that there might be a problem with the reconstruction of the face. The original fossil had been found in hundreds of pieces and was assembled over a six-week period by Alan Walker, Bernard Wood, and Richard Leakey's wife, Meave. The skull was far too large for an australopithecine. It cried out, *"Homo!"* However, the face had a bit of an australopithecine slant to it. Pictures taken before plaster was used to fill in the missing pieces reveal that the face of the fossil is rather free floating. It is attached to the skull only at the top, with nothing to stabilize the slant of the face. Further, the maxilla (upper jaw) is not attached to the rest of the face.

Others have also questioned the reconstruction of skull 1470. On several occasions, Richard Leakey protested that the skull was reconstructed in the only way possible. There were no other options. However, it seems that Leakey was not being straightforward. Roger Lewin, associated with Leakey on several projects, tells a different story regarding skull 1470.

> One point of uncertainty was the angle at which the face attached to the cranium. Alan Walker remembers an occasion when he, Michael Day, and Richard Leakey were studying the two sections of the skull. "You could hold the maxilla [upper jaw] forward, and give it a long face, or you could tuck it in, making the face short," he recalls. *"How you held it really depended on your preconceptions.* It was interesting watching what people did with it." Leakey remembers the incident too: "Yes, if you held it one way, it looked like one thing; if you held it another, it looked like something else."[9]

There is no question that bias intervened in the reconstruction of skull 1470. The face was given the larger slant off of the perpendicular to make it look more like a transitional form between primates and humans, especially when at the time of its reconstruction it was thought to be 2.9 million years old.

Bias is also obvious in the way famed artist Jay Matternes put "flesh" on the bones of skull 1470, as seen in the June 1973 issue of *National Geographic*. Matternes shows the possessor of skull 1470 to be a young black woman who looks very human except that she has an apelike nose. Human noses are composed of cartilage which normally does not fossilize, and the nose is missing on 1470. It is obvious that the purpose in giving the reconstructed skull 1470 woman an apelike nose was to make her look as "primitive" as possible. The decision of what kind of nose to give her was an entirely subjective one made by Matternes or his advisers. With a human nose, none would question the full humanity of that woman in *National Geographic*.

The very modern morphology and the very old date (2.9 m.y.a.) of skull 1470 presented an intolerable situation for human evolution. The ten-year controversy concerning the date of this fossil was finally "settled" in 1981, when the accepted date became 1.9 m.y.a. The account of this controversy, showing that the dating methods are not independent of evolution or independent of each other, is found in the appendix of this book. That case study of the dating of the KBS Tuff and of skull 1470 offers clear evidence that when the chips are down, factual evidence is prostituted to evolutionary theory.

Although there was no compelling reason why skulls 1470 and 1590 could not have been ascribed to some form of *Homo sapiens* (based on their morphology), they were grouped with the much smaller *Homo habilis* material from Olduvai Gorge. Skulls 1470 and 1590 helped to give status and acceptance to the *Homo habilis* taxon. Small cranial material similar to that found at Olduvai Gorge was also found at East Rudolf, such as KNM-ER 1805 and 1813. A chart in chapter 12 lists the major fossil material currently ascribed to the *Homo habilis* category.

There were, however, always some thoughtful evolutionists who were troubled by *Homo habilis*. Four problems with the taxon were obvious: (1) the reversal problem, going from gracile *habilis* to robust *erectus* and then back to gracile *sapiens*; (2) the juvenile nature of some of the postcranial material upon which the bipedality and the toolmaking ability of the taxon had been largely based; (3) the disparity of cranial volumes, with 1470 and 1590 being within the human range, while others, such as 1805, 1813, and O.H. 24, are far too small to be considered human; and (4) the fact that the postcranial material had not been found in direct association with the cranial material. Thus, evolutionists exhibited a deep faith that all of this material actually belonged in the same category.

The problem of *Homo habilis* now appears to be solved. In 1986, Tim White, working with Don Johanson and others at Olduvai Gorge, discovered a partial adult skeleton that has been designated Olduvai Hominid 62 and dated at about 1.8 m.y.a.[10] The cranium and teeth of O.H. 62 are very similar to the smaller *Homo habilis* skulls, 1805, 1813, and O.H. 24 and thus are properly classified with them. However, it was the *first time* that postcranial material had been found in unquestioned association with a *Homo habilis* skull. The surprise was that the body of this *Homo habilis* adult was not large, as *Homo habilis* was supposed to be. It was actually smaller than Lucy—just a bit more than three feet tall. Thus, we have strong evidence that the category known as *Homo habilis* is not a legitimate taxon but is composed of a mixture of material from at least two separate taxa—one large and one small. This new discovery also seems to remove the taxon *Homo habilis* as a legitimate transition between *afarensis* (or *africanus*) and *Homo erectus.*

Evidence that *Homo habilis* was made up of two distinct forms—a larger human form and a smaller nonhuman form—had been presented even before this discovery. Based on her work on endocranial casts of two *Homo habilis* fossils, 1470 and 1805, Dean Falk wrote:

> The evidence presented . . . shows that KNM-ER 1805 should not be attributed to *Homo.* . . . the shape of the endocast from KNM-ER 1805 (basal view) is similar to that from an African pongid, whereas the endocast of KNM-ER 1470 is shaped like that of a modern human.[11]

There is no compelling reason why the large *Homo habilis* material (skulls 1470 and 1590) cannot be classified as *Homo sapiens* based upon their morphology. While it is impossible to prove the association of skull 1470 with the leg bones, KNM-ER 1481, the probability is high that they both belong to the same type of individual. Although these leg bones are also dated at 1.9 m.y.a., they are virtually indistinguishable from modern human leg bones.

The small *Homo habilis* material, including KNM-ER 1805, 1813, O. H. 24, and O. H. 62, does not belong in the genus *Homo* and should never have been so classified. This material is best described as being australopithecine—variants of (or companion species to) *africanus* or *afarensis. Homo habilis* thus becomes the little man who isn't there.

It is difficult for evolutionists to accept the fact that *Homo habilis* is dead. In their report on the discovery of O. H. 62, Tim White and Don Johanson recommended that all of the *Homo habilis* material continue to be considered as one species. They then proposed a scenario that borders on the absurd.[12] Based on the punctuated equilibria model of human evolution, they suggest that there was one million years of stability with no evolutionary change whatsoever (from Lucy at about 3 m.y.a. to O. H. 62 at 1.8 m.y.a.). Then came a very rapid burst of evolutionary development leading to *Homo erectus* at 1.6 m.y.a. (based on the *Homo erectus* fossil skull, KNM-ER 3733, found at East Rudolf). However, the fossil skeleton, KNM-WT 15000, found on the other side of Lake Turkana (Rudolf), northern Kenya, indicates that at 1.6 m.y.a. *Homo erectus* adults were six feet tall. Johanson and White thus said that after more than one million years of remarkable stability, hominids almost doubled in size and had a significant skull morphology change in just 200,000 years. That is evolution with a vengeance!

That scenario is in contradiction to the facts of the fossil record. In chapter 12 we demonstrated that the *Homo erectus* fossils do not start at 1.6 m.y.a. but go all the way back to 2 m.y.a. *Homo erectus* fossils are contemporaneous with the entire *Homo habilis* material. Hence, neither *Homo habilis* as a taxon, nor any part of it, could be ancestral to *Homo erectus*.

Sober reflection is causing some evolutionists to recognize the possibility that the taxon *Homo habilis* is flawed. Richard Klein, in his chapter on *Homo habilis*, five times mentions the possibility that the *habilis* material may comprise several taxa.[13] However, by the manipulation of dates, as we mentioned in chapter 12, he implies that a part of the *habilis* material may still be ancestral to *Homo erectus*. The dates of the individual fossils show this idea to be impossible. *Homo habilis* is dead.

The Australopithecines

It is beyond the scope of this book to give extensive coverage to the australopithecine fossil record. It is also unnecessary. The australopithecines are simply extinct primates. The fact that true humans appeared in the fossil record before the australopithecines and lived as contemporaries with the australopithecines throughout all of australopithecine history reveals that the australopithecines had nothing to do with human origins. Australopithecine authority Charles Oxnard (University of Western Australia) concludes: "The genus *Homo* may, in fact, be so ancient as to parallel entirely the genus *Australopithecus*, thus denying the latter a direct place in the human lineage."[14] The composite fossil chart in chapter 16 shows this to be the case.

In spite of this evidence of contemporaneousness, *Australopithecus africanus* has again been slipped into the human family tree by many evolutionists. In chapter 5 we detailed the shifty history of that sly little taxon *africanus*. Since the 1950s, *africanus* had been considered a direct nonhuman ancestor. It got into trouble in 1973 because of evidence presented by T. C. Partridge that one of the fossils in that taxon, the Taung skull, was only about 0.75 m.y. old. For *africanus* to evolve into modern humans in just 0.75 m.y. is out of the question.

Evolutionists were mercifully delivered from dealing with that dating problem when Johanson and White proposed a new evolutionary family tree in which *africanus* was replaced by *afarensis* (including Lucy) as our direct non-human ancestor. In 1986, the discovery of WT 17000, the Black Skull, caused Johanson and White's tidy arrangement to come crashing down. The australopithecine family tree is now in confusion, with repercussions in the human branch as well. And slippery *africanus* has slipped back into the human family tree. (See charts in chapter 5.)

However, Partridge's evidence for the recent date for *africanus* has not been adequately addressed by evolutionists. Since he presented his evidence twenty years ago, evolutionists now feel free to ignore it. The more serious problem for *africanus* is that fossils identical to those of modern humans parallel the entire history of *africanus*. Thus, *africanus* cannot be our ancestor. Some evolutionists, such as Oxnard, are open enough to admit it. The charts in chapter 16 demonstrate this fact.

Recently some evolutionists have shifted in their thinking as to whether or not any of the australopithecines are legitimate human ancestors. Surveying one hundred years of paleoanthropology, Matt Cartmill (Duke University), David Pilbeam (Harvard University), and the late Glynn Isaac (Harvard University) observe: "The australopithecines are rapidly sinking back to the status of peculiarly specialized apes. . . ."[15]

The dramatic discovery of KNM-WT 17000, a super-robust australopithecine, dated at about 2.5 m.y.a., has brought chaos to the australopithecine family tree. The super-robust form (*boisei*) was originally thought to have evolved from *robustus* and to have become extinct at about 1 m.y.a. That super-robust forms were on the scene much earlier than *robustus* came as a shock. Turmoil surrounds both the East African and the South African robust australopithecines. R. A. Foley (Cambridge University) writes of them: "They are almost certainly not human. They are probably not robust, and they are possibly not even australopithecines. Worse still, it may be that they should not be treated as a group at all."[16]

Because of the increasing uncertainty regarding the robust australopithecines, more and more paleoanthropologists are beginning to refer to them by their older name, *Paranthropus*, indicating that they probably belong in a separate extinct genus.

The case for the australopithecines as human ancestors has been based on three claims made for them by evolutionists: (1) that they were relatively big brained; (2) that they were bipedal; and (3) that they appear in the fossil record at the relevant time. The most unique distinction between humans and animals is ignored by most evolutionists. It is that humans are created in the image of God. Only this spiritual dimension explains both our glory and our agony.

Besides that spiritual distinction, it is obvious from the charts in this book that the australopithecines do not appear in the fossil record at the relevant time. They are far too late. Furthermore, although brain organization is more important than brain size alone, the significant gap between the cranial capacities of the largest australopithecine and the smallest human, fossil or living, has not been bridged. There is not a smooth transition from nonhuman to human fossils in this regard.

The evidence for australopithecine bipedality is controversial. While there is strong evidence that australopithecine locomotion was significantly different from that of humans or other primates,[17] the issue is irrelevant. Bipedality does not prove a human relationship. The birds are bipedal, but no one suggests that they are closely related to humans. Evolutionists make much of the alleged australopithecine bipedality because to make a case for human evolution they must demonstrate the origin of bipedality from a primate stock. Unfortunately, the australopithecine "evidence" comes far too late in the fossil record. As shown by the Laetoli footprints (discussed in chapter 16), when the australopithecines first appear in the fossil record, true humans were already walking.

16

Fossil Failure
on a Grand Scale

THE *HOMO ERECTUS*, archaic *Homo sapiens*, and the Neandertal fossil categories have been discussed. Another category of major importance is the category consisting of anatomically modern *Homo sapiens* and *sapiens*-like fossils. Some fossils in this category are comparatively recent in date and rather easy to diagnose. There are also fossils in this category which are as much as 4.5 m.y. old. With these fossils, it is harder to make a specific diagnosis within the genus *Homo*, especially if they consist only of postcranial material. What we can say is that when we compare these older fossils with the comparable bones of modern humans, they are virtually identical. This means there are fossils that are indistinguishable from modern humans that extend all the way back to 4.5 m.y. on the evolutionary time scale. As we go back in time, the quantity of this fossil material decreases substantially, but the quality of the fossils presented here is excellent.

A chart of anatomically modern *Homo sapiens* and *sapiens*-like fossils is found in this chapter. Like the other charts, the vertical scale is expanded on the upper end of the time scale because that is where the vast majority of the fossils are located. This distortion is neces-

Anatomically Modern *Homo sapiens*-like Fossils
(Evolution Time Scale)

Many modern fossils
  between 30,000 y.a.
  and the present

T- Tools/artifacts in association
F- Evidence of use of fire
( ) Min. no. of fossil individuals

30,000 y.a.

Lake Mungo & Keilor remains, Australia[1] (4)
Cro-Magnon remains, France[2] (5) T
Niah skull, Borneo[3] (1)
Bacho Kiro mandibles, Bulgaria[4] (2)
Liujiang remains, China[5] (1)
Jebel-Qafzeh remains, Israel[6] (11) T

100,000 y.a.

Krapina A cranium, Yugoslavia[7] (1) T

Singa cranium, Sudan[8] (1)
Wadjak 1 & 2 skulls, Java, Indonesia[9] (2)
Skhūl 5 & 6 remains, Israel[10] (2) T
Border Cave skeletal remains, South Africa[11] (3)
Klasies River Mouth Caves skeletal remains, South Africa[12] (5) T F
Mumba Rock Shelter teeth, Tanzania[13] (1)
Omo-Kibish 1 & 3 skulls, Ethiopia[14] (2)
Fontechevade skulls, France[15] (2) T

250,000 y.a.

Pontnewydd mandible, teeth, Wales[16] (4) T
Rhodesian (Broken Hill/Kabwe) leg bones, Zambia[17] (2) T

500,000 y.a.

Java Man (*Pithecanthropus* 1) femur, Indonesia[18] (1)
Sondé tooth, Java, Indonesia[19] (1)

1 m.y.a.

1.5 m.y.a.

Koobi Fora footprints, Kenya[20] (1)
Gomboré IB-7594 (Melka-Kunturé) arm bone, Ethiopia[21] (1)
Olduvai Hominid 48 clavicle, Tanzania[22] (1)
KNM-ER 813 leg & ankle bones, Kenya[23, 24] (1)

KNM-ER 1470, 1481 cranium, leg bones, Kenya[25] (2)
KNM-ER 1472 femur, Kenya[26, 27] (1)

2 m.y.a.

KNM-ER 1590 cranium, Kenya[28, 29] (1)

Olduvai stone structure, Tanzania[30, 31]

4 m.y.a.

Laetoli footprint trails, Tanzania[32] (3)

Kanapoi (KP 271) arm fragment, Kenya[33] (1)

5 m.y.a.

Composite Fossil Chart
(Evolution Time Scale)

modern *H. sapiens*	archaic *H. sapiens*	*Homo erectus*	Australopithecines
Springbok Flats	Cape Flats/Dire Dawa	Mossgiel/Cossack	
Predmosti/Brno	Tangier/Eliye Sp.	Lake Nitchie	
Lake Mungo/Keilor	Dar es Soltan	Java Solo people	
Cro-Magnon/Niah	Haua Fteah/999	Kow Swamp/Cohuna	
Bacho Kiro/Liujiang	Florisbad/L. H. 18	Coobool/Talgai	
Jebel-Qafzeh	La Chaise/Azych	WHL 50/Témara	
Wadjak/Singa/Krapina	Krapina D/Eyasi	Narmada/Xujiayao	
Border Cave/Skhūl	Montmaurin	Cave of Hearths	
Klasies River/Mumba	Klasies River	Hazorea/L. H. 29/Mapa	
Fontechevade/Omo 1	Omo 2/Jebel-Irhoud	Hexian/Wadi Dagadlé	
	Bilzingsleben/Garba	DingCun/Changyang	
Pontnewydd	Castel di Guido	Rabat/Peking Man	
Rhodesian leg bones	Rhodesian Man skull	Sidi Abderrahman	
	Saldanha/Atapuerca	Kapthurin/Salé/Yunxian	
	Swanscombe/Steinheim	Jinniu Shan/Dali	
.5 m.y.a.	Bodo/Arago 21	Thomas 1&3/Arago	
Java Man femur	Vértesszöllös	Luc Yen/Lang Trang	
Sondé molar	Ndutu	Java Man skullcap	
		Peking Man/ O. H. 23	
		Lainyamok/Lantian	
		Ternifine/Yunshien	
		Java (Sangiran)	
	Petralona	Olduvai Hominid 12	Taung
	Mauer	Olduvai Hominid 28	
		Olduvai Hominid 22	
		Yayo	
		Yuanmou	
		Dmanisi	
		Olduvai Hominid 2	
1 m.y.a.		Lantian	
		Olduvai Hominid 29	
	Note	Olduvai Hominid 51	
	For more precise dating of fossils, consult the charts of the individual categories.	Gomboré II	
		Olduvai Hominid 9	
		Olduvai Hominid 36	
		Sambungmachan	
		Omo L-996-17/O.H.15	
		KNM-ER 992, 803	
		KNM-ER 3883, 737, 820	
		KNM-WT 15000	
Koobi Fora prints		KNM-ER 1808, 730	
Gomboré IB-7594		KNM-ER 3733, 1507	
KNM-ER 813/O.H. 48		SK-15, 18a, 18b	
Olduvai Structure		SK-84/SK-847	
2 m.y.a.		SKX-5020	
KNM-ER 1472		Java (Djetis)/Damiao	
KNM-ER 1590		KNM-ER 3228	
KNM-ER 1470, 1481			
3 m.y.a.			Lucy
Laetoli footprints			
4 m.y.a.			
Kanapoi humerus			
5 m.y.a.			

Vertical labels: Neandertals (archaic column, upper section); *Homo habilis* (flawed taxon); *A. Afarensis*; *A. robustus/boisei*; *A. africanus* sary to show all of the fossils in their time relationships. This distortion should be taken into consideration when studying the charts. Excluded from the chart of anatomically modern *Homo sapiens* are fossils that are dated younger than 30,000 years. They are irrelevant for our purposes.

The most important chart of all, the composite master chart, is also found in this chapter. This chart is a composite of all of the other charts in this book. On this composite chart the full impact of the contemporaneousness of the various fossil groups is revealed, especially the time relationship of the australopithecines to the human fossils. Because of the small size of the composite chart and the large number of fossils included on it, placement according to the date of the fossil is not quite as precise as it is on the charts of the individual fossil categories. The individual charts should be consulted for more precise dating purposes. (Remember that the author does not accept any of the evolutionist dates for these fossils.)

Because the lower portion of the chart of anatomically modern *Homo sapiens* fossils is the more significant (and the more controversial), we will deal specifically with items in that section. The fossil bones known as KNM-ER 1470, 1481, and 1590 have been discussed in chapter 15 and in the appendix. The morphology and size of these skulls together with the fully modern morphology of the leg bones (1481) indicate that true humans were on the scene in northern Tanzania at about 1.9 m.y.a. on the evolution time scale.

At the bottom of Bed I, the lowest bed at Olduvai Gorge, Tanzania, a circular stone structure was found that could only have been made by true humans. This object, fourteen feet in diameter, is considered to be the world's oldest man-made structure. Technically, it is an artifact rather than a fossil. Mary Leakey discovered it on the oldest of the occupation sites (or living floors) at Olduvai during the 1961-62 digging season.

Mary quickly realized that there was a pattern in the distribution of these stones, an intentional piling of stones on top of each other. The stones themselves are lava rocks that are not indigenous to what was a lakeshore when the structure was built. The several hundred rocks were brought from a source some miles away. Other stones on the occupation site are scarce and haphazardly scattered.

What was this structure? No one knows. Was it a habitation hut, a hunting blind, a weapons pile, or a shelter of some kind? The people of the Okombambi tribe in Southwest Africa construct such shelters today. They make a low ring of stones with higher piles at intervals to support upright poles or braces. Over these poles are placed skins or grasses to keep out the wind. Turkana tribesmen living in the desert of northern Kenya make similar shelters.

What staggers the mind is that the living floor where this structure was found was dated by evolutionists at 1.8 m.y.a. (K-Ar). Revisions[34] published in late 1991 indicate that it may be as old as 2 m.y. (Laser-fusion $^{40}Ar/^{39}Ar$). Mary Leakey also found the usual stone and bone waste, as well as Oldewan tools. These tools are normally considered by evolutionists to be very primitive. However, Mary reports that a similar type of stone chopper is used today by the remote Turkana tribesmen to break open the nuts of the doum palm.[35]

The conceptual ability required to make such structures, the physical ability to carry hundreds of large stones several miles, and the fact that similar structures are made by humans today constitute strong evidence that true humans were on the scene at Olduvai at about 2 m.y.a. on the evolution time scale.

Beginning in 1978, associates of Mary Leakey discovered a series of what appear to be human footprint trails at site G, Laetoli, thirty miles south of Olduvai Gorge, in northern Tanzania. The strata above the footprints has been dated at 3.6 m.y.a., while the strata below them has been dated at 3.8 m.y.a. (K-Ar). These footprint trails rank as one of the great fossil discoveries of the twentieth century.

Mary Leakey told the story in the April 1979 issue of *National Geographic*. She described the footprints as "remarkably similar to those of modern man."[36] Three parallel trails are seen, made by three individuals, with one individual walking in the footprints of another. The trails contain a total of sixty-nine prints extending a length of about thirty yards. More prints may yet be uncovered. The prints were made in fresh volcanic ash spewed out by Mount Sadiman to the east. A unique combination of circumstances caused these amazing prints to be preserved.

These footprint trails have produced a large body of literature. Virtually everyone agrees that they are strikingly like those made by modern humans. In spite of that fact, the evolutionist community has ascribed them to the Lucy-type hominid known as *Australopithecus afarensis*. This taxon includes mandibles found elsewhere at Laetoli by Mary Leakey as well as fossils found by Donald Johanson in the Afar region of Ethiopia. The assumption, based upon the somewhat similar ages of the fossils in the two different localities and the belief that *afarensis* was bipedal, is that the *afarensis* fossils represent the type of individuals who made the Laetoli footprint trails. Obviously, this is totally unprovable.

The specialist who has conducted the most extensive recent study of these footprints is Russell H. Tuttle (University of Chicago). He did so at the invitation of Mary Leakey. The footprint trails at Laetoli appear to have been made by individuals who were barefoot, probably habitually unshod. When Tuttle began his study, he discovered that very few studies have been done on habitually unshod peoples. Studies done on the footprints of shod people would not necessarily be applicable to the Laetoli prints.

As a part of Tuttle's investigations he observed the Machiguenga Indians in the rugged mountains of Peru, a habitually barefoot people. More than seventy individuals from ages seven to sixty-seven, both male and female, constituted his study. He concludes: "In sum, the 3.5-million-year-old footprint trails at Laetoli site G resemble those of habitually unshod modern humans. None of their features suggest that the Laetoli hominids were less capable bipeds than we are."[37]

Elsewhere, Tuttle writes: "In discernible features, the Laetoli G prints are indistinguishable from those of habitually barefoot *Homo sapiens*."[38] He is especially struck by the similarity of the Laetoli prints to those of the Machiguenga Indians: "Casts of Laetoli G-1 and of Machiguenga footprints in moist, sandy soil further illustrate the remarkable humanness of Laetoli hominid feet in all detectable morphological features."[39]

Not only did Tuttle reject the notion that the Laetoli footprints were made by *Australopithecus afarensis*, but he found that the former work on those footprints by J. T. Stern, Jr., and Randall L. Susman (State University of New York, Stony Brook) was flawed:

In any case, we should shelve the loose assumption that the Laetoli footprints were made by Lucy's kind, *Australopithecus afarensis*. The Laetoli footprints hint that at least one other hominid roamed Africa at about the same time.[40]

. . . my studies on the Laetoli footprints provide no support for the apish model of Stern and Susman, who, in fact, waffled from their initial position on the basis of undocumented rumors about faults in the casts that they had studied.[41]

If the Laetoli footprints are so much like those of modern humans, why would Tuttle talk about the existence of "one other hominid" in East Africa, one whose identity is totally unknown? Why not ascribe those footprints to humans? Tuttle is honest enough to give us the reason: "If the G footprints were not known to be so old, we would readily conclude that they were made by a member of our genus, *Homo*."[42]

The real problem—the only problem—is that to ascribe those fossil footprints to *Homo* does not fit the evolutionary scenario timewise. According to the theory of evolution, those footprints are too old to have been made by true humans. It is a classic case of interpreting facts according to a preconceived philosophical bias.

Interpreting the Laetoli footprints is not a question of scholarship; it is a question of logic and the basic rules of evidence. We know what the human foot looks like. There is no evidence that any other creature, past or present, had a foot exactly like the human foot. We also know what human footprints look like. But we will never know for sure what australopithecine footprints look like because there is no way of associating "beyond reasonable doubt" those extinct creatures with any fossil footprints we might discover. On the one hand, we have very positive identification: the human foot and the Laetoli footprints. On the other hand is the total absence of the kind of information needed to make any identification of those prints with australopithecines. Juries deal with that kind of problem continually. The human mind deals with that kind of logic every day. Were it not for the darkness evolution casts upon the human mind, there would be no question at all as to which category those Laetoli footprints should be assigned.

The fossil at the bottom of the anatomically modern *Homo sapiens*-like chart is an old friend, Kanapoi KP 271. We discussed that elbow bone in chapter 5. William Howells had the same problem with that fossil that Russell Tuttle has with the Laetoli footprints. According to evolution, that elbow bone is just too old to be human. Allow me to quote Howells again:

> The humeral fragment from Kanapoi, with a date of about 4.4 million, could not be distinguished from *Homo sapiens* morphologically or by multivariate analysis by Patterson and myself in 1967 (or by much more searching analysis by others since then). We suggested that it might rep-

resent *Australopithecus* because at that time allocation to *Homo* seemed preposterous, *although it would be the correct one without the time element.*[43]

Evolutionists refuse to call extremely old fossils by their proper names. The reason is to protect evolution theory. Hence, it is obvious we are dealing not with science but with something more like quicksilver. There may be many terms we could use in referring to the evolutionist's methodology, but my mother, who would not let me get by with very much, certainly had a name for it.

An Imaginary Conversation

Mr. C (Creationist): Why do you evolutionists continue to call the Kanapoi elbow bone, KP 271, an australopithecine? Howells originally called it *africanus*. Klein now calls it *afarensis*. Yet, computer analysis has determined many times that it could legitimately be considered fully human.

Mr. E (Evolutionist): I suppose it all depends on how much faith you put in computers to do that sort of thing.

Mr. C: Certainly a computer would be more objective than a human. We all tend to see things a bit differently, especially when our personal philosophy is involved. Lord Zuckerman severely criticized those who felt they could diagnose a fossil just by "eyeballing" it.

Mr. E: Well, you know what I think of Zuckerman!

Mr.C: But seriously, wouldn't the computer diagnosis be more reliable?

Mr. E: Not necessarily. This can be a rather subjective area also. I'd want to see the work myself, take my own measurements, crunch the numbers, and do my own evaluating.

Mr. C: But there is a vast amount of data out there that you accept without personally checking it out. Why are you suspicious only of data that is contrary to evolution?

Mr. E: That's not the point. The evidence for evolution is massive. Evolution is universally accepted by scholars. When you get one little item that's out of line, it deserves to be looked at with a jaundiced eye.

Mr. C: When are you going to look at it?

Mr. E: What was that?

Mr. C: I asked you when you were going to allow your jaundiced eye to look at that particular piece of evidence?

Mr. E: Oh, I don't think that's necessary. My colleagues have done a good job.

Mr. C: But this could be important. In dealing with evidence, one bit of solid evidence can overthrow a large amount of circumstantial evidence. The fossil is small, all right. But it seems to be of good quality, especially for that age.

Mr. E: That's my whole point. Are you going to let one scrap of bone at 4.5 m.y.a. challenge one hundred years of scientific research and progress?

Mr. C: Granted, KP 271 stands alone. How many fossils like KP 271 in the 4.5 m.y. time frame do you feel it would take to cause you to take a second look at human evolution?

Mr. E: You creationists have a marvelous ability to live in the past. The battle is over. Give it up. Get on with life. Human evolution is a fact. It has been established on solid scientific evidence. Sure, there are questions about the mechanisms. But the fact of human evolution is established. You are letting your fundamentalist religion mess up your science.

Mr. C: If several more fossils were found in the 4.5 m.y. time frame that were indistinguishable from modern humans, would that cause you to take a second look at your position?

Mr. E: That's an improper question. Obviously, there's something wrong with KP 271. It's an anomaly. Someday we'll figure it out. Right now we're more interested in the origin of modern humans. We're moving on. We're not rehashing the old chestnuts.

Mr. C: Let's move the discussion out of the realm of human fossils. Suppose you discovered an old manuscript that claimed to be written by the old bard himself, William Shakespeare. However, in the manuscript, one of the characters mentions turning on the *radio*. Would that alert you that something might be wrong? Or would the word *radio* have to appear about sixteen times in the manuscript before you bothered to look into the matter?

Mr. E: Your illustration is off base. There is no relationship between the evaluation of ancient manuscripts and the evaluation of human fossils. The methodology is totally different.

Mr. C: The principle is the same. In historical research, if something appears in the wrong time frame, it deserves to be looked at. Even one item should spark your attention, but I don't sense that your attention has been sparked. What about the Laetoli footprints at 3.75

m.y.a.? Every evolutionist recognizes how similar those are to human barefoot prints.

Mr. E: Of course, they are. That fits perfectly with human evolution. Bipedality came first. Then came the enlargement of the brain.

Mr. C: The *afarensis* fossils seem to be partly arboreal. Their feet weren't exactly like ours. Their toes were somewhat curled, and the ends of the toes were pointed. Tuttle notes a definite difference.

Mr. E: That's a matter of interpretation! Besides, *afarensis* had to make those tracks. There was no one else around to make them.

Mr. C: That's begging the question. That's what we're trying to prove. You can't arbitrarily call those tracks *afarensis* tracks, and then say that there wasn't anyone else around. If humans made those tracks, there were humans around.

Mr. E: You obviously don't know how science works. You can't bring your fundamentalist religion into science. One thing we know for sure is that we humans *weren't* around at that time. We couldn't be, because we evolved from *afarensis* or some form like him. It is this mixing of science and religion by folks like you that has caused the crisis in science education in our nation. I suppose you believe in a flat earth, too!

The Big Picture

Up to this point, we have been painting with a broad brush. We have been concerned with the big picture. The facts of the big picture are that first, fossils that are indistinguishable from modern humans can be traced all the way back to 4.5 m.y.a., according to the evolution time scale. That suggests that true humans were on the scene before the australopithecines appear in the fossil record.

Second, *Homo erectus* demonstrates a morphological consistency throughout its two-million-year history. The fossil record does not show *erectus* evolving from something else or evolving into something else.

Third, anatomically modern *Homo sapiens*, Neandertal, archaic *Homo sapiens*, and *Homo erectus* all lived as contemporaries at one time or another. None of them evolved from a more robust to a more gracile condition. In fact, in some cases (Neandertal and archaic *Homo sapiens*) the more robust fossils are the more recent fossils in their respective categories.

Fourth, all of the fossils ascribed to the *Homo habilis* category are contemporary with *Homo erectus*. Thus, *Homo habilis* not only *did* not evolve into *Homo erectus*, it *could* not have evolved into *Homo erectus*.

Fifth, there are no fossils of *Australopithecus* or of any other primate stock in the proper time period to serve as evolutionary ancestors to humans. As far as we can tell from the fossil record, when humans first appear in the fossil record they are already human. It is this abrupt appearance of our ancestors in morphologically human form that makes the human fossil record compatible with the concept of Special Creation. This fact is evident even when the fossils are arranged according to the evolutionist's dates for the fossils, although we believe the dating to be grossly in error. In other words, even when we accept the evolutionist's dates for the fossils, the results do not support human evolution. The results, in fact, are so contradictory to human evolution that they effectively falsify the theory. This, then, is the big picture.

The Local Picture

There is a second approach to the human fossil record. We could call it the local picture. Although it is independent of the time element (the dates for the fossils), it confirms the big picture. It involves situations where different types of fossils are found at the same place geographically and at the same level stratigraphically. According to their morphology they should be placed in two different evolutionary categories. Since this approach is totally independent of the dating methods, it acts as a control. It confirms what we said previously about the lack of evidence for evolution in the human fossil record. If two different types of fossil humans are found at the same place and at the same level, it falsifies human evolution. The date of the geologic stratum in which they are found does not matter. The date is irrelevant, and the method is time-independent.

A chart in this chapter shows specific cases where fossils belonging in two different evolutionary taxa are found in the same place and at the same level. Although the chronological arrangement on this chart is unnecessary, it is done to show that local contemporaneousness exists throughout the entire alleged history of human evolution. It is independent confirmation of all that we have said thus far.

Local Contemporaneousness of Various Fossil Categories

Date y.a.	Locality	anat. modern *Homo sapiens*	Neandertal	archaic *Homo sapiens*	*Homo erectus*	Ref.
10,000	Coobool Crossing, Australia	many moderns			a number of erectus-like	44
30,000+	Willandra Lakes (Lake Mungo), Australia	many moderns			WHL-50 and others	45
50,000–100,000	Tabun Cave, Mount Carmel, Israel	Skhūl 5 & 6 cave close to Tabun	Tabun remains close to Skhūl			46
80,000	Czechoslovakia	Mladec Cave	Sipka Cave			47
100,000	Klasies River Mouth Caves, South Africa	Mixed— modern and archaic		Mixed— modern and archaic		48, 49
100,000	Krapina, Yugoslavia	Krapina A skull	Many individuals	Krapina D skull		50, 51
105,000	Skhūl Cave, Mount Carmel, Israel	Skhūl 5 & 6 individuals	Skhūl 2, 4, 7 & 9 individuals			52
130,000	Omo River, Ethiopia	Omo 1 & 3		Omo 2		53
130,000+	Eyasi Lake, Tanzania	Mumba Rock Shelter		Eyasi skull		54
150,000	Kabwe, Zambia (Broken Hill Mine)	2-3 modern individuals		Rhodesian Man skull		55
450,000	Arago, Tautavel, France			Skull 21	Mandible 13, Innominate 44	56
500,000	Trinil, Java, Indonesia	Sondé tooth			Java Man skullcap	57
500,000	Trinil, Java, Indonesia	Java Man femur			Java Man skullcap	58
1.64 m.y.a.	Koobi Fora, area 103, Kenya	Koobi Fora footprints			KNM-ER 730, 737, 1808	59, 60
1.64 m.y.a.	Koobi Fora, area 104, Kenya	KNM-ER 813			KNM-ER 3733	61, 62
1.64 m.y.a.	Koobi Fora, Ileret area, Kenya (near area 103,4)	KNM-ER 813, Koobi Fora prints			KNM-ER 803, 820, 1507 & 3883, WT 15000	59, 61, 62
1.9 m.y.a.	Koobi Fora, Kenya	KNM-ER 1470, 1481, 1590			KNM-ER 3228	63, 64

The Fossils Have Failed

The human fossil record, like the fossil record in general, has failed to furnish evidence for evolution. Evolutionists, understandably, are reluctant to admit it. One way they now handle the problem is to claim that the fossils really are not important. Indicative of this new trend is an incredible quotation by Mark Ridley (Oxford University): ". . . no real evolutionist, whether gradualist or punctuationist, uses the fossil record as evidence in favor of the theory of evolution as opposed to special creation. This does not mean that the theory of evolution is unproven."[65] The heading of Ridley's article reads: "The evidence for evolution simply does not depend upon the fossil record."[66]

Ridley then presents us with a classic case of revisionist history. Just as the Chinese government slaughtered hundreds of prodemocracy students in Beijing's Tiananmen Square and then revised history by calmly announcing that such a thing never happened, Ridley tells us not only that the fossils are unimportant as evidence for evolution but that from Darwin's time on they never were important:

> The gradual change of fossil species has *never* been part of the evidence for evolution. In the chapters on the fossil record in the *Origin of Species* Darwin showed that the record was useless for testing between evolution and special creation because it has great gaps in it. The same argument still applies.[67]

Now allow me to quote Darwin:

> But just in proportion as this process of extermination has acted on an enormous scale, so must the number of intermediate varieties, which have formerly existed, be truly enormous. Why then is not every geological formation and every stratum full of such intermediate links? Geology assuredly does not reveal any such finely graduated organic chain; and this, perhaps, is the most obvious and serious objection which can be urged against the theory. The explanation lies, as I believe, in the extreme imperfection of the geological record.[68]

Ridley claims that the fossil evidence has never been a part of evidence for evolution and that Darwin recognized that the fossil evidence did not discriminate between creation and evolution because of the gaps in the fossil record. In contrast, Darwin said that the number of intermediate forms fossilized in the rocks must be enormous, that they have not been found, that their absence is the most serious objection one could have against his theory, and that they have not been found only because of the imperfection of the fossil record. Darwin felt that the fossil evidence was so important, and the lack of transitions was such a serious threat to his theory, that he devoted an entire chapter in the *Origin* to "The Imperfection of the Geological Record." Ridley has forgotten his Darwin.

For a hundred years evolutionists paraded the fossils they had found as evidence for evolution. They promised more and better fossils in the future, hoping that luck and the tooth fairy would validate their hopes. In the early 1970s, when it became obvious that we had a more than adequate sampling of the fossil record, the grim reality dawned that those transitional fossils were not to be found. The punctuated equilibria model of evolution was then invented to explain why they were not found. However, it is imperative to emphasize that the punctuated equilibria model does not remove the *need* for transitional fossils. It just explains why those transitions have not been found. Certainly, the punctuated equilibria theory is unique. It must be the only theory ever put forth in the history of science which claims to be scientific but then explains why evidence for it cannot be found.

Some authorities in the field are honest enough to admit that the fossil evidence for human evolution is a problem. Although their number is not legion, it does include some very respected names:

David Pilbeam: "There is no clear-cut and inexorable pathway from ape to human being."[69]

David Pilbeam on whether man evolved from gibbons, chimps, or orangutans: "The fossil record has been elastic enough, the expectations sufficiently robust, to accommodate almost any story."[70]

Mary Leakey on the constructing of evolutionary family trees: ". . . in the present state of our knowledge, I do not believe it is possible to fit the known hominid fossils into a reliable pattern."[71]

J. S. Jones and S. Rouhani: "The human fossil record is no exception to the general rule that the main lesson to be learned from paleontology is that evolution always takes place somewhere else."[72]

Robert Martin: "So one is forced to conclude that there is no clear-cut scientific picture of human evolution."[73]

The popular myth is that the hominid fossil evidence virtually proves human evolution. The reality is that this evidence has been a disappointment to evolutionists and is being de-emphasized. In actuality, the human fossil evidence falsifies the concept of human evolution. The Bible, the Word of the living God, clearly declares that humans were specially created. The human fossil evidence is completely in accord with what the Scriptures teach.

17

Remember Baby Fae?

A COMMON MYTH endorsed by many Christians is that the subject of human origins is unimportant. Human destiny is the vital issue. In sincerity they ask, "Does it make any difference whether one believes in creation or evolution as long as one trusts in Christ as Savior, is born again, and is going to heaven?"

Certainly I would be the first to admit that insuring one's destiny is far more important than knowing one's origin. However, the two issues are inseparably related philosophically, theologically, and biblically. One can be born again and still believe in evolution. However, to continue to believe in evolution reveals either a lack of biblical understanding or the employment of a mental gymnastics akin to a controlled schizophrenia. If that statement seems disgustingly dogmatic as well as uncharitable, let me explain.

Do you remember Baby Fae? Baby Fae was the infant born prematurely on October 14, 1984, in Barstow, California. She had a fatal heart defect known as hypoplastic left heart syndrome, a condition in which the left side of the heart is so underdeveloped that it is virtually useless. In a dramatic transplant operation that made medical history and world headlines, Dr. Leonard Bailey, a pediatric cardiac surgeon at Loma Linda University Medical Center, inserted the walnut-size heart of a female olive baboon into Baby Fae's chest cavity in a desperate and daring attempt to save her life. For twenty-one days the world hoped and prayed and cheered for Baby Fae—until she died.

Loma Linda University Medical Center is one of more than sixty hospitals in the nation operated by the Seventh-Day Adventist Church. It has an excellent reputation for pediatric heart surgery and has the largest intensive care unit for newborn babies in California. It also has a solid reputation for medical care and research. One of the classic studies showing the link between smoking and lung cancer was carried out at Loma Linda. The school specializes in training medical missionaries.

Dr. Leonard Bailey is a devout Seventh-Day Adventist. That denomination is strongly creationist and has been a leader in the creationism movement. After graduating from college, Bailey studied medicine at Loma Linda University Medical School and took further surgical training at Toronto's top-flight Hospital for Sick Children. He then returned as a staff member at Loma Linda University Medical Center. He was forty-one years old and chief of pediatric heart surgery when he operated on Baby Fae.

Adult donor hearts usually come from brain-injured persons. Such injuries seldom occur in infants. In fact, infant hearts for transplant are so seldom available that transplants into very young children are rarely attempted. Bailey has devoted his career to trying to help victims of hypoplastic heart. It was in Toronto that he first considered using cross-species transplants (called xenografts) as a way of getting around the extreme shortage of infant hearts available for transplant. He performed more than 150 cross-species transplants on sheep, goats, and baboons, such as grafting lamb hearts into baby goats.

Bailey had little encouragement in his pioneer research. He found it difficult to get his research papers published. The procedure was so novel that prestigious journals would not give consideration to his articles. "People didn't understand why I thought this was important," he said. "They weren't watching the babies die that I was." Nor could he get federal funding. Ultimately, a research fund was set up by twenty physicians at Loma Linda who contributed part of their own salaries each month until more than one million dollars had been raised for the project.

When the announcement was made that the historic surgery had been performed, there was a surprising amount of criticism from the medical community. Had the parents of Baby Fae been informed that the chances of success were very small? Yes. Had the parents been informed of other options? Yes. Had Bailey ever done a heart transplant on humans before? No. (However, they did not ask that question of South African surgeon Christiaan Barnard when he performed the world's first human heart transplant). Had Bailey published on heart transplants? No. His articles had not been accepted because they were not thought to be important, but he had published at least thirty articles on other aspects of pediatric heart surgery.

Bailey's research had caused him to believe that xenografts were more likely to be successful on infants because their immune systems were not as fully developed. Furthermore, the advent of a new drug, Cyclosporine, not having the side effects of other drugs used to suppress the baby's rejection mechanism, gave Bailey hope that the time was ripe for this type of surgery. Yet, he was criticized for not looking for a human heart and for performing the surgery on such a tiny infant. Bailey found it hard to understand why people would question a procedure that was designed to try to save an otherwise hopeless infant.

There was a second area from which criticism of Bailey's surgery came. It was the most stunning of all. It was from animal-rights activists and organizations such as People for the Ethical Treatment of Animals (PETA), the largest animal-rights group in the country. Animal-rights activists carried picket signs outside Loma Linda Medical Center with messages such as: "Ghoulish tinkering is not science," "Stop the madness," "They are prolonging Fae's suffering," and "Christiaan Barnard refused to kill any more primates."

Because of the unique nature of the surgery, animal-rights activists were especially vocal. It was the first time many people became aware of this growing movement in our nation, a movement that began to gain momentum in 1981. It soon became obvious that the chief concern of the activists was not that Baby Fae might be suffering but that a healthy baboon had been killed to provide a transplant heart for her. Bailey was shocked. Having grown up with a Christian worldview, he was surprised to learn that our nation is drifting so far toward a non-biblical view of human life.

Famed heart expert Dr. Michael DeBakey, when asked what he would say to those who object to taking an animal's life to obtain an organ for human transplant, expressed the traditional Christian worldview:

It's a matter of values. It depends on whether you put a higher value on a human life. I don't think one should simply destroy animals for trivial reasons, but when it may save the life of a human being, I would put a higher value on the life of a human being.[1]

It *is* a matter of values. And the values of our nation and of the world are changing. Those who protested the surgery on Baby Fae were marching to the tune of a different drummer. They were protesting that the baboon's rights had been violated. Philosopher Tom Regan (North Carolina State University) put it this way: "No one has the right to violate somebody else's rights." He goes on to explain: "The baboon has rights. The baboon doesn't exist for us; the baboon exists for the baboon."[2]

There are more than a dozen animal-rights groups in the United States. They are becoming highly visible because many movie, television, and radio personalities contribute time and money to these causes. Even some of the more moderate humane societies have been taken over by animal-rights activists. Each of these activist groups has its own political agenda.

PETA, with over 250,000 members, a seven-million-dollar-plus annual budget, and a Washington headquarters staffed by over sixty full-time employees, has a very aggressive and comprehensive agenda. "The mission is to stop all use of animals by society—not only as bio-medical-research subjects but as food, clothing, sport, and pets.[3]

Some animal-rights groups are akin to terrorist organizations that advocate violence, including threats on human life. Their reasoning is that just as it took violence to abolish human slavery, violence is also necessary to free enslaved animal species.[4] Between 1981 and 1990, animal-rights activists committed seventy-one illegal acts against research facilities, with damage estimated at ten million dollars.[5]

There is a relationship between animal-rights groups and some environmentalist groups in that they tend to share a common philosophy. However, since these groups cover a broad spectrum of belief, agendas, and methods, it is possible to speak about them only in the most general terms. Only a few of them advocate the use of violence to achieve their ends.

The common philosophical thread weaving through these groups is that the natural universe has intrinsic rights equal to those of human beings. Some limit those rights to the other primates. Others extend those rights to all animals that are believed to have feelings, going all the way down the ladder of life to somewhere between a shrimp and an oyster. The most radical activists extend those rights to the plants and to the rocks. One theologian writes: "Human beings transgress their divine authority when they destroy or fundamentally alter the rocks, the trees, the air, the water, the soil, the animals—just as they do when they murder other human beings."[6]

Some animal-rights groups equate "killing broiler chickens in slaughterhouses to the Holocaust perpetrated by the Nazis."[7] More and more we hear the charge that "meat is murder," and that "man is the tyrant species."[8]

Since this thinking is relatively new in our society, we can well ask where this concept of animal rights comes from and what are its philosophical roots. Putting all of nature on equal footing smacks of pantheism, and pantheism is growing rapidly in our country because of the New Age movement. One-fourth of Americans are said to believe in reincarnation, a doctrine basic to Hinduism and Buddhism. The New Age movement is actually just a Madison Avenue version of Hinduism coupled with the only other popular philosophy in our nation which equates humans and animals—evolution. While it is difficult to prove that evolution is largely responsible for this equating of human and animal rights, it is more than coincidence that all of the animal-rights advocates who have expressed themselves publicly on the subject are evolutionists.

According to evolution, it is merely the "luck of the draw" that man has evolved a big brain with its ability to think, to reason, and to subdue other animals. Had certain mutations not happened in our ancestors and instead happened in the ancestors of the chimpanzees, we might be where they are and they might be where we are. According to evolution, I am only quantitatively better than the other animals (bigger brain), not qualitatively better (created in the image of God). Hence, I have no ethical right to use my superiority, achieved by chance, to violate the rights of other animals who through no fault of their own did not evolve the same abilities. This essential equality of all life has been best expressed by PETA's national director, Ingrid L. Newkirk, in her famous slogan: "A rat is a pig is a dog is a boy!"[9]

I have tremendous admiration and respect for the animal-rights activists. I salute them for their honesty, integrity, courage, and for their clear, logical thinking. Of all the evolutionists in the world, they alone have dared to take the philosophy of evolution to its logical conclusion and to publicly take their stand for it. Other evolutionists are wanting in logic or courage, or both. For if evolution is true, the animal-rights activists are absolutely right. You and I are just animals, and we have no right to trespass on the rights of other animals. So, although I believe that the animal-rights activists are dead wrong, I salute them. They are more noble than their fellow evolutionists. While all Christians should have compassion for animals because they are a part of God's creation, there is a qualitative difference between humans and all animals.

Is God an Evolutionist?

The question of our origins is fundamental, because what we are is determined by where we came from. Why did Dr. Leonard Bailey work so hard to save Baby Fae? Because as a creationist he believes that she was created in the image of God. Why did the animal-rights advocates protest the surgery? Because they believe that Baby Fae was just an animal, the product of the evolutionary process, and as such she had no right to live at the expense of the life of a baboon.

Two diametrically opposing worldviews are in conflict. The critical nature of this conflict has been largely obscured by two misunderstandings, both having to do with Charles Darwin. The first misunderstanding concerns Darwin's religious attitudes. The second concerns the nature of his accomplishments.

Many think of Darwin as a religious man. He studied for the ministry at Cambridge. His wife, Emma Wedgwood (daughter of Josiah Wedgwood, maker of Wedgwood china), seems to have been an evangelical Christian. Also, Darwin refers in the *Origin* to creation and to God. There are even reports that toward the end of his life Darwin experienced the new birth. Unfortunately, all of this gives a false impression.

It was parental influence that caused Darwin to study for the ministry. Even in his early years Darwin's personal commitment to orthodoxy seems to have been minimal. Emma's influence upon Darwin was basically one of restraint. He never shared her faith but often softened expressions of his opposing views out of respect for her. His references to creation in the *Origin* were highly critical of that doctrine. His references to God were only those of a theist, not those of an orthodox Christian. Darwin's spiritual journey was actually a journey in reverse. He moved from an anemic orthodoxy in his early years to a nonorthodox theism in his middle years to agnosticism in his senior years. It is inaccurate to ever describe Darwin as an atheist. However, the influence of evolution upon him was fatal. It would not be difficult to show a relationship between his commitment to evolution and his rejection of biblical Christianity, although it is clear that he was quite ignorant of the biblical Christianity he rejected.

Reports of Darwin's alleged conversion have been common in some evangelical circles for many years. Not only was Darwin said to have experienced the new birth, but he was also reported to have repudiated his writings on evolution. I have personally seen tracts making these claims. As much as we might wish that these reports were true, I am sorry to say that they are actually false. Since none of Darwin's biographers report any such experience and there are no other records of it, about the only way to verify such an event would be through his personal correspondence.

In 1919, Darwin's son Francis published a biography of his father together with a collection of his father's correspondence.[10] A study of the letters written by Charles Darwin between the time of his alleged conversion and the time of his death clearly reveals that he experienced no such change of mind and heart. Not only is there a total lack in these letters of any reference to a conversion to Christ, there are many positive expressions of his continued faith in evolution. As far as can be determined, Darwin remained an evolutionist and an agnostic to the day of his death. It is a misunderstanding to think there is evidence from Darwin's life that evolution is compatible with biblical Christianity.

The second major misunderstanding of Charles Darwin concerns the nature of his accomplishments. While it is commonly thought that his major work was an attempt to put the concept of evolution on a solid scientific foundation, that was only a secondary matter. His scientific evidence for evolution was not that impressive. Darwin's major accomplishment was not in the area of science but in the area of philosophy. Allow me to quote one of the world's leading biologists, Ernst Mayr (Harvard University):

> Nothing could be more wrong than the immemorial assertion: "Darwin was no philosopher."

> One must grant Darwin's opponents the validity of two of their objections. First, Darwin produced embarrassingly little concrete evidence to back up some of his most important claims. This includes the change of one species into another in succeeding geological strata, or the produc-

tion of new structures and taxonomic types by natural selection. More importantly, . . . Darwin . . . left . . . totally unanswered certain serious problems like the origin of life and of new genetic variation. . . .[11]

Not only was Darwin's contribution primarily in philosophy, it was a philosophy bent on a specific mission: to show that creation is unscientific. The most extensive research into Darwin's religious attitudes and motivations has been done by historian Neal C. Gillespie (Georgia State University). He begins his book with this comment: "On reading the *Origin of Species*, I, like many others, became curious about why Darwin spent so much time attacking the idea of divine creation."[12]

Gillespie goes on to demonstrate that Darwin's purpose was not just to establish the concept of evolution. Darwin was wise enough not to stop there. Darwin went for the jugular vein. Darwin's master accomplishment was to convince the scientific world that it was unscientific to believe in supernatural causation. His purpose was to "ungod" the universe. Darwin was a positivist. This is the philosophy that the only true knowledge is scientific knowledge; no other type of knowledge is legitimate. Obviously, to accept that premise means to reject any form of divine revelation. Darwin accomplished one of the greatest feats of salesmanship in the history of the world. He convinced scientists that it was unscientific to deal with God or creation in any way. *To be scientific, they must study the world as if God did not exist.*

Ernst Mayr recognized the true significance of what is called the Darwinian revolution:

> . . . the Darwinian revolution was not merely the replacement of one scientific theory by another, as had been the scientific revolutions in the physical sciences, but rather the replacement of a world view, in which the supernatural was accepted as a normal and relevant explanatory principle, by a new world view in which there was no room for supernatural forces.[13]

In all of this, it is important to realize that Darwin was not an atheist. He did not exterminate God. He just evicted God from the universe which God had created. All that God was allowed to do was to create the "natural laws" at the beginning. From then on, nature was on its own. With God out of the picture, evolution fell into place rather easily, since evolution seemed to be the only viable alternative to Special Creation. Allow me to say it again. The establishment of evolution was only a secondary accomplishment of Darwin's. His major accomplishment was to "ungod" the universe. With that accomplished, evolution became a "natural."

We are now getting down to basics. The real issue in the creation/evolution debate is not the *existence* of God. The real issue is the *nature* of God. To think of evolution as basically atheistic is to misunderstand the uniqueness of evolution. Evolution was not designed as a general attack against theism. It was designed as a specific attack against the God of the Bible, and the God of the Bible is clearly revealed through the doctrine of creation. Obviously, if a person is an atheist, it would be normal for him to also be an evolutionist. But evolution is as comfortable with theism as it is with atheism. An evolutionist is perfectly free to choose any god he wishes, as long as it is not the God of the Bible. The gods allowed by evolution are private, subjective, and artificial. They bother no one and make no absolute ethical demands. However, the God of the Bible is the Creator, sustainer, Savior, and judge. All are responsible to him. He has an agenda that conflicts with that of sinful humans. For man to be created in the image of God is very awesome. For God to be created in the image of man is very comfortable.

Evolution was originally designed as a specific attack against the God of the Bible, and it remains so to this day. While Christian Theistic Evolutionists seem blind to this fact, the secular world sees it very clearly. Darwin's associate, Thomas Huxley, saw it clearly over one hundred years ago, when he wrote:

> The doctrine of Evolution, therefore, does not even come into contact with Theism, considered as a philosophical doctrine. That with which it does collide, and with which it is *absolutely inconsistent*, is the conception of creation, which theological speculators have based upon the history narrated in the opening of the book of Genesis.[14]

Things have not changed. Darwin is very much alive. Sir William Paton (University of Oxford) wrote in 1986:

> Perhaps it is not so much the possible existence and human awareness of some higher being that can most worry a scientist, but the idea of that being interfering with the natural course of events, with "natural law."[15]

Evolutionists believe that evolution embraces the entire history of the universe. If Huxley's god doesn't even come into contact with evolution (the history of the universe), it is obvious that this absentee god is not the God of the Bible. Paton's "higher being" is forbidden to interfere with natural law. Thus, "natural law" is really Paton's god, for it is the highest power. Any similarity between Paton's natural law and New Age pantheism's "the force be with you" is not accidental. The New Age gods are the natural fruit of evolutionary theory. The present spiritual battle in our society is not between theism and atheism. It is between the God of the Bible and the gods of evolution and pantheism.

Gods in Black and White

How important is the doctrine of creation to the Christian concept of God? Donald K. McKim (University of Dubuque Theological Seminary) expresses it well:

> Since God as Creator is the explanation for the existence of the world and for human existence, it is the activity of creation that establishes our deepest and most essential relation to God: as Creator and thus Lord. The doctrine of God as Creator, then, is perhaps the most basic conception of God that we know.[16]

Christians working in pantheistic and polytheistic regions of the world have always been concerned about presenting the God of the Bible in such a way that people would realize that he is not just one of many gods. The danger was that people might accept him as just another god added to the many they already worshiped. Christians have found that the best way to present the God of the Scriptures in all of his uniqueness is to present him as "Creator of heaven and earth" (Gen. 14:19). Since non-Christian religions seldom have a doctrine of creation, this doctrine serves as a foundation upon which to build biblical concepts. Many Christians don't always think about how much the concept of creation shapes our concept of God—concepts such as God's power, glory, majesty, beauty, holiness, loving care, personality, and distinctness from his creation (Rom. 1:18-20).

What does the theory of evolution imply about the nature of God? For years we have been told that evolution is just a scientific theory and that it does not interfere with our faith in God. Even Darwin talked about the wonder of God's using evolution as his method of creation. Few realized that evolutionists were talking about a different god. Creationists have insisted that the nature of God implied by the theory of evolution is a far cry from the God revealed in the Bible. Finally, a respected philosopher of science, David Hull (Northwestern University) has admitted what creationists have been saying all along. Since Darwin gained many of his insights while visiting the Galapagos Islands off the coast of Ecuador, Hull refers to the god of evolution as the god of the Galapagos:

> Whatever the God implied by evolutionary theory and the data of natural history may be like, He is not the Protestant God of waste not, want not. He is also not a loving God who cares about His productions. He is not even the awful God portrayed in the book of Job. The God of the Galapagos is careless, wasteful, indifferent, almost diabolical. He is certainly not the sort of God to whom anyone would be inclined to pray.[17]

The gods implied by the concept of evolution are ugly—both in their nature and in their behavior.

Having It Both Ways

There are those who say that we can be both a Christian and an evolutionist. This is true if (1) we place more faith in human theories of origins than we are willing to place in the Word of God, and (2) we refuse to take evolution to its logical conclusions. Paleoanthropologist John T. Robinson (University of Wisconsin, Madison) is typical of many Theistic Evolutionists. He claims that there are really two different concepts of evolution: (1) the *scientific theory* of evolution, which a Christian is free to accept and still maintain his faith, and (2) the *philosophy* of evolution, which is atheistic and is in diametric opposition to Christian faith.[18] In other words, we can accept the scientific theory of evolution while rejecting its philosophical consequences and implications. To think this way is to play games with our own minds.

Ideas always have consequences. It is barbaric butchery to attempt to sever data from its philosophical and theological implications, as Theistic Evolutionists do. Robinson may be able to do it, but thousands of people who have rejected Christ and the Bible because of evolution are not so mentally constituted. Unfortunately, they are more logical. Robinson had a Christian background in South Africa, and he has used mental gymnastics to "believe two contradictory things at the same time," to quote a phrase from Alice in Wonderland. In light of Genesis 2:21-25 (the creation of Eve) and Exodus 20:8-11 (the fourth commandment), there is no way to make the Bible speak with an evolutionary lisp. It defeats the mind to think of ways in which God could have revealed the concept of creation more clearly than he has already done in the Scriptures.

We smile at the statement attributed to Mark Twain: "Faith is believing what ain't so!" Actually, to have faith is to accept reports of matters that we have not personally verified. As such, the normal person exercises faith hundreds of times each day. We act on faith in virtually everything we do. To personally verify everything before we act on it would leave no time for anything else. Faith is so basic to everyday life that we do not even recognize it for what it is.

It is because we accept the testimony of our fellow humans so readily that makes it a serious matter to question the testimony of God. The apostle John wrote: "We accept man's testimony, but God's testimony is greater because it is the testimony of God" (1 John 5:9).

It is the task of the wise to take complex concepts and make them simple. Unfortunately, many who claim to be wise have taken God's very straightforward statements about his nature and his creation and made them very foggy. The biblical doctrine of creation with all of its implications is ours to accept or reject. It is not ours to change.

What Are You Worth?

The deeper question is the meaning and value of human life in an evolutionary context. Much has been written about the beauty of life moving upward over time and the magnificence of the human condition resulting from evolution. No one has expressed it with more optimism than has the famous evolutionist, George Gaylord Simpson:

He [man] stands alone in the universe, a unique product of a long, unconscious, impersonal, material process, with unique understanding and potentialities. These he owes to no one but himself, and it is to himself that he is responsible. He is not the creature of uncontrollable and undeterminable forces, but his own master. He can and must decide and manage his own destiny.[19]

Unfortunately, there is a dark side to this utopian picture. It is hard to believe that Simpson and other evolutionists have not been aware of

this dark side, but only recently have evolutionist philosophers seriously addressed it: the implications of evolution as to the intrinsic value of
human life. Philosopher James Rachels (University of Alabama, Birmingham), in his 1990 book *Created from Animals* is one of the few evolutionists who has dared to tackle the problem:

> . . . Darwinism leaves us with fewer resources from which to construct
> an account of the value of life. Traditional theorists could invoke man-
> kind's divine origins and special place in God's plan, as well as the idea
> that human nature is radically different from animal nature. Using these
> notions, they could devise a robust account of the sanctity of human life
> and its consequent inviolability. A Darwinian must make do with
> skimpier materials. With the old resources no longer available, one
> might well wonder whether we are left with enough to construct a viable
> theory.[20]

As an evolutionist, Rachels then bravely attempts to build a case for
the value of human life if humans are not in a category distinct from
animals (that is, not created in the image of God).[21] It is a pathetic
struggle, because, as he himself says, evolution leaves him with so little
to work with. He declares that traditional (read "biblical") morality
has placed too much value on human life, and he strives for "a more
modest view of the value of life."

> We have found no reason to support a policy of distinguishing, in prin-
> ciple, between the kind of consideration that should be accorded to hu-
> mans and that which should be accorded to other animals.[22]

Rachels' new ethic is called moral individualism. It is essentially the
idea that the value of human life is based on the value placed on it by
the person who has it. In other words, the value of my life is determined primarily by the value I myself place on it. Rachels seems unaware that there is already a name for this philosophy: self-
centeredness.

Rachels also seems oblivious to the fact that his philosophy is not
new but has been practiced throughout human history by all who have
rejected the value God places on human life. All humans tend to value
themselves. The weakness of the evolutionary ethic is its inability to
provide a strong motive for one human to place value on the life of another human. A person has no obligation to do so; he chooses to do so.

And if he chooses not to do so, there is no accountability, responsibility, or compulsion. Evolutionary ethics has all the power of a toothless tiger. This type of ethics is causing the crisis in ethics in our society. In reality, it is not an ethic; it is a lack of ethics.

Because Rachels quickly runs out of ingredients, he cooks a very thin philosophical soup. True to his evolution ethics, he justifies suicide, euthanasia, and abortion on his way to developing a philosophy for the value of animal life. In fact, he seems much more at home in emphasizing the value of animal life than he does the value of human life. In trying to value human life he was bound to fail. Evolution has stacked the cards against him.

To value individual human life in an evolutionary scenario is a contradiction in concepts. In evolution, the individual has no value. It is the population gene pool that alone has value, because it is out of that gene pool that the alleged new species will develop as evolution proceeds over time. Since there are over five billion individuals making up the human gene pool today, it does not take a rocket scientist to figure out how much value one individual has. For all practical purposes, the value of the individual in an evolutionary scenario is zero.

An incredible contrast is seen in the value Jesus Christ places on the individual human being: "What good is it for a man to gain the whole world, yet forfeit his soul? Or what can a man give in exchange for his soul" (Mark 8:36-37)?

Christ makes the amazing statement that an individual human soul (person) is worth more than all the real estate and mineral wealth of Planet Earth. The value of Planet Earth is beyond calculation, almost infinite. Furthermore, Jesus Christ, as God, gave his life on the cross to save from sin each person who comes by faith to him. Thus, the value God places on each individual is the value of Planet Earth plus the value of Jesus' life. To whom is the individual worth so much? To God. This is the key to the value of human life. It is the value placed on each person by his Creator/Redeemer. Because the Creator/Redeemer has valued my neighbor as well as me, and because I am responsible to my Creator/Redeemer, I am compelled to value my neighbor as myself. This valuation by the Creator/Redeemer is the only solid basis for an ethical system and for the intrinsic value of human life. In the final judgment, those who have rejected Christ and his ethic will understand for the first time how much their own quality of life on earth was made richer because of the value God has placed on each human life. Unfortunately, then it will be too late.

The true difference between evolution and creation is best determined by the value each places on the individual human life. In evolution, the value of the individual human life is practically zero. In creation, the value of the individual human life is virtually infinite. It would not be stretching it too far to say that the difference between evolution and creation is the difference between zero and infinity. Our being created in the image of God—the very thing Darwin tried to erase—makes the difference.

There is a myth that the question of origins is not important. Even some Christians embrace this myth. It may be possible to study the world and life without considering origins, but it is not *natural* to study the world and life without considering origins. God himself has decided that the question of origins is important (Gen. 1-2; Rom. 1:18-20). In reality, the question of origins has a direct bearing on our nature, our destiny, our concept of God, and our philosophy of life. It is without question one of the most profound matters an individual must face. It is the difference between zero and infinity.

Remember Baby Fae.

18

Is the Big Bang a Big Bust?

To go from human origins to the origin of the universe is a jump of magnificent and absurd proportions. Yet, it must be made. The biblical doctrine of human creation is inseparably linked in the Genesis account to the creation of the universe. The two accounts stand or fall together.

Evolutionists often accuse creationists of being content with refuting evolution without presenting positive evidence for creation. If there are cases where that is done, the criticism is valid. While creation and evolution seem to be the only viable alternatives for the origin of humans (or the origin of anything), it is technically true that to disprove evolution does not automatically prove creation. Logically, there is the possibility of a third, as yet unknown, alternative.

Our work thus far has been more than just negative evidence against evolution. We have shown that the nature of the fossil record is such as to effectively falsify the concept of human evolution. On the other hand, since humans have been humans as far back as one can go in the fossil record with no evidence of evolutionary ancestors, that fact constitutes positive evidence that we were created. However, just as philosophers of science recognize that science cannot prove anything in the absolute sense, so creationists recognize that the fossil record is incapable of furnishing absolute proof of creation. Proof for creation must be of a different sort, namely, a divine revelation that is historically accurate. This is what Genesis claims to be.

As to Genesis, two questions are foundational: first, in our scientific age, is the Genesis cosmology so hopelessly antiquated that only intellectual lightweights would bother with it? and second, which logically follows, even if creation were true, does Genesis give us a reliable account of it? In this chapter we will deal with the Genesis cosmology. The next chapter will deal with Genesis as serious history.

There are two facts that the average modern feels he knows about the universe: first, that it began in a fireball known as the Big Bang, and second that the universe is about fifteen billion years old, give or take a few billion years. Even if one wished to take the early chapters of Genesis literally, most feel it is out of the question. Science has "proven" that it did not happen that way.

The popular attitude toward Genesis today is one of the following: (1) Genesis is among thousands of tribal myths dealing with creation, none of which are to be taken seriously; (2) Genesis was written primarily as a polemic against polytheism and is not concerned with the origin or the age of the universe, or (3) Genesis only intends to tell us the who and the why of creation, while the task of science is to tell us the when and the how of creation.

The result of these attitudes is that many sincere evangelicals approach Genesis with a hermeneutic which if applied to the Gospel of John or the Epistle to the Romans would be the death of the very gospel in which they profess to believe. They feel that Genesis cannot be interpreted literally lest it conflict with the facts of science. In the broad spectrum of contemporary theology, the attitude toward the literal interpretation of Genesis ranges from that of a mild embarrassment all the way to the attitude seen at Harvard: "Literalism is, perhaps, the only sin at the Div[inity] School, where gay students dance openly with their lovers at a social in the Refectory."[1]

A Cosmology Primer

The idea that the universe is fifteen billion years old and that this age gives validity to some evolutionary or developmental cosmology is widely believed but is totally incorrect. It is a classic case of getting the cart before the horse. Let me explain.

One of the great astronomers of the twentieth century, Sir Arthur Eddington, expressed this basic principle of cosmology:

For the reader resolved to eschew theory and admit only definite observational facts, *all* astronomical books are banned. *There are no purely observational facts about the heavenly bodies.* Astronomical measurements are, without exception, measurements of phenomena occurring in a terrestrial observatory or station; it is only by theory that they are translated into knowledge of a universe outside.[2]

Eddington said that (1) there are no undisputed observational facts about the universe, including measurements of age and distance, and (2) any knowledge we have of the universe is based on our theories of the universe. Obviously, our knowledge is true only if our theories are true. (Eddington was not talking about the Solar System so much as about the universe beyond. Many do not realize that the farthest direct age/distance measurement we can make in the universe is limited to about three hundred light years, done by triangulation using the diameter of the earth's orbit as a baseline. All age/distance measurements beyond that are indirect, and are based on assumptions which may or may not be valid.)

It has been many years since Eddington wrote those words. We have now placed men on the moon and sent unmanned space probes to a number of the planets. It is tempting to think that the limitations Eddington spoke of regarding astronomical information no longer hold true.

In a lecture by the man who virtually pioneered radio astronomy, Sir Bernard Lovell (Jodrell Bank Observatory, England), a question was asked about the quasars. Lovell replied:

If you ask me how far away those objects are [and hence how old], then the answer is the extraordinary one that *you cannot calculate the distance unless you know what cosmological model applies to the universe.* The distance is so much on the Big Bang model, so much on the Steady-State Theory, and it has another value if the constants in the cosmological equations are different and the universe is in a cyclical condition.[3]

For some reason, popular thinking has completely reversed a basic principle of cosmology. The popular mind believes that the age of the universe has been independently established at about fifteen billion years. This in turn validates some form of the Big Bang model. In reality, the time scale of fifteen billion years is a *result* of the Big Bang model. That time scale is actually built into the model, and that time scale is valid only if the Big Bang model is true in the absolute sense. Let me put it another way. It is unscientific to claim that the universe is fifteen billion years old. An accurate scientific statement would be that the universe is fifteen billion years old only if the Big Bang cosmology represents the true condition of the universe. Hence, one does not falsify the Big Bang model by falsifying the time scale. It is the other way around. The time scale logically crumbles if the Big Bang model is falsified. It is ironic that although the Big Bang model is experiencing serious challenge in the scientific community, few are questioning the time scale. The much-hailed support for the Big Bang based on variations in the microwave background, announced in May 1992, is a trivial matter considering the fatal scientific flaws that can be marshalled against the theory. The euphoria will be short lived.

Lovell mentioned three different cosmological models and stated that each had a different time scale. He further stated that one cannot calculate the cosmological distance (and hence the age of the universe) unless one knows which cosmological model applies to the universe. This is certainly contrary to what most people believe to be the case.

Cosmologist Robert Wagoner (University of Washington) expressed this same fact when he said that the time scale of the universe is dependent upon which theory of gravity one builds into his cosmological model. Referring to the age of the universe in the Big Bang model, he explained:

> This time-scale is given by the theory of gravity. If we invoke some other theory of gravity, such as the Brans-Dicke theory, the time-scale would be much different and the nuclear production would be much different. It [the time-scale] is also sensitive to the density of the universe.[4]

Wagoner said that the time scale (age) is dependent upon the theory of gravity, about which we know little. He also said that it is based on the density of the universe, about which we know very little. (It is a matter of intense debate in current cosmology.) Lovell had said that the age/distance factors are also dependent on the cosmological constants, about which we know virtually nothing. Yet, Lovell declared that we cannot determine the age of the universe until we know precisely which cosmological model applies to the universe. If that is the case, science can tell us nothing *with certainty* regarding the age of the universe.

The implications of what Eddington, Lovell, and Wagoner expressed are staggering. First, if our knowledge of the universe is based upon the cosmological models we ourselves build, then it is obvious that we are dealing in circular reasoning. We get out what we put in. If our cosmological model gives us an age of fifteen billion years, it is because we built that type of model in the first place. We know that with computers it's "garbage in, garbage out." The same is true of cosmological models.

A second implication is that if we must know the condition of the universe before we can get the correct measurements, and yet we need those measurements to determine the condition of the universe, we have a catch-22 situation that may be scientifically impossible to solve. Thus, the universe seems to have a degree of built-in inaccessibility.

Third, in cosmology there does not seem to be an infallible way of discriminating between valid and invalid data without an outside reference. Whether one agrees with the Genesis cosmology or not, in the light of the inaccessibility of authoritative information about the universe, it seems logical that a loving God would give us independent information about origins.

A fourth implication is that if the time scale in cosmology is always a by-product of the cosmological model, then the validity of the cosmological model is the factor to be addressed, not the time scale. In the Genesis model, the determining item is not the six-day time scale but the ability of an infinite, all-powerful God to create the universe in a brief time span and in the relatively recent past. If such a God is fully able, then a six-day creation in the relatively recent past (which a straightforward reading of Genesis would suggest) is a reasonable and logical by-product of the model. To accept the Genesis concept of creation but reject the Genesis time-scale is not a viable option. Ronald Youngblood (Bethel Seminary West) comments on the literal six-day creation concept:

> After all, it seems to represent what the text of the Bible actually says, at least when interpreted literally. It also seems to dovetail best with the biblical concept of an all-powerful God, who "spoke, and it came to be" (Ps. 33:9).[5]

Since time, distance, and age measurements in the universe are not absolute and independent, a fifth implication is that there is a heavy philosophical and theological element in cosmology. If a person rejects the Genesis cosmology, it is because his preference is toward a cosmological model more in keeping with his personal worldview. Thus, cosmological models can be more theological statements of the persons holding them than they are reflections of the true condition of the universe. A cosmological model is really just an expression of how the person holding it looks at the universe. These models can be clever ways for unregenerate humans to display their disinterest in God, ways to push God out of his universe. However, they give the appearance of sophistication because of the alleged scientific evidences used to buttress them. In light of the extreme subjectivity of all cosmological models, the Genesis cosmology is certainly a respectable, legitimate, and logical one. It has internal consistency and furnishes satisfying answers to questions which other cosmologies do not even address.

A sixth implication is that since each cosmology is a self-contained entity with its own time scale as a by-product, the practice of some Christians of mixing a Genesis creation concept with a Big Bang time scale reveals a failure to comprehend substantive elements as opposed to derived elements in cosmology. The result of such a mixture— Progressive Creationism or the Day-Age theory—is not just a mixing of apples and oranges. It is a mixing of apples and walruses.

The Light Breaks Through

While these statements may seem bizarre amid the barrage of scientific pronouncements regarding the age of the universe, there are times when the light breaks through. Sometimes that light comes from unexpected sources. In this case it comes from one of the world's leading solar astronomers, John Eddy (High Altitude Observatory, Boulder).

The setting was a symposium entitled "Time in Full Measure," held at Louisiana State University, Baton Rouge, on April 13, 1978. The sessions were reported in *Geotimes* by Raphael Kazmann (Department of Civil Engineering, LSU). The purpose of the symposium reveals that the age of the earth and various geologic structures that make up the crust of the earth are not just academic questions.

The College of Engineering . . . was interested in methods of establishing the age of geologic formations as part of solving the engineering problems of providing permanent containment of radioactive and other noxious, but nonradioactive, wastes, and in determining the long-term stability of formations underlying dam sites or potential nuclear-fueled steam-electric generating plants.[6]

John Eddy, opening the morning session, dropped a bombshell. It was an amazingly candid statement regarding our ability to measure the age of the sun. Kazmann reported it in *Geotimes*:

> There is no evidence based solely on solar observations, Eddy stated, that the Sun is 4.5-5 x 10^9 years old. 'I suspect,' he said 'that the sun is 4.5 billion years old. However, given some new and unexpected results to the contrary, and some time for frantic recalculation and theoretical readjustment, *I suspect that we could live with Bishop Ussher's value for the age of the earth and sun. I don't think we have much in the way of observational evidence in astronomy to conflict with that.*' Solar physics now looks to paleontology for data on solar chronology, he concluded.

> He [Eddy] concluded that astronomy, as an observational science, can say nothing about chronology as far back as 4.7 x 10^9 [4.7 billion] years.[7]

Since our sun is just an ordinary star, and galaxies are just assemblages of stars, what Eddy said about the sun would seem to apply to all of the stars and galaxies.

The reason Eddy's remarks are so striking is that he did what creationists have urged scientists to do for many years. He made a clear distinction between the facts about the sun (what can actually be observed) and the solar theories. Because this distinction is seldom made, people have the mistaken idea that the sun really is 4.5 billion years old and that the universe really is 15 billion years old.

The honesty Eddy displayed is rare. First, he stated that there is nothing in the *observations* of the sun to enable us to discriminate between the 4.5 billion-year age the evolutionists have given the sun and the date of 4004 B.C. the sixteenth-century archbishop Ussher gave as the date for creation, including the sun. (Recent Creationists now go back as much as 10,000 to 15,000 years for the creation event.) Eddy's statement is so amazing that I must repeat it. There is no way to determine *from the observations* whether the sun is just thousands of years old or 4.5 billion years old. The cosmological model one adopts determines the sun's age.

Eddy also confessed that there is not much by way of *observational* evidence in astronomy (he did not limit it to solar observations) to conflict with a very brief age for the earth and the sun. This is an incredible statement when one considers the scorn heaped on Recent Creationists for believing that the universe is young. Eddy is, of course, expressing his personal belief in cosmic evolution when he states that he suspects that the sun is 4.5 billion years old.

Further, Eddy stated that as a solar astronomer he could live with a very recent date. The only thing to be done would be "frantic recalculation and theoretical readjustment." Whereas evolutionists have continually told us that the evidence *demands* a vast age for the sun, we discover that this is not the case.

Next was the revelation that solar physics gets its time scale from paleontology, the fossils. Creationists have long declared that the ultimate basis for the long time span was the evolutionist's interpretation of the fossil record. Evolutionists, on the contrary, have claimed that the long time span was based on a number of independent confirmations. Here is one case (a most unlikely case) where there is a clear connection. Who, of the uninitiated, would have suspected that the age of the sun was based on the evolutionist's estimate of the age of fossil corals? This is another illustration of the fact that in cosmology the time scale comes from the model, not from independent information.

The John Eddy story does not end there. On April 26, 1984, the philosophy department of the University of Colorado, Boulder, sponsored a symposium on creationism. The main speaker was Douglas J. Futuyma (at the time editor of the journal *Evolution* and associate professor of ecology and evolution at the State University of New York, Stony Brook). Futuyma had just published his book *Science on Trial*.[8] The significance of the title can best be expressed by a quotation by Ashley Montagu (Princeton University) on the back cover of the book: "The attack on evolution, the most thoroughly authenticated fact in the whole history of science, is an attack on science itself."

The chairman of the philosophy department at the University of Colorado invited me to respond to Futuyma's lecture. Because John Eddy is well known in Boulder, I used his statement from the *Geotimes* article in my response. As soon as I concluded, five young men descended on me like buzzards at a road kill. Since they dispensed with the amenities of an introduction, I can only guess that they were friends or associates of John Eddy or graduate students in astronomy at the university.

"John Eddy would *never* say such stupid things as you claim he said!" the first one screamed.

The second one continued, "You have struck a new low in intellectual dishonesty!"

A third one confined himself to telling me what he thought about my mother.

I did not respond. I simply reached into my briefcase and handed them copies of the *Geotimes* article. As they read, the change in the expressions on their faces (from rage to deepest confusion) was beyond description. They left mumbling that they were going to see John Eddy about this immediately.

I don't know if John Eddy has a problem. But I have one. Why would a person make statements in Baton Rouge that he would not make in Boulder? Or more specifically, why would a person make statements in Baton Rouge so *contrary* to what he says in Boulder that when his Baton Rouge statements are quoted in Boulder, the one quoting him is accused of lying? There is a reason.

The age of the earth and the sun is foundational to the creation/evolution issue. A recent creation demands a supreme being who created the universe. A recent creation also destroys *any* possibility of evolution. But that is not all. The scientific area may be one of the *least* important areas this issue impacts. The age of the earth and the sun have profound implications in theology, philosophy, anthropology, and ethics, to name just a few areas. In fact, there are few areas of life that would be untouched. That's why it is such an emotional issue. It is also a political one; jobs have been lost over it. The ultimate irony is that Recent Creationists probably have more acceptance today in the academic circles of the (former) Soviet Union than they have in the academic circles of the United States. Our nation truly has "forgotten Joseph."

The question still remains, What does Eddy know about the sun that would cause him to bring up Archbishop Ussher's date in the first place? It concerns the solar neutrino crisis. (The Baton Rouge conference had a session on that problem, also.) Many in the scientific establishment know about this crisis, for it has been in the scientific journals for about thirty years. However, few people know about the implications of this crisis for the age of the sun. Eddy does. It involves some very elusive objects called solar neutrinos.

A neutrino is an atomic particle so small that it can penetrate even the densest matter—a particle about as small as any in the universe. (There are three types of neutrinos, but we will avoid technicalities.) A neutrino is as close to nothing as something can be and still be something. Actually, the neutrino was not discovered; it was "invented." It was invented in 1932 when atomic scientists needed a "nothing particle" to make their equations come out even. The discovery in 1955 of the real thing represents one of the brilliant epochs of science, of which there are many.

Later, Raymond Davis (University of Pennsylvania) developed a unique trap deep in the Homesteak Mine, Deadwood, South Dakota, to detect neutrinos coming from the sun and test the theory of stellar evolution. The experiment has been going on now for more than twenty-five years. The results so far are a failure for the evolutionary theory of the sun (and stars), and may be a support for recent creationism.

Writing of the failure of the Davis experiment to confirm the theory of stellar evolution, Roger K. Ulrich (University of California, Los Angeles) confessed it with masterful understatement:

> . . . the theory of stellar evolution is challenged in a very fundamental way. This discrepancy between the theory and observation raises the possibility that the theory is incorrect, so that our conclusions based on the theory can no longer be accorded complete confidence.[9]

The implications of this problem for stellar evolution are so serious that some evolutionists now falsely claim that the problem has been solved. Some esoteric and unverified ideas are suggested for the solution, but current journal articles reveal that the problem is still far from solution.[10]

A recent Nature article explains why the solar neutrino experiment (actually, there are several experiments going on) is unique:

> The experiment is important because neutrinos are the only known particles to reach Earth directly from the solar core. Thus they provide an observational basis for testing the complex theoretical framework describing the thermonuclear interactions that power the Sun.[11]

Because neutrinos are believed to be the only particles reaching the earth from the core of the sun, this type of experiment represents the only *observational* evidence that tests the theory of solar and stellar evolution regarding how and why the sun shines. We can understand why these matters are important to astrophysicists, but the literature often describes the problem as a crisis. There is more here than meets the eye.

Although astronomers know what the deeper problem is, it is almost never mentioned, even in the scientific literature. One of the few times the real problem has been hinted at was in an article by astronomer Jessie L. Greenstein. Regarding the discrepancy between theory and observation in the solar neutrinos, Greenstein wrote:

> This large discrepancy suggests a radical modification of fundamental physics or drastic changes in the theory of the solar interior (an error in central temperature caused by wrong opacity theory, or *a low helium content*).[12]

Greenstein gave two possible sources of error: (1) fundamental physics or (2) the theory of the solar interior. Of the two, the second seems the more likely. We know a great deal about fundamental physics because of four centuries of investigation. We know virtually nothing about the solar interior. The possible source of error in the theory of the solar interior was said to be in the estimate of the central temperature of the sun. Again, Greenstein gave two possibilities: (1) wrong opacity theory or (2) low helium content. Logic can suggest which of the two is the more likely source of error. Opacity theory has to do with the insulating qualities of gases—in this case, ionized gases. While we know little about the sun's interior, we know a great deal about gases through experimentation in the laboratory, even at very high temperatures. An error here could be expected because of many uncertainties, but not necessarily a massive one. However, the error in the experiments has been described as a massive one. The low helium content of the solar interior is the more likely source of the massive error. Of all of the possibilities, this is the one area about which we know virtually nothing.

One thing we think we know about the sun is that it shines by converting hydrogen to helium. Hence, the helium content of the sun is dependent upon how long the sun has been shining. If we were to build a creationist model of the sun, one of the first things we would build into our model would be a low helium content based on the belief that the sun has been shining only thousands of years rather than billions of years. Yet, this would possibly help solve the problem. We would have a model which would conform to these observations. The creationist model predicts what the solar neutrino experiment suggests is the true situation. The cause of the massive error also becomes obvious. To impose a five-billion-year evolution model upon a sun that has been shining only thousands of years is to invite a massive error.

Of all of the attempts over a thirty-year period to resolve the solar neutrino crisis the one possibility not even remotely considered by evolutionists is a recent age for the sun. The evolutionist's age for the sun and the universe represents his holy of holies; no trespassing is allowed. That's why the implications of the solar neutrino crisis are treated with such a holy silence. The solar neutrino experiments do not prove the Genesis cosmology. But they do indicate two things: (1) that the accepted age for the sun is far from certain, and (2) that to take the Genesis cosmology and its time scale seriously is not an absurd idea.

The sun is our nearest star. It is only 8.3 light-minutes away. Yet, we do not know for sure how or why it shines. If we know so little about a star that is just eight light-minutes away, is it wise to accept uncritically an age for the universe based on the interpretation of astronomical objects that are said to be 15 billion light-years away? Does that not say something about the quality of the information upon which that interpretation is based?

I stated earlier that the fallacious practice of Progressive Creationists and others of mixing a Genesis creation concept with a Big Bang time span represents an improper synthesis of two mutually exclusive cosmologies. There is a second reason why this mixing is illegitimate. As far as I know it is seldom addressed.

In all forms of the Big Bang model, all matter was initially concentrated at one point. The Big Bang explosion propelled this matter outward in all directions. The expansion of the universe which astronomers postulate today is simply the residual expansion from that initial explosion. The matter expands at the cosmic speed limit—the speed of light. Hence, in the Big Bang model, there is always a direct relationship between time and distance. At the beginning, time equals zero ($T=0$) and distance equals zero ($D=0$). After the Big Bang, time increases in direct proportion to distance. Vast distances translate to immense time or age. To say that a galaxy is ten billion light-years away is also to say that it is ten billion years old.

However, in the Genesis cosmology there is no relationship between time and distance. When God spoke the universe into existence with the power of his word, it was fully developed immediately. In the beginning (T=0), the distance (D) was already immense. *Distance is not a measure of time or age because there is no time-distance relationship.* This is why the mixing of a Genesis creation concept with the Big Bang time scale produces absurdities.

One of the fallacies of the Big Bang cosmology is based upon this time-distance relationship. Because the light from distant galaxies appears to be red-shifted, astronomers believe that the universe is expanding. Although the idea of an expanding universe is just one of several possible interpretations of the red-shift of light, this expansion in itself is not necessarily unbiblical. What is unbiblical is the extrapolation of this expansion backward fifteen billion years to an alleged Big Bang. The validity of this extrapolation can be put to an observational test. When we look out in space we look back in time. If the universe has been expanding for fifteen billion years, we would expect to see a difference in the spatial density of the galaxies. Galaxies far out should be closer together than galaxies close to our own local galaxy (the Milky Way system), because they are closer in time to the Big Bang. However, we do not see these differences even when measured within the parameters of the Big Bang cosmology. Sir Bernard Lovell commented on this fact:

> But it is a misfortune for the observer, as I emphasized earlier on, that we do not find any change even in what we call the spatial density of the galaxies. We find that the number of galaxies per given volume of space, per cubic megaparsec, as we say, is the same at the distance of four billion light years as it is in the region around our own system.[13]

The implication of this observation is quite fascinating. It means that if the universe is expanding, *it has not been expanding very long.* If the universe were fifteen billion years old, the spatial density of the galaxies would be very different close in as compared to far out.

The purpose of this chapter is to show that even from the perspective of finite human observation, Genesis 1 is a worthy and respectable cosmology. It is a most unpopular one, to be sure. But truth is never determined by popular vote. It must be significant that while the currently popular theories are violently opposed to the Genesis cosmology, the observational evidence, tiny as it is, supports Genesis rather decently.

In all that I have said, my intention is not to disparage modern cosmology. Just the opposite is true. Modern cosmology is one of my favorite areas of study. It is probably the most dynamic area of science today. It is also full of conflict. As far as the methodology of science is concerned, this conflict is a very healthy sign. Although evolutionists continually accuse creationists of not understanding how science works, we fully understand that this conflict is one way science sharpens and clarifies its insights.

Modern cosmology is also full of uncertainty. In fact, cosmologist Michael Turner (University of Chicago) has recently coined this rule: "In a cosmological theory, you can't invoke the tooth fairy twice."[14] The wild ideas being generated by this uncertainty are what make modern cosmology such a fascinating and exciting study. But if anyone is so mentally constituted as to think that modern cosmology can cast authoritative light on Genesis 1, he either has read very little in modern cosmology or he has a very improper understanding of God's ability to accurately communicate truth.

Some say that it was impossible for humans to know about the origin of the universe until we had the tools of the twentieth century. If it were impossible for the human race to have accurate information about the origin of the universe until a breed of secular twentieth-century cosmologists came on the scene, does that not say something about the quality of God's revelation? It does not suffice to say that it was not necessary for humans to have information about the origin of the universe until now. We are not talking about our need to know, but about God's ability to communicate.

Herbert Dingle is a famous British philosopher who has challenged Einstein's special theory of relativity. Astronomer William K. Hartmann, quoting Dingle, said that while most people believe that we live in an evolving system, "yet this is not self-evident." After praising the literal Genesis creation cosmology for its unity, harmony, consistency, and explanatory value, the Dingle quotation concludes: "It would be a good discipline for those who reject it [the Genesis cosmology] to express clearly their reasons for such a judgment. . . ."[15]

19

Genesis:
The Footnotes of Moses

THE PRECISE CIRCUMSTANCES of the composition of the Book of Genesis have been a matter of continual interest for Bible scholars. Since there is strong internal as well as external evidence that Moses wrote Exodus, Leviticus, Numbers, and Deuteronomy, and since the Pentateuch is considered to be a unit, the approach of most conservative scholarship has been that Moses wrote Genesis also.

However, nowhere does Scripture say that Moses actually wrote the narratives or the genealogies of Genesis. There is no statement in Genesis referring to Moses as its author as there clearly is in the other books of the Pentateuch. Not even Christ or the apostles say that Moses actually wrote or spoke the words they quote from Genesis. While accepting the Mosaic authorship of Genesis, conservative scholars have not detailed the means by which Moses received his information.

There are three possible means by which the Book of Genesis was composed under the inspiration of the Holy Spirit: (1) Moses received his information by direct revelation, (2) Moses wrote Genesis utilizing material passed on by oral tradition, or (3) Moses wrote or compiled Genesis utilizing earlier written documents. Literary items in Genesis make the first possibility seem unlikely. The second possibility also seems remote because of the probability of information being lost or degraded in oral transmission. The third option seems most likely. The majority of evangelical scholars accept some version of the third view but give few details.

Ancient Writing

One of the arguments used by critics of the past century in their attack on the historicity and integrity of Genesis was that the art of writing went back only to the time of David, about 1000 B.C. Hence, no portion of Genesis could have been in written form before that time. It is now known that they were not only wrong but very wrong. By the 1930s our museums were rich with cuneiform writing on clay tablets dating back to 3500 B.C. Excavations of the royal archives at Ebla, in northwest Syria, possibly dating as far back as 2700 B.C., reveal that writing at that early date was commonplace. Whereas it was not necessary in that era for the average person to know how to read and write, writing was readily available to everyone through a class of professionals known as scribes. In fact, the ancient Sumerians, Babylonians, and Assyrians seemed unwilling to transact even the smallest items of business without recourse to a written document. This characteristic is dramatically seen at Ebla.

It may surprise some to learn that a clear reference to writing is found in Genesis 5:1: "This is the written account of Adam's history." This suggests that the art of writing was known within the lifetime of Adam, which could make writing virtually as old as the human race. To a creationist, this is not surprising. It is obvious that at the time of their creation, Adam and Eve knew how to speak. Yet, language is incredibly complex, and no one understands its origin. The ability to write is in the same magnitude of complexity as the ability to speak. Since God created our first parents with the ability to speak, it is reasonable to suggest that he created them with the ability to learn to write as well. A naturalistic, evolutionary origin of language stretches credulity.[1]

Cuneiform writing became the system used by all civilized countries east of the Mediterranean—Assyria, Babylonia, Persia—and by the Hittites, who are mentioned seven times in Genesis beginning at Genesis 15:20. Cuneiform writing consists of a series of wedge-shaped impressions (*cuneia* means "wedge") made in plastic clay. The Hebrew word for "to write" means "to cut in" or "to dig." Abraham, Isaac, and

Jacob all would have written in cuneiform. Cuneiform was not a specific language but a method of writing on clay tablets which embraced many languages and dialects.

The clay of the Euphrates Valley is remarkable for its fineness, as fine as well-ground flour. The scribes would mix into the clay a bit of chalk or gypsum to keep the tablets from shrinking or cracking. They were then dried in the sun or in a kiln. These clay tablets, next to stone, are the most imperishable form of writing material known. It is possible that the two tablets upon which God wrote the Ten Commandments were actually clay tablets (Exod. 32:15, 16). The western Asian archeological record suggests that virtually everything written before Abraham left Ur of the Chaldees, and much after that, was written on clay tablets in cuneiform.

As early as 2350 B.C. clay envelopes were used for private clay tablet correspondence and sealed with a private seal. A reference to this seal is found in Job 38:14, believed to have been written before the time of Abraham. Judah carried a seal with him and gave it to Tamar (Gen. 38:18, 25). Joseph was given Pharaoh's seal-ring (Gen. 41:42) which enabled him to act in an official capacity on behalf of Pharaoh.

Although papyrus was the common writing material in Egypt, cuneiform writing was understood, as the Tell-el-Amarna tablets, found in Egypt in 1888, reveal. Among these clay tablets were letters, dated about 1400 B.C., from Palestinian officials to the Egyptian government—all written in cuneiform.

Those who do not consider the early chapters of Genesis to be reliable history use oral transmission as the explanation for those chapters of the book. However, it is absurd to think that God would entrust his eternal Word to the fragile memory of humans. Scripture teaches the opposite. In Deuteronomy 31:19-21, Moses was given a song to teach to the people. He was specifically commanded to write it down so that it would not be forgotten. God said that forgetting is what the people were disposed to do. Obviously, God has little faith in oral transmission.

The Structure of Genesis

All scholars agree that the most significant and distinguishing phrase in Genesis is: "these are the generations of . . . ." Commentators of all theological schools divide the book around that phrase, which is found eleven times in Genesis (2:4, 5:1, 6:9, 10:1, 11:10,

11:27, 25:12, 25:19, 36:1, 36:9, 37:2). The translators of the Septuagint (the Greek Old Testament) regarded that phrase as being so significant that they gave the book its name after that term. *Genesis* is the Greek equivalent of the Hebrew *TOLeDOT*, "generations."

It is common for ancient records to begin with a genealogy or a register documenting close family relationships. Because several of the *TOLeDOT* phrases are followed by genealogies in Genesis, scholars have almost universally assumed that the *TOLeDOT* phrase serves as an introduction to the section which follows. Hence, the major sections of the Book of Genesis have been made to begin with the *TOLeDOT*. Since the person named in the *TOLeDOT* does not figure prominently—if at all—in the narrative which follows, the word has taken on the meaning of "descendants" ("these are the descendants of . . . .").

Yet, the lexicon defines *TOLeDOT* as "history, especially family history" or something associated with origins. This would mean that the term is concerned with ancestors rather than descendants. It also suggests that the phrase looks back to the preceding narrative rather than looking ahead to what follows.

The first use of *TOLeDOT* in Genesis 2:4 ("these are the generations of the heavens and the earth") clearly establishes that this reference at 2:4 is back rather than ahead. There is simply nothing following Genesis 2:4 that deals with "the heavens and the earth." Many commentators recognize that here *TOLeDOT* looks back even though they interpret the other occasions where it is used as looking ahead. They fail to see that Genesis 2:4 is the key, and that all of the *TOLeDOT* phrases refer back to the previous material.

James Moffatt, in his translation of the Bible, actually lifted the *TOLeDOT* phrase out of Genesis 2:4 and transferred it to Genesis 1:1 for it to serve as an introduction to chapter 1 of Genesis. Other liberal writers have stated that it was out of place at Genesis 2:4, or that it was put there by a compiler merely to serve as a transition.

The Colophon and Mesopotamian Writings

In 1936, P.J. Wiseman wrote a book entitled *New Discoveries in Babylonia About Genesis*. Wiseman seems to have found the key that unlocks the details of the authorship of Genesis. His thesis is that there are internal clues in Genesis that reveal how it was written; that the actual authors of Genesis were Adam, Noah, the sons of Noah, Shem,

Terah, Ishmael, Isaac, Esau, Jacob, and Joseph; that the authors, other than Joseph, probably wrote in cuneiform on clay tablets; and that Moses, utilizing these records, was the redactor or editor of Genesis rather than its author.

Wiseman's work was recently edited and reissued by his son,[2] Donald P. Wiseman, a noted evangelical scholar. The younger Wiseman was assistant curator of Western Asian antiquities at the British Museum, and later professor of Assyriology at the University of London. He is also general editor of the Tyndale Old Testament Commentary series. Donald Wiseman endorses his father's work, as does R. K. Harrison, professor of Old Testament, University of Toronto, who has incorporated it into his monumental *Introduction to the Old Testament*.[3] Although P. J. Wiseman is often cited by evangelical scholars, his remarkable insights into the composition of Genesis are not well known by the evangelical community.

Wiseman asked the question, How was information recorded and how were documents formulated in ancient Mesopotamia, which was the geographical context of much of the book of Genesis? The heart of Wiseman's contribution to the problem of the formulation of Genesis was his insight in identifying the *TOLeDOT* phrases in Genesis with ancient Mesopotamian colophons. A colophon is a scribal device placed at the conclusion of a literary work written on a clay tablet giving—among other things—the title or description of the narrative, the date or occasion of the writing, and the name of the owner or writer of the tablet.

It is not surprising to the student of ancient eastern customs that many of their literary habits were precisely the opposite of our own. For instance, the Hebrews commenced their writing on what to us is the last page of the book and wrote from right to left. In ancient Mesopotamia (Iraq) it was the end and not the beginning of the tablet which contained the vital information regarding date, contents, and ownership or authorship. This custom was widespread and persisted for thousands of years.

Perhaps the most striking aspect of the colophon practice was that the name in the colophon was the name of the owner or writer of the tablet. Sometimes the owner would also be the writer. However, if a person was not able to write, he would hire a scribe to do the writing for him. The scribe would not include his own name, but the name of the one who had hired him—the owner of the tablet. Thus, it is impossible to overemphasize the importance of the colophon at Genesis 5:1: "This is the written account of Adam's ancestry." Not only does the Hebrew word *sepher* mean "book" or "a complete writing," but the presence of Adam's name suggests that it was a written account *owned or written by Adam*, not just a written account *about* Adam. Genesis 2:4 to 5:1 gives evidence of being a firsthand, eyewitness account of the experiences of Adam, possibly written by him on a clay tablet.

Derek Kidner (Tyndale House, Cambridge) understands the impact of the Hebrew word *sepher* at Genesis 5:1.

> The opening, *This is the book* . . . , seems to indicate that the chapter was originally a self-contained unit ('book' means 'written account', of whatever length), and the impression is strengthened by its opening with a creation summary, and by the set pattern of its paragraphs.[4]

However, Kidner rejects Wiseman's theory that Genesis contains a series of colophons in which the names given are the names of the original writers or owners of the tablets: ". . . by insisting on a complete succession of named tablets the theory implies that writing is nearly if not quite as old as man."[5]

At the risk of being thought a bit naive, one could ask what is wrong with the art of writing being nearly as old as the human race. Here is where preconceptions enter in. If one believes that God created humans directly as humans, there is nothing at all wrong with the idea. The problem is really with Kidner. He is a Theistic Evolutionist. His philosophy demands that humans not be that intelligent that early in their history. Therefore, Adam could not have known how to write. It's rather gracious of Kidner to allow Adam to be able to speak. This is not the only occasion where Kidner forsakes solid biblical exegesis because of his preconceived notions about origins.

The colophon also included the date or the occasion of the writing. It is easy for us in the twentieth century to miss this fact because we date our writings by the calendar. Not so the ancients. The creation account (Gen. 1:1–2:4) is dated "in the day that the Lord God created the heavens and the earth" (Gen. 2:4). "In the day" equals "when" and implies that the creation account was written very close to the actual time of creation, not centuries later. In Genesis 37:1, 2, Jacob dated his tablet as having been written "when he lived in Canaan." Although by our standards that phrase is not precise, it does reflect a specific period in Jacob's life. Before that time he spent many years working for Laban in Haran. After that period he lived with Joseph in Egypt until his death. Leviticus (although probably not written on clay tablets) is dated as having been written when Moses was "on Mount Sinai" (Lev. 27:34), and Numbers was dated as having been written when Moses was "on the plains of Moab" (Num. 36:13). This type of dating was accurate enough for the people of that era, considering the nature of their society.

The use of colophons persisted almost unchanged for over three thousand years in ancient Mesopotamia and elsewhere. Colophons are found in the Ebla tablets in northwest Syria (2700 B.C.), in the Akkadian texts from Ras Shamra (1300 B.C.), and continued at least until the time of Alexander the Great (333 B.C.). Colophons are not unknown today. In one of my English Bibles, at the end of the Epistle to the Romans, is this statement: "Written to the Romans from Corinth, and sent by Phoebe, servant of the church of Cenchrea." Readers of *Time* and *Newsweek* will recognize that many of the major news articles have the name of the author and the place of writing at the end of the article, such as: "Written by Susan Smith in Washington." These are all suggestive of colophons.

The Authors of Genesis

The internal evidence suggests that Genesis was written on a series of clay tablets as follows:

Genesis 1:1–2:4 Origin of the heavens and the earth. No author is given. Wiseman suggests that the author was God himself, who wrote it as he wrote the Ten Commandments, probably on clay tablets. According to its date, as given in the text itself, it was written very soon after the act of creation.

Genesis 2:5–5:2 Tablet written by or belonging to Adam.

Genesis 5:3–6:9a Tablet written by or belonging to Noah.

Genesis 6:9b–10:1 Tablet written by or belonging to the sons of Noah.

Genesis 10:2–11:10a Tablet written by or belonging to Shem.

Genesis 11:10b–11:27a Tablet written by or belonging to Terah.

Genesis 11:27b–25:19a Tablets written by or belonging to Ishmael and Isaac.

Genesis 25:19b–37:2a Tablets written by or belonging to Esau and Jacob.

It is significant that the last colophon is at Genesis 37:2a. From Genesis 1 to 11, the Mesopotamian setting and local color are very obvious. From Genesis 12 to 37:2a, that influence persists. Abraham came from Ur of the Chaldees, Isaac sent back there for his wife, and Jacob got his wife from Haran and worked there for many years. However, from Genesis 37:2b to the end, the setting and local color change dramatically. We are now in Egypt. This section has a strong Egyptian flavor and was probably written by Joseph on papyrus or leather; hence it is without colophons. (Colophons are only associated with clay tablets.)

Strengthening the arguments presented thus far is the fact that in every case the person named as the owner or writer of the tablet could have written the contents of that tablet from his own personal experience. It is also significant that in every case, the history recorded in the various tablets ceases just prior to the death of the person named as the owner or writer of the tablet.

The Role of Moses

All of the tablets could have come to Moses in the way that family records were normally handed down. Nothing would have been more precious to the patriarchs than their family histories and genealogies. It is possible that there were many sets of these tablets and that each member of a patriarchal family had his own set. Of all of the personal items that Noah would have taken on the ark, the family histories would have been considered by him the most precious and most worthy of preservation.

Because of his education in the household of Pharaoh, Moses had the finest scholarly training of that day. He would have known how to read the languages of the cuneiform tablets as well as Egyptian. Cuneiform writing was well-known in Egypt because of Egypt's relationship with Mesopotamia. Moses' task would have been first of all to organize the book—under the guidance of the Holy Spirit—into a unified whole. The use of previously written documents in no way does violence to the concept of verbal plenary inspiration. Luke also tells us that he used previously written documents (Luke 1:1-4). It is reasonable to assume that each of the original writers of the tablets was guided by the Holy Spirit as well. By retaining the colophons, Moses clearly indicates the sources of his information. Just as a scholar documents his sources today with footnotes or endnotes, so Moses documented his sources of information by the colophons. These colophon divisions, based on the different sources, constitute the framework of the Book of Genesis.[6]

Moses' second task would be translation. Any tablets written in Mesopotamia would have had to be translated into Hebrew. If this translation had not been done before Moses' time, Moses would have been qualified to do it. Joseph's records, if written in Egyptian, would also have had to be translated into Hebrew by Moses.[7]

The third major task for Moses as the redactor or editor would be to bring place names up to date for the Israelites of the exodus. Geographic names change, and this updating is seen clearly in Genesis 14:2, 3, 7, 8, 15, and 17. This tablet, written in Abraham's day, had in it many geographic names that had become obsolete in the over four hundred years between Abraham and Moses. It is indicative of Moses' deep regard for the sacred text that he did not remove the old names but just added an explanatory note telling of the new names. Such notations are also seen at Genesis 23:2, 19, and 35:19. Genesis 23:2 and 19 also indicate that these notations were made before the Israelites entered Canaan, since Moses had to state where these places were. Had the Israelites already been in the land, these notations would not have been necessary.

Several passages indicate the antiquity of the tablets Moses had in his possession. In Genesis 16:14, regarding the well or spring to which Hagar fled, Moses added this note: "It is still there, between Kadesh and Bered." Genesis 10:19 is one of the most important evidences of the great antiquity of the Book of Genesis. This passage, part of Shem's tablet, had to have been written before the destruction of Sodom and Gomorrah. Since these cities were destroyed (Gen. 19), never rebuilt, and their very location was forgotten, it is obvious that this tablet telling of the settlement of clans near those cities had to be written while those cities were still standing.

Implications of the Evidence

The implications of this evidence for the origin of Genesis are staggering. Rather than Genesis having a late date, as is universally taught in nonevangelical circles, it implies that Genesis 1-11 is a transcript of the oldest series of written records in human history. This is in keeping both with the character of God and with the vital contents of these chapters. It is reasonable to expect that the first humans created by

God would have had great intelligence and language capabilities, and that God would fully inform them as to their origin.

This research also confirms the idea that the Genesis creation and Flood accounts are the original accounts of these events and were not derived from the very different and polytheistic Babylonian accounts.[8] It also supports the fact that monotheism was the original religious belief and not a later evolutionary refinement from an earlier polytheism.

This research further serves to falsify the widespread idea that Genesis 1 and Genesis 2 give conflicting accounts of creation. It also suggests that the higher critical theories on the composition and date of Genesis are factually bankrupt.

Just as God has not left us in doubt about our destiny, so he has not left us in doubt about our origin. We have the footnotes of Moses.

20

Adam and the Evangelical

SOME PEOPLE feel that the early chapters of Genesis have little informational value for the student of origins or for the historian. On the other end of the spectrum are those who feel that the early chapters of Genesis are completely reliable in terms of cosmology and history. A third group consists of those evangelical creationists who believe in the full inspiration of Scripture (including the early chapters of Genesis) but who also accept the validity of the humanly devised geologic column and its time scale. Those evangelicals have a problem with Adam.

Specifically, the problem is twofold: (1) What is the relationship of Adam to the fossils that give every evidence of being human and are dated all the way back to 4.5 m.y.a.? (2) Since archeologists claim that Genesis 4 and 5 present a picture of what they call the Neolithic (New Stone Age), where in Genesis is the Paleolithic (Old Stone Age) or the Mesolithic (Middle Stone Age) in relation to Adam?

Those who believe that Genesis was written for other than historical purposes are not troubled by these questions. We who accept the Genesis framework of history, believing that the commonly accepted geological and archeological time spans are founded on invalid assumptions, are interested in these questions but not troubled by them. The evangelical creationist who accepts both the integrity of Genesis and the integrity of these geological and archeological time frames must develop a synthesis between what appear to be two mutually exclusive views of human history.

These evangelicals have sought to resolve the conflict in one of two ways: (1) the old earth–old Adam view, in which the biblical Adam is dated early enough to precede all of the fossils deemed to be human, or (2) the old earth–recent Adam view, in which the biblical Adam is dated at about 10,000 y.a., and the fossil material in question represents "pre-Adamites" who may or may not have been human in the biblical sense of the term. Although both of these views are sincere attempts at reconciliation, both have serious biblical, theological, and paleontological flaws. Further, neither view seems to be an effective apologetic for the Christian faith. (We are not discussing here the issue of evangelicals who believe in Theistic Evolution. The arguments against naturalistic evolution are equally valid against Theistic Evolution, and have been addressed elsewhere in this book.)

The Old Earth–Old Adam View

The old earth–old Adam creationist view was catapulted into popularity in the evangelical community through the late Bernard Ramm's book *The Christian View of Science and Scripture*, published in 1954.[1] Although much of Ramm's scientific data was obsolete when the book was published,[2] and there are internal inconsistencies in his system, his book has been used almost continuously as a text in a number of respected Christian colleges and universities.

Recent Creationists sometimes consider old earth–old Adam creationists as just closet evolutionists. There are many similarities in the two positions. However, Ramm makes the following distinction between any form of creation and any form of evolution. Evolution, he says, is change from within the organism, based on mutations. Creation is change from outside the organism, based on direct acts of God.[3] While Ramm's distinction is valid, it is not the only one nor the best one that can be made. Creation, when applied to the work of God, has historically and traditionally been considered to be a sudden act or event. Evolution has always been considered to be a process involving much time. Since old earth–old Adam creationists seem to have a massive amount of process mixed with a tiny bit of event in their system, they should not be surprised that their commitment to creationism has been questioned.

The problem is that the word *creation* has been used in such sloppy ways. So many people with different concepts of origins have appropriated the term to themselves that the term no longer conveys much information. However, when used of God's creative activity, the biblical usage seems quite clear. If the days of Genesis 1 are literal days, there is no question that biblical creation involves event, not process. It was the injecting of vast amounts of time into Genesis 1 that changed the concept of creation from event to process. But since Genesis 1 is the battleground, one cannot appeal to it without being accused of begging the question.

However, there is evidence external to Genesis 1 that ought to settle the issue. "For in six days the LORD made the heavens and the earth, the sea, and all that is in them" (Exod. 20:11). At the risk of being called simplistic, one cannot imagine how God could have said more clearly that creation in Genesis 1 is event, and does not involve a process. The setting of this Exodus passage is the Ten Commandments. We are to work six days and rest one day, because God in creation worked six days and rested one day. Although old earth–old Adam creationists have tried very hard to blunt the force of this passage, Terence E. Fretheim (Luther Northwestern Theological Seminary) states its significance:

> . . . the references to creation in Exodus are not used as an analogy—
> that is, your rest on the seventh day ought to be like God's rest in cre-
> ation. It is, rather, stated in terms of the imitation of God or a divine pre-
> cedent that is to be followed: God worked for six days and rested on the
> seventh, and therefore you should do the same. Unless there is an exac-
> titude of reference, the argument of Exodus does not work.[4]

Another clear passage is Hebrews 11:3: "What is seen was not made out of what was visible." This passage is usually considered to refer to the original *ex nihilo* (out of nothing) creation of the universe. Creation as event seems to be demanded, since it is difficult to go from nothing to something in stages.

There are also parallels to the original creation in the creation miracles of our Lord. Creation miracles include the turning of water into wine (John 2:1-11), the feeding of the five thousand (Matt. 14:15-21), the feeding of the four thousand (Matt. 15:32-38), and the raising of Lazarus from the dead (John 11). These (and other miracles of creation) all involve event rather than process. A future analog would be the translation and the resurrection of believers at the Lord's coming. Here the Scriptures state that these things will take place "in a flash, in the twinkling of an eye" (1 Cor. 15:52). Thus, those who define Special Creation as event seem to be on the side of the angels, at least in this case. It remains for old earth–old Adam creationists to demonstrate any biblical evidence where God's creation activity *clearly* involves a process.

Bernard Ramm says his old earth–old Adam view includes: (1) the pictorial-day theory of the Genesis days, (2) the moderate theory of concordism, and (3) Progressive Creationism.[5] By the pictorial-day theory he means "that creation was *revealed* in six days, not *performed* in six days."[6] By moderate concordism he means that since the events in the days of Genesis 1 are not in agreement with the geological record, those days probably overlap and are best considered as being partly chronological and partly topical in their arrangement. By Progressive Creationism he means that creation, while it involved some event, was largely a process or development guided by natural law over time—specifically 4-5 billion years for Planet Earth. Short of a nuclear explosion, it is hard to imagine a more effective demolition job on Genesis 1 than Ramm has proposed.

Ramm does hold to a number of fiat creative acts, one being the creation of humans. However, he also believes that "a crass literalistic interpretation [of human creation] with its literal anthropomorphisms is . . . objectionable to good exegetical taste."[7] He seems to say that creation in Genesis 1 was on the taxonomic family level,[8] or higher. Since both australopithecines and humans are in the family *Hominidae,* this could imply that Adam was an australopithecine, or that Adam was the stem from which both the australopithecines and modern humans developed (evolved?). Writing in 1954, Ramm felt that the human race (Adam) was about 500,000 years old.[9]

Inconsistencies are endemic to the old earth–old Adam viewpoint. One of them is the concept of radiations or development after the initial creative acts of God. Ramm claims that any development or radiation in his system is on the horizontal scale. That is, a large number of varieties developed from an original created form without that radiation involving increased complexity. However, when Ramm's book was published, the human fossils dated at 500,000 y.a. (Ramm's age for Adam) were considered to be very primitive both physically and mentally. Hence, to go from Adam to today's humans would demand a vertical development (evolution) utilizing new genetic information. Recent Creationists accept horizontal or lateral radiation but deny the possibility of vertical development through natural processes or natural law, because it would require the acquisition of new genetic information which God would have to create.

Ramm is seldom very clear on the details of his system, but because he has so few creation events compared to a vast amount of development, it is impossible for his system to explain the present world without invoking vertical radiation (evolution) even though he disclaims that type of radiation.[10] Hence, in spite of his sincerity, Ramm has not presented a legitimate creation scenario for life in general or for human origins in particular. Although he claims to be a creationist, his system is actually a mixture of creation and evolution, with the heavier emphasis on evolution.

Theologian Carl F. H. Henry confirms that in the origin of humans, Progressive Creationism (one form of the old earth–old Adam view) involves a degree of evolution because it goes beyond mere development or horizontal radiation. Henry gives an accurate definition of that view:

> *Progressive creation*: God immanently supports and directs an extensive *evolution* of species but also acts transcendently at special stages of the *evolutionary process* to create the main biological taxonomical orders of being; man may be dependent physically on *intermediate manlike forms* but in distinction from the primates he is specially made in God's image.[11]

Other old earth–old Adam creationists have attempted harmonizations similar to that of Ramm. Like Ramm, many of them appeal to an essay, "Primeval Chronology," written by the famous Princeton Old Testament scholar William Henry Green (1825-1900). Robert C. Newman (Biblical School of Theology) includes Green's essay in the appendix of his book *Genesis One and the Origin of the Earth* published in 1977.[12] Green examined the chronologies of Genesis 5 and 11:

> . . . we conclude that the Scriptures furnish no data for a chronological computation prior to the life of Abraham; and that the Mosaic records do not fix and were not intended to fix the *precise* date either of the Flood or of the creation of the world.[13]

Newman's purpose in quoting William Henry Green is to suggest that if a solid conservative scholar such as Green believed that the early chapters of Genesis allowed for an old earth–old Adam position, that position cannot be all bad. However, quoting Green in today's discussion on the age of Adam is a classic case of quoting out of context.

William Henry Green lived at a time when Archbishop Ussher's date for creation, 4004 B.C., was popular. It was based upon the belief that the chronologies of Genesis 5 and 11 were strict father-son relationships. The sciences of archeology and paleontology were in their infancy. When Green first wrote his essay in 1863, only the original Neandertal fossils had been reported. When Green revised his essay for publication in 1890, only the Neandertal fossils from Spy, Belgium, could be added to the list of reported discoveries. The dates for these Neandertalers were unknown and are still relatively unknown. The scientific community of that day estimated that these fossils were thousands of years older than Ussher's creation date of 4004 B.C. Hence, they posed a problem for creationists. Green addressed this problem. Using biblical data, he concluded that the gaps in the Genesis chronologies allowed for these fossils to be post-Adamic.

Green's argument seems as valid today as when he first presented it. However, Green's words, as well as quotations from B. B. Warfield and James Orr, are used by old earth–old Adam creationists in a totally different context from that which those men intended. For instance, James Orr (writing in 1904) believed that the human race originated about 12,000 to 15,000 B.C.[14] That age is about the maximum age many Recent Creationists today place on the human race. Yet, James Orr is used by old earth–old Adam creationists to support ages of hundreds of thousands or even millions of years for human origins. William Henry Green was thinking of expanding Genesis 5 and 11 just a few thousand years. Yet Robert Newman, using Green as his primary justification, states that many of the promises of Jesus regarding his return could be better understood ". . . if mankind had lived in sin for tens of hundreds of thousands of years before his [Christ's] first coming."[15] Newman's time span is a bit fuzzy, but ten times one hundred times one thousand equals one million. Since each of his terms is in the plural, the minimum time that Newman seems to suggest for man's living in sin (Adam to Christ) is two million years.

Newman bases the legitimacy of this vast time span from Adam to Christ on Green's evidence for gaps in the Genesis chronologies.

Twenty names are in the Genesis 5 and 11 chronologies from Adam to Abraham, with a maximum of nineteen gaps. If we assume there is a gap between each of the names, the gaps would have to be over 100,000 years each to support Newman's chronology. However, we know that many of the names in those chronologies represent true father-son relationships, such as Adam being the father of Seth and Noah being the father of Shem. That means that in the few cases where there are legitimate gaps in the genealogies, those gaps would have to cover periods of over 200,000 years each to satisfy Newman's suggested chronology. If old earth–old Adam creationists stretched the Genesis chronologies to include all of the fossil material that can legitimately be called human (as Newman and others have suggested), they would have to place Adam at 4.5 m.y.a., making the gaps in the Genesis chronologies well over 500,000 years each. The absurdity of this type of interpretation of Genesis ought to be obvious to all. The impropriety of using Green, Orr, and others of a past era to justify that type of interpretation ought also to be obvious.

Appearing also in 1977 was the book *Creation and the Flood*[16] by old earth–old Adam geologist Davis A. Young (Calvin College). Rejecting Theistic Evolution, Young claimed full allegiance to creationism and to the inspiration of Scripture. He is the son of famed Old Testament scholar the late Edward J. Young of Westminster Seminary. Davis Young's harmonization has had profound influence in evangelical circles, partly, I suspect, because he comes from a family whose commitment to orthodoxy is unquestioned.

Young embraced the Day-Age theory, in which the days of Genesis 1 are not considered normal twenty-four-hour days but are instead considered to be historical time periods (ages) of indeterminate length. The term *day* is figurative, but the six days represent historical events taking place over time. Like Ramm, Young saw creation largely as a process.

Since the sequence of events in Genesis 1 is quite different from the sequence geologists interpret in the geologic column, Young achieved concordance by assuming there is a large degree of overlap in the events of each of the days (ages). The recorded activities of each of the days (ages) were the *major* events of that day, not the total events of that day. For instance, most of the plants would have appeared on day three, but some plants could have first appeared on other days. Most of the marine animals and birds would have appeared on day five, but some could have first appeared on other days. By invoking the Day-Age theory with overlap, Young claimed that the ". . . overall correspondence is by no means exact or perfect, but it is surprisingly close."[17]

Like Ramm, Young had many inconsistencies in his system. Although he claimed to be a creationist, the evolution Young denied with his right hand he accepted with his left hand. He allowed for the possible evolution of life from nonlife, and for the possible evolution of the horse, and other animals, from a more generalized stock.[18] Young also used the complexity of the rocks as evidence that the creation of those rocks had to be processes rather than events. But he rejected the one mechanism—a world-wide Flood—which could explain the complexity of those rocks. He did accept the sudden creation of some life forms even though those life forms are far more complex than the rocks he insisted could only be formed by a long process because of their complexity.

Geologist Richard Bambach (Virginia Polytechnic Institute and State University) points out an inconsistency in Young's thinking that seems to be basic to all creationists who accept the evidence for an old earth. What is the rationale, he asks, for accepting the scientific evidence for an old earth while rejecting the scientific evidence for evolution, when the evidence for both is often of the same nature and has the same presuppositions? Bambach claims that Young adjusts his religion to fit the old-age evidence while using his religion to reject the evolution evidence.[19]

Young was emphatic regarding the creation of the first humans and the entrance of sin into the human family through the sin of one man. Using Warfield, he said that the time span of the early chapters of Genesis could be ". . . tens or hundreds of thousands of years. We simply do not know."[20] He speculated, and then concluded:

> Could he [Adam] have looked like a Neanderthal Man? Even more daring, could Adam have been a creature like Australopithecus?
>
> It is probably virtually impossible for the Christian to identify, from the fossil record, the time when special creation occurred.[21]

In the tradition of Ramm, Newman, and Young, geneticist Pattle P. T. Pun (Wheaton College) published his 1982 work, *Evolution: Nature and Scripture in Conflict?*[22] Pun holds the Day-Age theory with the days (ages) overlapping. He also believes that creation is primarily a process. He feels that ". . . humans have existed for hundreds of thousands of years prior to Christ's first coming."[23] What these humans looked like, Pun feels, depends on how we define humans. If we define humans in terms of cranial shape and tool-making ability, the first humans may have been the five-million-year-old australopithecines.[24] (Pun is mistaken on the dates, the alleged human cranial characteristics, and the tool-making abilities of the australopithecines.)

In a recent radio interview, Pun leaned heavily on the geological interpretations of Davis Young. He again suggested that the australopithecines were human.[25] Here we see the absurdities in the old earth–old Adam position. Not even evolutionists put the australopithecines in the genus *Homo*. Pun simply has not done his homework on the australopithecine fossils. Like other old earth–old Adam creationists, Pun's scenario for Adam is the impossible dream.

In speculating that the australopithecines might have been the originally created "humans" or "Adam," neither Ramm, Young, nor Pun addresses the question of where the qualitatively new genetic information came from that would allow the tiny-brained australopithecines to "develop" into the big-brained humans of today. To imply that qualitatively new genetic information can originate by natural processes is to embrace the central concept of evolution. Whether these writers realize it or not, they are embracing evolution and calling it creation. It is a serious omission not to address this problem and thus give the impression that extensive horizontal radiations from a limited number of created organisms can account for the biological world as we know it today.

Although I have cited only a small part of the old earth–old Adam literature, there is an interesting practice in much of it that demands comment. When old earth–old Adam creationists place Adam prior to the human fossils to make Adam the head of the human race, in most cases the antiquity that must be assigned to Adam is seriously underestimated. This practice is so consistent in the literature that one wonders if there is a psychological factor at work. The old earth—old Adam creationist goes back in time as far as he dares, all the time subconsciously knowing that he is doing Genesis an injustice. Thus, he does not go back far enough to actually include all of the fossil material that can legitimately be considered human.

Let me give one illustration of this practice, a practice that is epidemic in the old earth–old Adam literature. Anthropologist James O. Buswell III (then at Trinity College, Deerfield, Illinois) wrote in *Christianity Today* that those who hold the old earth–old Adam position ". . . insist that Adam must have been created before the earliest of those forms that, by both anatomical and cultural evidence, may be interpreted unequivocally as human and as geologically ancient, according to the findings of human paleontology."[26]

After discussing the recent-creation date for Adam, Buswell then referred to himself as among ". . . those who find acceptable an earlier date for Adam's creation, say, hundreds of thousands of years ago."[27]

However, at the time Buswell wrote those words (1975), human fossils and artifacts had been found that were dated well beyond one million years, and some human fossils and artifacts were considered to be close to two million years old. Even allowing for the interpretive problems which are always present, it is hard to believe that Buswell seriously thought that he had harmonized Genesis and the human fossil record by dating Adam "hundreds of thousands of years ago." Certainly no evolutionist with a knowledge of the human fossil record would have bought that type of harmonization.

Of all the old earth–old Adam creationists in recent years, Davis A. Young has been the most influential. He is an excellent communicator, and not many geologists have been as outspoken in their creationist convictions as he has been. His two books on the Day-Age harmonization of science and Scripture have been extensively quoted by others in support of the old earth–old Adam position. Hence, many were shocked when he recently repudiated his former approach to harmonization. He told about it in a science symposium, "The Harmonization of Scripture and Science," held at Wheaton College, March 23, 1990. One cannot help but admire his honesty and openness. One cannot help but wonder if other old earth–old Adam creationists have not sensed the same discomfort with the Day-Age and Progressive-Creation positions that he has experienced. An extended quotation of Young's remarks is appropriate.

> The Day-Age hypothesis insisted with at least a semblance of textual plausibility that the days of creation were long periods of time of indeterminate length, although the immediate context implies that the term *yom* for "day" really means "day." Having devised a means for allowing Genesis 1 to be in harmony with an ancient planet, Day-Age advocates

needed to demonstrate that the sequence of creative activities of Genesis chapter 1 matched the sequence of events deciphered by the astronomers and geologists. Well, Day-Agers outdid themselves in constructing impressive correlations. Of course, these correlations . . . all differed from each other. While a fairly convincing case could be made for a general concord, . . . specifics of these correlations were a bit more murky.

There were some textual obstacles the Day-Agers developed an amazing agility in surmounting. The biblical text, for example, has vegetation appearing on the third day and animals on the fifth day. Geology, however, had long realized that invertebrate animals were swarming in the seas long before vegetation gained a foothold on land. This obvious point of conflict, however, failed to dissuade well-intentioned Christians, my earlier self included, from nudging the text to mean something different from what it says. In my case, I suggested that the events of the days overlapped. Having publicly repented of that textual mutilation a few years ago, I will move on without further embarrassing myself.

Worse yet, the text states that on the fourth day God made the heavenly bodies after the earth was already in existence. Here is a blatant naked confrontation with science. Astronomy insists that the sun is older than the earth. How do Day-Agers worm out of this? The usual subterfuge involves the suggestion that the light originally visible on earth was sunlight that was obscured and diffused by the thick atmospheric mists that began to dissipate with the separation of the waters on the second day. Not until the fourth day, then, had the mists thinned to the point where the sun became visible from the earth.

After examining other techniques at harmonizing Genesis with contemporary models of earth history, Young confessed, "Genius as all these schemes may be, one is struck by the forced nature of them all. While the exegetical gymnastic maneuvers have displayed remarkable flexibility, I suspect that they have resulted in temporary damage to the theological musculature."

Young then concluded what Recent Creationists have been stating for years. "Interpretation of Genesis 1 through 11 as factual history does not mesh with the emerging picture of the early history of the universe and of humanity that has been deciphered by scientific investigation. Dickering with the biblical text doesn't seem to make it fit the scientific data. . . . Concordistic harmonizations have generally failed. I don't particularly like to admit that because I have been a concordist much of my life."[28]

Davis Young is correct in saying that harmonization based on the old earth–old Adam position has failed. Because he does not recognize that his data has been placed in a philosophic framework alien to Genesis, he now has nowhere to go. He is suggesting that ". . . the Bible may be expressing history in nonfactual terms. . . ." There is a name for nonfactual history: fiction. However, Young clearly does not intend to imply that. Hence, his words convey no information. We see the frustrations of a man who is utterly sincere in wanting to maintain biblical integrity but is unable to extricate himself from the man-made philosophic framework of earth history.

Old earth–old Adam creationists recognize the need to maintain the clearly taught biblical doctrines of (1) the unity of the human race in Adam (Acts 17:26), (2) the individual, Adam, as the first created human (Gen. 1-2; 1 Cor. 15:45), and (3) the entrance of sin into the human family through one man, Adam (Rom. 5:12-21). However, their position fails because the acceptance of an old earth forces them to deal with Scripture in a contrived manner and to stretch the Genesis chronologies to absurd lengths to maintain Adam as head of the race.

The Old Earth–Recent Adam View

Old earth–recent Adam creationists seek to honor the equally clear implication of Genesis that Adam lived in relatively recent times. But their acceptance of an old earth and the antiquity of the human fossils presents a problem for them, also. These creationists seek to solve the problem by suggesting that the human fossils that are dated earlier than the Neolithic (the New Stone age, starting at about 10,000 y.a.) represent "pre-Adamites" who may or may not have been human in the biblical sense of the term. Since there is no biblical evidence whatsoever for the existence of such creatures, this biblical silence gives the old earth–recent Adam creationists freedom to indulge in flights of fantasy regarding the nature and status of those alleged beings.

Old Testament specialist Ronald Youngblood holds to an old earth, and believes that the days of Genesis 1 are a literary framework or device similar to such devices found in other Near Eastern literature. Hence, those days have no relationship to time. However, he feels that Genesis demands both the instantaneous and recent creation of Adam. This presents the problem of identifying the human fossil material.

> Adam and Eve, the first "man" and "woman" in the Biblical sense of those terms, date back to a few tens of thousands of years ago at best.[29]

> Whatever hominids (Cro-Magnon, Neanderthal, and earlier) may have existed prior to the time of Adam, they had only animal intelligence and were not bound to God in a covenant relationship.[30]

The Cro-Magnon people made the paintings on the cave walls of France and Spain. Even their stone tools are works of art. They also made fishhooks. The earlier Neandertals buried their dead with flowers. Their rugged stone tools brought down the mightiest game. Is all this just "animal intelligence"? Youngblood reasons that since chimpanzees can be taught to use sign language, and since language is a mark of culture, that shows that the Cro-Magnon and Neandertal people had only animal "culture." In other words, the cultural activity of the hominid fossils does not prove that they were human in the biblical sense of the term. But Youngblood uses an improper analogy. Studies show that there is an immense gap between this ape language and human language.[31] The gap is qualitative, not just quantitative. There is no acquisition of culture by these animals. On the other hand, every kind of evidence that we have a right to expect from the fossil and archeological record indicates that the Cro-Magnon and Neandertal peoples were human in the same ways that we are human. Furthermore, if death is the result of Adam's sin, why did those pre-Adamites die? What happened to those Adam look-alikes when Adam appeared on the scene? Youngblood does not address these problems.

Gleason L. Archer, Jr. (Trinity Evangelical Divinity School) also accepts an old earth and a recent Adam. He believes that one cannot hold to the accuracy of Genesis and yet stretch the date for Adam back to 200,000 y.a. or beyond.[32] Archer deals with the fossil material, such as *Pithecanthropus (Homo erectus)*, Swanscombe Man, Neandertal, and Cro-Magnon, as follows:

> . . . it seems best to regard these races as all prior to Adam's time, and not involved in the Adamic covenant. We must leave the question open, in view of the cultural remains, whether these pre-Adamite creatures had souls (or, to use the trichotomite terminology, spirits).[33]

Since Romans 5:12-21 clearly teaches that all humans following Adam must be his literal descendants, there cannot be a genetic relationship between Adam and the pre-Adamites. Thus Archer conveniently disposes of them: "They may have been exterminated by God for reasons unknown prior to the creation of the original parent of the present human race."[34]

Archer not only creates human races without biblical authority, he destroys human races without biblical authority. Since he recognizes that those beings might have had souls (or spirits), we are not dealing with just animal death before the fall of man into sin, but possibly with human death before the Fall. If sin first entered the world through Adam, was God justified in exterminating those pre-Adamic "humans" when they apparently had not committed any sin? Is this not a problem that could involve the righteousness of God? Archer is silent on these matters.

Both Youngblood and Archer exclude those pre-Adamites from a covenant with God, a covenant God is said to have made with Adam. This covenant, known in Reformed theology as the covenant of works, is more a theological than a biblical covenant. However, even if such a covenant is legitimate, it is irrelevant. Any covenant God would have made with Adam would be based upon the fact that Adam, as distinct from the animals, was created in the image of God (the *imago Dei*). Hence, the real issue is not whether or not those fossil individuals (considered by Youngblood and Archer as pre-Adamites) had a covenant relationship with God, but whether or not they possessed the *imago Dei*. Their cultural attainment would suggest that they did.

John R. W. Stott (All Souls Church, London) is another old earth–recent Adam creationist. He wrote:

> But my acceptance of Adam and Eve as historical is not incompatible with my belief that several forms of pre-Adamic "hominid" may have existed for thousands of years previously. These hominids began to advance culturally. They made their cave drawings and buried their dead. You may call them *homo erectus*. I think you may even call some of them *homo sapiens*, for these are arbitrary scientific names. But Adam was the first *homo divinus*, if I may coin the phrase, the first man to whom may be given the Biblical designation "made in the image of God."[35]

For someone who accepts an old earth, Stott underestimates notoriously the age of the human fossils. Furthermore, he misapplies the concept of the divine image. It is obvious that something of the divine image was retained after the Fall (Gen. 9:6). It also seems that something of the divine image is restored at salvation (Col. 3:10). Most evangelical theologians recognize that the divine image somehow involves man's rational powers. Two of the greatest theologians in the history of the church, John Calvin[36] and Jonathan Edwards,[37] believed that the *imago Dei* originally had two components: (1) a moral component having a disposition toward holiness and love, and (2) a natural component of reason and will. The moral component was lost in the Fall. The component of reason, although affected by the Fall, was retained. Based on Romans 1:18-20, Edwards further concluded that our human reasoning ability is qualitatively like God's, although infinitely smaller quantitatively.

Whether or not one agrees with Calvin and Edwards, it is difficult to exclude culture as one of the expressions of the divine image. Anthropologists know that human culture is partly made up of how we relate to God and to each other. It is the divine image that also allows us, like God, to be creative, to conceptualize, to deal in abstractions, and to be artistic. These things separate all humans from all animals; without them we would not be human. Since those fossil humans (Stott's pre-Adamites) had human bodies and human culture, for Stott to claim that they did not have the divine image demands more omniscience on his part than I suspect he has. Some old earth–recent Adam creationists now recognize that they must grant the *imago Dei* to their pre-Adamites in spite of all the problems it creates.

Because Derek Kidner is a Theistic Evolutionist, he would not normally be included in these discussions. However, his scenario is not quite that simple. Furthermore, some old earth–recent Adam creationists are impressed with his solution to the pre-Adamite problem. In addressing the subject of origins in his commentary on Genesis, Kidner speaks of "creation by evolution."[38] Creation by evolution is a contradiction in terms. It is this sloppy and imprecise use of words, words which normally convey precise concepts, that is responsible for much of the confusion in this area.

"God initially shaped man by a process of evolution,"[39] Kidner states. When God breathed the spiritual component into one of those evolved creatures, Adam was created. Eve, however, was apparently the product of Special Creation. The divine image and the effects of the Fall were then extended outward to include all other pre-Adamites who had evolved.

> Yet it is at least conceivable that after the special creation of Eve, which established the first human pair as God's viceregents (Gen. 1:27, 28) and clinched the fact that there is no natural bridge from animal to man, God may have now conferred His image on Adam's collaterals, to bring them into the same realm of being. Adam's "federal" headship of humanity extended, if that was the case, outwards to his contemporaries as well as onwards to his offspring, and his disobedience disinherited both alike.[40]

Kidner's inconsistency in suggesting that God allowed Adam's body to evolve and then directly created Eve's body needs no comment. But since Kidner's interpretation of the pre-Adamites means that not all humans have descended from Adam, he must explain that ". . . the unity of mankind 'in Adam' and our common status as sinners through his offense are expressed in Scripture in terms not of heredity but simply of solidarity."[41]

It is unclear what Kidner means by our "solidarity" in Adam in contrast to our being genetically related to Adam. He does not amplify it. Considering the seriousness of the doctrinal issue here, he certainly owes his readers an explanation. He recognizes that he has a problem with Genesis 3:20, which states that Eve is the "mother of all living," but he attempts to explain it away. He ignores Acts 17:26: "From one man he [God] made every nation of men."

At times biblical interpretation raises questions of ethics. Is it ethically proper to interpret a passage of Scripture in such a way that the result is exactly the opposite of what the passage appears to say? Some years ago, the late J. Barton Payne wrote an article entitled, "Hermeneutics As a Cloak for the Denial of Scripture." It happens. Payne had seen it happen, and he was concerned. Not only does it happen, but it is legal. In evangelical circles, it is not politically correct for one to say that he does not believe a certain Scripture. But he can say, "I don't interpret it that way," and then by interpretation change the meaning completely. Not only can that be an accepted procedure, it is sometimes considered a mark of scholarship and sophistication. Genesis 3:20, Acts 17:26, and Romans 5:12-21 state that all humans are genetically descended from Adam and Eve. Kidner states that some humans were not. He does it by interpretation. It looks rather like denial to me.

Another harmonization strategy utilized by old earth–recent Adam creationists is to suggest that Genesis 1 depicts the creation of mankind in general (an acceptable translation of *Adam*) and that the creation of the literal Adam came later. An innovative variation of this strategy was suggested recently in a paper read before the Evangelical Theological Society by John H. Walton[42] (Moody Bible Institute). Walton emphasized that he was not convinced that the theory he was presenting was right, but he was intrigued by the way it seemed to offer solutions to the problems. I deal with his ideas because these strategies have been suggested by other evangelicals also.

The basic concept is that there were two phases to God's creative activity. Genesis 1:1-2:3 is phase one. Genesis 2:4 begins phase two. Between those two phases is a time period of indeterminate length, possibly as much as millions of years. Walton felt that this theory would make the "hybrid Day-Age theory" unnecessary. He was impressed with the "one size fits all" ability of his theory to encompass everything from the literal-day approach to the framework-hypothesis position.

In phase one, which includes the seven-day sequence, the people God created were primitive and would be equated with those found in the human fossil record. Those people would have the divine image, because that is mentioned in Genesis 1. Since the Bible states that there was no sin until Adam's fall, those people could be viewed as unblemished. However, there is evidence of violence in the fossil record, so it would be better to consider those people as not being morally responsible, in the same way that babies today have the divine image but are not considered to be morally responsible. Although primitive, those people had souls and were loved by God. They also had the potential for moral responsibility as they matured. Death was normal for people and animals in phase one of creation.

Phase two of creation begins at Genesis 2:4. A key point is Walton's translation of Genesis 2:5-6: "Now no cultivated shrub had yet appeared on the earth, and no cultivated grasses had yet sprouted because God had not yet sent rain on the earth, and there was no man *available* to till the ground." The phase-one primitive people were not able to carry on agriculture but were the hunter-gatherers of the archeological record. In phase two God inaugurated civilization and created

Adam and Eve. Walton felt that this scenario fits well with the Neolithic revolution and its domestication of plants and animals. The animals God created in phase two may have been an entirely different set compared with those created in phase one.

Walton also suggested that God created Adam in a preadolescent state. Phase one was the baby stage of mankind. Although death was natural in phase two, also, God had provided the Tree of Life so that Adam and Eve could avoid death. The Tree of the Knowledge of Good and Evil was actually a wisdom tree. It was God's intention that Adam and Eve would gradually grow into maturity. By eating of the wisdom tree against God's will, Adam and Eve "grew up too fast." They became sexually aware and experienced lust. Although they were not cursed for eating of the tree, they were expelled from the garden. By not having access to the Tree of Life, they eventually died. Our condition is the same as theirs. We also die because we do not have access to the Tree of Life.

Marriage was a part of the phase-two creation. Although the primitive people of phase one were still living when Adam and Eve were created, there was no intermarriage between the phase-one people and the descendants of Adam and Eve. He suggested that the phase-one people perished in the Flood, if they had not passed off the scene before that time.

Walton emphasized that his paper was tentative, and its purpose was simply to suggest new strategies for harmonization. However, as with all other old earth–recent Adam views, there are serious problems with these suggestions. The concept of death seems to be the Achilles heel of all old earth–recent Adam scenarios. When Walton was asked why the pre-Adamites died, he replied that they were created to die. But the idea that death is an integral part of God's creation does not square with the biblical declaration that death is an enemy: "The last enemy to be destroyed is death" (1 Cor. 15:26). Furthermore, death is not a part of the original creation, but is an intruder imposed upon humans after creation was completed. "For since death came through a man, the resurrection of the dead comes also through a man. For as in Adam all die, so in Christ all will be made alive" (1 Cor. 15:21, 22).

The humans of phase one and phase two were not that different, Walton feels. Adam was created just a bit more mature. However, for biblical reasons, he insists that the two populations were reproductively isolated. The idea that two rather similar populations in the same general area would not intermarry or mix genetically is an extremely naive one. It would be almost unparalleled in human history. There is only one way under those conditions to insure complete genetic isolation. That would be to suggest that the two populations represented two different biblical "kinds." It goes without saying that Genesis knows nothing of two different "kinds" of humans.

Further problems with Walton's proposals involve carnivorous animals before the fall, and God's declaring in Genesis 2:3 that he is finished with creation only to have him begin again at Genesis 2:4. Furthermore, if the Fall consisted only in man's getting quickly what God had intended that he get slowly (maturity), then the penalty seems far harsher than the crime. Walton's scenario shares the weaknesses of the other old earth–recent Adam positions. It has a wrong view of death, a wrong view of the Fall, and a wrong view of the effects of the original sin.

While I have not discussed all of the old earth–recent Adam proposals, the ones I have discussed have been representative. On balance, the old earth–recent Adam position seems to be even more flawed Biblically, doctrinally, and paleontologically than the old earth—old Adam position.

Concluding Observations

The harmonization problems in both the old earth—old Adam position and the old earth–recent Adam position stem from the acceptance of the alleged evidence for an old earth. Some concluding observations are in order.

First, in the extensive body of old-earth creationist literature that discusses both the old-Adam and recent-Adam positions, one sees throughout an exceptionally high level of trust in the validity of the humanly devised geological age system and the archeological time span. In fact, the degree of faith and trust placed by old-earth creationists in these humanly devised systems is frightening. Nowhere in this literature does there seem to be the slightest doubt that these ages are accurate, so accurate that Genesis must be accommodated to fit them. Just as Davis Young did not realize how much he was nudging Scripture to make it say something it was not saying until he had stepped back and reevaluated his position, is it possible that these sincere old-earth creationists actually, perhaps without realizing it, place more trust in these humanly devised systems than in the Scriptures themselves?

Second, an almost universal characteristic of the old-earth (both old-Adam and recent-Adam) creationist literature is a lack of expertise in dealing with the human fossil material. I have yet to find an exception. While these creationists are extremely capable in their areas of expertise, it is obvious that they are out of their element when they address the subject of the human fossils. Their errors both in fact and in understanding are so basic and so obvious as to endanger the acceptance of their harmonizations with some of the very people—the scientific establishment—to whom their harmonizations are addressed.

Third, the faith placed in the radiometric dating methods by old-earth creationists involves their trust in data beyond their ability to evaluate. Some years ago, I presented a paper at an Evangelical Theological Society meeting challenging the validity of the long-term radiometric dating methods. My paper was in response to an article in *Christianity Today* by a noted evangelical anthropologist. He had urged all creationists to accept these dating methods and thus bring harmony to the creationist movement. This anthropologist was in the audience when I presented my paper. In the discussion period, he admitted that he did not know enough about the details of radiometric dating to refute my paper. Yet, in his article he had stated that the human fossil and cultural evidence could be interpreted "unequivocally" as being geologically ancient, based on the dating methods.

I have had world-famous geologists tell me that they did not know the details of the radiometric dating methods or the basic assumptions upon which they were based. Yet, they were very dogmatic about the fact of an old earth. When people speak dogmatically on matters beyond their expertise, they are either expressing a deep faith or a deep bias, or both.

It is not uncommon for evangelical Bible scholars and theologians to accept uncritically the idea of an old earth without having the scientific background to properly evaluate the evidence. A 1990 volume on theology states:

> Recent-creationist attempts to undermine the results of the several scientific methods of dating are insufficient to discount these methods entirely. The data for scientific dating are drawn from many different sources and show a significant degree of agreement.

> With all the work that has been done by the Creation Research Society, this organization has yet to make a sufficiently conclusive case for the

hypothesis that the Flood of Noah's day was the efficient cause of the bulk of the fossils and geological formations everywhere around the world.[43]

I know the two authors well. They are both outstanding evangelical theologians. However, the issue is not their theological credentials; the issue is their scientific credentials. They have made judgments on scientific matters far beyond what their expertise in science would allow. The authorities they cite for their acceptance of the scientific dating methods are Davis Young and Robert Newman, neither of whom are authorities in the dating methods. The authority they cite for rejecting the Noachic flood as the major cause of the geological work and fossilization we see in the crust of the earth is Richard Bube, a Theistic Evolutionist with no expertise in geology, hydrology, or paleontology. This practice of making judgments in areas beyond one's expertise is becoming epidemic among those who reject Recent Creationism. Can it be that evangelicals at times express strong opinions in areas where they have little expertise, expecting their words to carry weight because of their accomplishments in other areas where they do have expertise?

A fourth observation has to do with appeals to authority in science. Science at its best is a noble enterprise. One of its basic rules is that a scientist is not to appeal to authority to make his point. The facts are supposed to speak for themselves. The reason for this rule is that scientific truth is not arrived at by majority vote or by opinions of experts. I once heard a lecture by Peter Bergmann, the world-famous physicist from Syracuse University. In his early years he had been an associate of Albert Einstein. He apologized in his lecture for mentioning that Einstein agreed with the particular point he was making, because, he said, we scientists are not supposed to appeal to authority. An appeal to authority can be a subtle (or not-so-subtle) form of intimidation. It can be a way of pressuring one to side with the majority or with the experts. It implies that the majority or the experts are always right, which in the major events in the history of science has proven not to be the case. We all know that an appeal to authority in science is very wrong, very illogical, and very effective.

No one uses the appeal-to-authority argument more often than do evolutionists. Slightly behind evolutionists in the use of this argument are old-earth creationists. They are right when they say that far more people in the intellectual and scientific establishment believe in an old earth than believe in a recent earth. However, there is something humorous about old-earth creationists using that argument. As creationists, they themselves are also in the minority. If they were to follow their own advice, they would become evolutionists.

Old-earth creationists use the appeal-to-authority argument so often and with such force that one can be forgiven for wondering if there might be a psychological factor at work. It is hardly a secret that Recent Creationists are *persona non grata* in the intellectual world. Thus, there could be the temptation to accept the old-earth position because it gives them intellectual respectability, even though accepting that position may require their endorsing data beyond their ability to evaluate. It is a truism that the world practically worships the Recent Creationists of the past (Isaac Newton and the rest) while despising the Recent Creationists of the present.

My last observation is of a concept practically every old-earth creationist utilizes called the Double-Revelation theory. Simply put, it is the idea that God has given us two books: the book we call Scripture, and the "book" of nature. Since God has given both revelations, the assumption is that these two revelations should agree or harmonize. In fact, Recent Creationists are often accused of dishonoring God by insisting upon an interpretation of Genesis that forces Genesis out of harmony with the accepted "facts" of science. As logical as the Double-Revelation theory seems, and as universally as it is held, it is foolishness nonetheless.

It is helpful to understand the roots of the Double-Revelation theory. Historian Gale E. Christianson (Indiana State University) tells us about it in his book on the life and times of Isaac Newton.

In the midseventeenth century, science did not have the financial support from government, industry, and the universities that it has today. Only a small group of people of independent means could devote their full time to it. Before the term *virtuoso* was applied to musicians it was applied to a group of natural philosophers in England who believed that they could formulate a rational understanding of the universe based on immutable mathematical laws. Nature was looked upon as a great machine running without external aids. The virtuosi believed in God, but their worship was more through the wonder of his creation than through the words of Scripture. Their concept of creation did not come from Genesis but from the Great-Chain-of-Being philosophy.

These virtuosi sensed a problem between the precision of the mathematical laws they had worked out for nature versus what they felt was the inability of words, even the words of Scripture, to accurately describe nature. Hence, they developed the Double-Revelation theory that Scripture was supreme in matters of faith and ethics, but the precise language of science was superior to describe nature. Christianson continues:

> . . . the virtuosi contended that when God gave the Scriptures to man He necessarily employed the language of the common people. Biblical language had to conform to the daily experience of a pre-scientific Hebrew culture and hence contains certain unscientific propositions. . . .

> The virtuosi advocated that man, in his striving to know the physical nature of the universe, begin with God's *work* rather than his *word*. The teachings of the Bible must remain paramount in questions of human conduct and morality, but in dealing with nature the impersonal language of science must take precedence over the common tongue.[44]

This unwarranted removal of the biblical teaching on nature from the full inspiration and inerrancy of the rest of Scripture has a modern counterpart. Old-earth creationists put it this way: "The Bible gives us the 'who' and the 'why' of creation, and science tells us the 'when' and the 'how' of creation."

The Double-Revelation theory is foolishness. It is foolishness because while science can instruct us well in matters of the natural world, it cannot instruct us at all on the events of creation. Creation was a one-time occurrence, a singularity. While science thrives on experimentation and repeatability, it cannot handle singularities. It has no way of studying things that happened only once in the past. Creation is simply outside the domain of science. That is why God had to reveal the details to us. The fact that the scientific establishment has made pronouncements on creation reveals its arrogance. It is unfortunate that some creationists have listened to them.

The Double-Revelation theory is also foolishness because of the nature of biblical truth in contrast to the nature of scientific truth. Whether one agrees with it or not, the Bible claims to be truth in the absolute sense, including its statements about nature. On the other hand, philosophers of science are unanimous in recognizing that science does not—in fact, cannot—traffic in absolute truth. All scientific truth is relative. What strange twist of logic would cause us to think that absolute truth and relative truth can be or should be harmonized? In fact, if they could be harmonized, would that not be elevating the relative truth of today to absolute status? Since the average life of a scientific theory is less than twenty years, a harmonization today guarantees a lack of harmony tomorrow. Logic would tell us that a true harmonization cannot take place until Christ returns in glory and all truth is illuminated by the truth of God. Until then, the normal thing to expect is a degree of discord, not harmony. In other words, the basic supposition of the Double-Revelation theory is utterly false.

It should not be difficult to see that when old-earth creationists must manipulate Scripture, as we have shown in this chapter, and then must further deny a worldwide Flood (as virtually all of them do), we are not dealing with harmony. We are dealing with discord. Although they certainly do not intend it to be taken this way, there is implicit in their position the idea that God did a very poor job of revealing creation and early earth history.

Some say that Genesis is not really clear. The problem is that Genesis is *too* clear. If Genesis were unclear, it could be adapted to an old-earth or evolutionary scenario quite easily. The fact that so much manipulation is necessary to do so indicates that Genesis has its own unique cosmological agenda.

There is a degree of humor in the fact that while some creationists who love God and his Word cannot see the problem, an evolutionist Marxist, Stephen Jay Gould, has no difficulty in recognizing that the proper reading of an inerrant Bible shows both a recent creation and a miraculous worldwide Flood.[45]

Appendix:
The Dating Game

A VERY POPULAR MYTH is that the radioactive dating methods are an independent confirmation of the geologic time scale and the concept of human evolution. This myth includes the idea that the various dating methods are independent of one another and hence act as controls. The methods appear so impressive that many creationists have accepted them as evidence that the earth is very old. Perhaps the best way to expose this myth for what it is—science fiction—is to present a case study of the dating of the East African KBS Tuff strata and the famous fossil KNM-ER 1470, as recorded in the scientific journals, especially the British journal *Nature*.

Richard Leakey, son of famed paleoanthropologists Louis and Mary Leakey, was just twenty-three when he borrowed a helicopter and first visited the rich fossil deposits east of Lake Rudolf (now Lake Turkana) in northern Kenya. The year was 1967. He was so stunned by what he saw that he immediately organized an expedition to that area to search for hominid fossils. The result is the permanent base known as Koobi Fora, which has produced some of the most striking hominid fossil discoveries in the entire history of paleontology.

The most important fossil discovered there is one that for all its fame has never been given a respectable name. It is KNM-ER 1470 (Kenya National Museum—where it is housed; East Rudolf—where it was found; and 1470—the museum acquisition number). Skull 1470 is very modern in appearance but was originally believed to be about 2.9 million years old. This conflict between its modern appearance and its ancient age presented a serious challenge to all currently held theories of human evolution. It precipitated a conflict over the dating of the fossil which lasted ten years.

One of the early geologists to work with Richard Leakey at East Rudolf was a young woman from Yale University, Kay Behrensmeyer. In seeking to unravel the geology of the area, she discovered a layer of volcanic ash or tuff that turned out to be crucial in the dating of the fossils and the artifacts found in association with it. The spot where she first located this tuff became known as the Kay Behrensmeyer Site. This volcanic tuff has become known ever since as the KBS Tuff. Richard Leakey, in a Denver lecture, laughingly remarked that Kay Behrensmeyer is the only woman in the world who has had a volcanic ash named after her.

If the KBS Tuff were located anywhere else, no one would give it a second thought. However, at East Rudolf it is of utmost importance. First, although human fossils and artifacts cannot be directly dated radiometrically, the KBS Tuff can be. It contains radioactive potassium 40, which decays to argon 40. Second, artifacts (tools) have been found in close association with the KBS Tuff. The assumption is that the date of the tuff gives an estimate of the age of the stone tools. Third, hundreds of *Homo* and australopithecine fossils have been found either above or below the KBS Tuff. The date of the KBS Tuff thus becomes a maximum age for fossils found above it and a minimum age for fossils found below it. Of all the fossils found in association with the KBS Tuff, skull 1470 is the most important.

The KBS Tuff is part of the Koobi Fora Formation, a sequence of sediments about three hundred feet thick that crop out on the eastern shore of Lake Rudolf. These sediments cover an area about fifty miles north and south along the shore and extend about twenty miles east of the lake. The KBS Tuff itself is only about three feet thick. Two other thin layers of volcanic tuff lie above the KBS (the Karari and the Okote Tuff) and another lies below it (the Tulu Bor Tuff). The bulk of the sediments in which the fossils are found are not volcanic tuffs. The four tuffs are like the floors of a four-story building, with the fossil-bearing sediments lying between them where the rooms of the building would be.

Although the KBS Tuff is volcanic in origin, it is not a primary air-fall tuff. That is, it was not deposited directly on the land when it was ejected from the volcano. Lake Rudolf was much larger at that time. Some ash fell into the lake and then made its way to the lake bottom.

Some was carried by rivers into the lake. Thus, the KBS Tuff has been transported by and deposited from water. For this reason it has a great deal of foreign material in it, making it very difficult to get pure samples for dating.

The first attempt to date the volcanic rock layer known as the KBS Tuff was a feasibility study done in 1969, well before the discovery of skull 1470. Richard Leakey supplied rock samples to F. J. Fitch (Birkbeck College, University of London) and J. A. Miller (Cambridge University) who were recognized authorities in potassium-argon (K-Ar) dating. Many species of mammals had been found below the KBS Tuff, as well as australopithecine fossils and human artifacts. It was imperative that these discoveries be placed in their proper chronological setting.

In their report in *Nature*,[1] Fitch and Miller first commented on the many possible sources of error in dating. "One of the most intractable of these," they said, "is the possible presence of extraneous argon derived from inclusions of pre-existing rocks."[2] To check for this extraneous argon, they first dated the raw rocks as they were originally submitted by Leakey. Their analysis gave dates from 212 to 230 million years of age. "From these results it was clear that an extraneous argon age discrepancy was present. . . ."[3]

The first question an outside observer would ask is, How did they know? The answer is that the associated fossils told them so. In spite of our being assured that the dating methods constitute an *independent* confirmation of evolution, the associated fossils had already determined the outside limits for dates that would be "acceptable." Based on their alleged evolution, the australopithecine and other mammalian fossils found beneath the KBS Tuff had determined that the rocks should be somewhere between two and five million years old. Anything beyond that was obviously the result of extraneous argon.

Dates of 212 to 230 m.y.a. were notoriously far off. These dates would place the KBS Tuff in the Triassic period of the Mesozoic era, which is early dinosaur times. Hence it was obvious that these dates were wrong. Without the associated fossils, however, there would be no way for a geologist to know if these were "good" dates or "bad" dates. Under other circumstances and without the fossils to guide them, geologists could well have accepted these dates as "good" dates. When fitting rock layers into their proper sequences over large geographic areas, it is evolution and the fossils that guide the geologists.

To compensate for this obvious error in dating the KBS Tuff, Fitch and Miller stated: ". . . it would only be possible to date this tuff by careful extraction of undoubtedly juvenile components for analysis."[4] In other words, Fitch and Miller then proceeded to remove from the whole-rock samples those components of the rock which they believed were "undoubtedly" juvenile or young, that showed no sign of weathering or alteration. The observer can be forgiven if he asks another question, How do they know for sure which components of the rock are undoubtedly young?

Thus began the long process, based upon evolutionary and other philosophical assumptions, by which the geochronologist manipulates or "massages" the data to guarantee that he gets a "good" date. I want to stress that the geochronologist does this in absolute sincerity. He is so committed to evolution and its attendant age demands that he believes implicitly that he removes error from his data to arrive at truth. The obvious subjectivity in it escapes him. It is a perfect illustration of circular reasoning in an experimental frame of reference. The experimenter manipulates the data to guarantee that he gets the result that is "needed." In computer language, it's "garbage in, garbage out."

Fitch and Miller requested new samples from Leakey containing "fresher" pumice lumps and feldspar crystals. Experiments were conducted on the pumice and the feldspar crystals separately, using three different processes: K-Ar age determination, ^{40}Ar-^{39}Ar total degassing, and ^{40}Ar-^{39}Ar age spectrum. Fitch and Miller concluded that the KBS Tuff was "very close to 2.6 m.y. (2.61 ± less than .26 m.y.)."[5] This figure of 2.61 m.y.a. was widely published in both the scientific and popular press. Richard Leakey stated that 1470 was found below rock that was "accurately dated"[6] and "securely dated"[7] at 2.6 m.y.a.

In 1972, before skull 1470 was discovered (or at least before it was announced), Vincent Maglio (Princeton University) published in *Nature*[8] a chronology of the hominid-bearing sediments east of Lake Rudolf which included the KBS Tuff. His work was based upon the vertebrate faunas. Several animal species seemed to show a significant degree of change through the stratigraphic sequence. This was interpreted as a rapid rate of evolution. Obviously, rapidly changing lineages make for more precise correlations than those that change slowly or not at all. The lineages were of two species of pig (suid) and one species of elephant. Although there were some problems, Maglio's dates for the sediments were somewhat compatible with the radiometric date arrived at by Fitch and Miller, and were considered at the time to confirm their date.

In 1974, a third chronology of the area was published in *Nature*[9] by Brock (University of Nairobi, Kenya) and Isaac (University of California, Berkeley). The study was based on the paleomagnetism of the deposits below the KBS Tuff utilizing 247 samples. They stated their conclusions for the group of fossils including skull 1470 as follows: "An age of 2.7 to 3.0 Myr for this group is strongly indicated."[10] Since this date referred to the sediments that skull 1470 was actually found in, and the KBS Tuff dated at 2.61 Myr lies above the fossil, it seemed to represent a "bulls-eye" for the correlation of the various dating methods. The heading of the article stated that their measurements "provide a valuable check on other dating methods."[11] Later they said that because the isotopic and paleomagnetic ages were consistent, ". . . this independent evidence greatly strengthens our proposed chronology."[12]

However, Brock and Isaac also made the following comment:

The correlations shown in Figure 4 are not fully independent, and rely partly upon K.Ar and faunal evidence as well as upon the basic polarity data.

The starting point for the correlation is the age of 2.61 ± 0.26 Myr obtained by Fitch and Miller from selected sanidine crystals from pumice specimens from the KBS Tuff.[13]

This comment indicates that the correlation by Brock and Isaac was not as independent of the other dating methods as they claimed it to be.

Also in 1974, Anthony Hurford (Birkbeck College, University of London) attempted to date the East Rudolf sediments using still another method: fission-track dating involving uranium. His purpose was to check out an unpublished study by Fitch and Miller that suggested that vast portions of the East Rudolf sediments had been changed or altered by volcanic heat or hot ground water around 1.75 m.y.a., causing partial or complete overprinting of the apparent ages obtained from them. Overprinting erases or obscures previous events.

Hurford's conclusion regarding his fission track specimen:

The specimen has either suffered no thermal annealing or that it has been totally aneated at 1.8 Myr.

As this tuff is within the Kubi Algi Formation and is stratigraphically below the 2.6 Myr KBS Tuff, the second alternative is accepted as the correct interpretation.[14]

He agreed with Fitch and Miller that the sediments had been altered
at about 1.75 to 1.8 m.y.a. One could be excused for asking why the
annealing of the lower sediments at 1.8 m.y.a. did not call into question the KBS Tuff date of 2.6 m.y.a.

A study of Hurford's methodology illustrates how dogma finds its
way into science. He started by referring to the date of the KBS Tuff
as a "firm date." Apparently the date became firm because he felt that
it was supported by the fossil and paleomagnetic evidence. He did not
mention that the fossil correlation was only of the most general sort
and that the paleomagnetic date was based on the radiometric date. In
spite of an obvious need for caution, Hurford's acceptance of the KBS
Tuff date became the benchmark on which he based his fission-track
conclusions.

It seems, however, that Hurford set up a strange scenario. In the
absence of clear physical evidence to the contrary, the Kubi Algi Formation, which is below the KBS Tuff, would be older than the KBS
Tuff, because it was laid down first. It is hard to understand how the
Kubi Algi Formation could have experienced an overprinting at 1.8
m.y.a., and then later on have the KBS Tuff laid down on top of it
with a date of 2.61 m.y.a. If, however, they were laid down in their
present sequence and then some sort of thermal event overprinted
the Kubi Algi at 1.8 m.y.a., how do we know that the KBS Tuff
wasn't affected as well? It would certainly seem to compromise the
"firm date" of the KBS Tuff. At any rate, it is clear that the various
dating methods are related, and the dates obtained are not independent of one another.

Late in 1974, Fitch, Miller, and associates published the results of
their revised study confirming their original dating of the KBS Tuff at
2.61 ± .26 m.y.a. They also reported a broad scatter of apparent ages
from ten different samplings ranging from 0.52 to 2.64 m.y.a. Referring to the other studies, they stated: "The compatibility of independent evidence is a very strong argument for accepting the chronology
now proposed for East Rudolf."[15] However, we have seen from the
other studies that they are not independent but were linked to the original radiometric date by Fitch and Miller.

By late 1974, two years after skull 1470 had been presented to the world, the KBS Tuff had been dated five different times by four different dating methods. The alleged compatibility of the four different methods would seem to make all of this a geologist's dream. What better proof could one want for the reliability of the various dating methods to furnish independent confirmation of the dates for the fossil material? Because 1470 was found below rock dated at 2.61 m.y.a. and above rock dated at 3.18 m.y.a., skull 1470 was estimated to be an incredible 2.9 million years old. Richard Leakey had found the world's oldest fossil belonging to the genus *Homo*. On the surface all seemed serene.

However, under the surface paleoanthropology was seething in ferment. Skull 1470 with its estimated date of 2.9 m.y.a. presented the evolutionary world with an intolerable situation. Richard Leakey did not exaggerate when he declared: "Either we toss out this skull or we toss out our theories of early man."[16] The problem was quite simple. The theory of human evolution did not allow for a skull so modern in morphology to be that old. It was absolutely predictable to those of us who watched these matters unfold that something would have to give. Only three things could happen to relieve the stress that the theory of human evolution was experiencing: (1) the date for 1470 could be revised; (2) 1470 could be assigned to the most distant and primitive form of *Homo*; or (3) 1470 could be reevaluated and designated an australopithecine. Actually, all three of these solutions happened in one way or another. The date was eventually revised, the fossil was assigned to the category *Homo habilis*, and some—including one of Richard Leakey's close associates, Alan Walker—said that 1470 was actually an australopithecine. As these revisions took place, paleoanthropologists heaved almost audible sighs of relief.

Richard Leakey, however, continued to fight for the original date. Although he was committed to evolution and was aware of the problem the date for skull 1470 presented for evolution, his situation was somewhat different. He was considered the discoverer of skull 1470. (Actually, it was discovered by Bernard Ngeneo, a member of his team.) No one will care if you discover the oldest fossil broccoli, but if you are fortunate enough to discover the oldest fossil human, the world will beat a path to your door. The acclaim and prestige such a person receives is beyond belief. Human fossils work a very special kind of magic. Richard Leakey needed this magic. He was only twenty-

eight when skull 1470 was discovered, and he had had no formal college training. He learned paleoanthropology at the feet of his parents,
Louis and Mary Leakey. Some paleoanthropologists have never forgiv-
en him for entering the field by a different door. If skull 1470 was 2.9
million years old, he had discovered the oldest member of the genus
Homo. If 1470 was not that old, he would lose that distinction. The
problems that 1470's age would pose for evolution were not as vital to
him as the status 1470 would give him in establishing him in the field
of paleoanthropology. Hence, he resisted any lowering of the date for
1470.

While Fitch and Miller were busy confirming their original results,
still another study was already under way by G. H. Curtis and his associates (University of California, Berkeley). They used conventional
K-Ar dates on pumice from three separate areas of the KBS Tuff.
They claimed to distinguish two tuff units. One from areas 10 and 105
gave an age of 1.6 m.y.a. and the other from area 131, where skull
1470 was found, gave an age of 1.82 m.y.a. These dates were considerably younger than the dates the five previous studies had reported.

Commenting on the broad scatter of results Fitch and Miller had
obtained earlier, they gave this explanation:

Contamination by ancient bed rock material during the reworking of the
tuffs was suggested to account for the anomalously old dates, whereas
subsequent alteration, 'overprinting,' of the pumice fragments used for
dating, by alkaline-rich and possibly heated ground water may explain
the anomalously young dates by partial loss of radiogenic argon.[17]

Since the whole point of their exercise was to establish the age of the
KBS Tuff, the question again must be asked, "How did they know that
the older dates or the younger dates were anomalous? Anomalous with
reference to what? It was obvious that it had already been determined
what the "proper" age should be. How was this determined? By the
concept of evolution. The age of the KBS Tuff and of skull 1470 must
be lowered.

The Curtis article challenged the validity of the ^{40}Ar-^{39}Ar technique for this particular dating situation and criticized the methodology of Fitch and Miller. It further stated that ". . . older pumices may
also be present in the KBS Tuff horizon which could account for the
2.61 Myr date reported by Fitch and Miller."[18] Criticizing the samples
used by Fitch and Miller, the dating method employed by Fitch and

Miller, and the laboratory technique of Fitch and Miller left little more to be said.

All of the above-cited articles spoke of the great difficulty in getting rock or crystal samples that were not altered, weathered, or derived from older rock. Curtis et al. explained at length their efforts to extract from the whole-rock samples the portions that were suitable for dating. However, Fitch and Miller also went to great lengths to extract suitable samples. The question arises, How does one know when one has good samples for dating? The only answer to that question is that "good" samples give dates that are in accord with evolutionary presuppositions. "Bad" samples are the ones that give dates not in conformity with evolution—a classic illustration of circular reasoning.

Curtis et al. also mentioned the factor that would ultimately determine the date of skull 1470: the evolution of the pigs.

> [When some palaeontologists compared fauna associated with] the KBS Tuff in East Rudolf with those of other, supposedly well calibrated localities, the reliability of the date of 2.61 Myr for the KBS was questioned. Although Maglio found that the morphology of elephant fossils fit with a 2.5 Myr date, Cooke and Maglio, in 1972, pointed out that fossil pigs from below the KBS Tuff horizon at East Rudolf seemed to correlate best with those from beds dating close to 2 Myr in the Omo River area to the north in Ethiopia.[19]

Notice that elephant evolution fit the older date, but pig evolution fit the younger date. The pigs would ultimately win. This does not support the idea of concordant results that evolutionists talk about. Perhaps that was why Maglio left geology to study medicine.

It is fascinating to see that Curtis et al. claimed authority for their dating results because of the high degree of correlation within their study. But that same claim was made for the older date. Also, five different dating projects involving four different dating techniques all supposedly agreed on the older date within reasonable margins of error.

The 28 October 1976 issue of *Nature* contained not one but two dating projects for the KBS Tuff by two different methods. These two methods seemed to agree on an older date for the tuff and hence for skull 1470.

The first of these studies was by Fitch, Miller, and P. J. Hooker (Cambridge University). They first recalculated the results of their 1969 work and told why:

Developments in the analytical techniques of ^{40}Ar-^{39}Ar dating since then enable recomputation of the results obtained, using, in addition, a more accurate value for the constant of proportionality (J) used in the 1969 experiments.[20]

Recalculating with 1969 rock samples and utilizing both the K-Ar and the ^{40}Ar-^{39}Ar methods gave them a revised estimate for the age of the KBS Tuff of 2.42 m.y.a. Calculating with rock samples obtained in 1971-73 and using only the ^{40}Ar-^{39}Ar technique gave a minimum age for the KBS Tuff of 2.4 m.y.a. The close correlation of these two dating efforts by different radiometric techniques gave them confidence in the accuracy of their results. Because of the inherent difficulties of obtaining proper samples for dating, they also looked to other methods for support. This support came in an accompanying article in the same issue of *Nature* involving fission-track dating of zircon crystals. We will consider that article later.

Fitch, Miller, and Hooker acknowledged the controversy that was raging around the dating of the KBS Tuff. That controversy was largely because of stone tools found in it and one very human-looking fossil found below it.

Over the past five years, opposition to the acceptance of a 2.5-Myr age for the KBS Tuff has come from three sources: first, archaeologists and palaeoanthropologists disturbed by the consequent antiquity of hominid fossils and stone tools found close to or associated with the KBS Tuff; second, palaeontologists reporting apparent misfits between the faunal sequences at East Rudolf and elsewhere; and third, from a small programme of conventional total fusion K-Ar age determinations on East Rudolf pumice samples undertaken at Berkeley.[21]

I will deal with these matters in reverse order. The "small programme" at Berkeley is a reference to the work of Curtis et al. who dated the KBS Tuff at 1.6 and 1.82 Myr. The flaw in that date is quite obvious to Fitch, Hooker, and Miller:

. . . K-Ar apparent ages in the range 1.6-1.8 Myr obtained from the KBS Tuff by other workers are regarded as discrepant, and may have been obtained from samples affected by argon loss.[22]

This exercise can appropriately be named the dating game. Since yours is obviously the correct date, those who arrive at a younger date had samples that obviously had experienced argon loss. A date older than yours can be explained if you declare that those samples had obviously inherited excess argon from older rock. How does one refute that kind of logic?

The second problem, "apparent misfits between the faunal sequences," will eventually be settled rather arbitrarily by a victory of the pigs over the elephants. Where else but in the world of science fiction could such a confrontation of pigs and elephants have such unlikely results?

It is the third problem that is most revealing. It involves archeologists who are "disturbed by the consequent antiquity of hominid fossils and stone tools found close to or associated with the KBS Tuff." "Disturbed" seems a strange word to describe scientists who are supposed to let the facts speak for themselves. I would think that words like *interested, amazed*, or *intrigued* would be far more appropriate. *Disturbed* sounds like they felt threatened. They were. The whole concept of human evolution was on the line. This was the real issue behind a controversy that raged for ten years over some ash out of a volcano in East Africa.

One more item needs to be mentioned. Fitch, et al. commented that the Berkeley group reported "scatter" in their dates ranging from 1.5 Myr and 6.9 Myr. Fitch et al. reported their own scatter in apparent ages ranging from .5 Myr to 2.4 Myr. In some cases the scatter was interpreted as overprinting events. In other cases, "naughty" crystals were removed to give results more appropriate to the overriding principle behind it all—human evolution.

The other article in that 28 October 1976 issue of *Nature* was written by Hurford, Gleadow (University of Melbourne, Australia), and Naeser (U. S. Geological Survey, Denver). It was about fission-track dating of zircon crystals found in the KBS Tuff. They began with a rather remarkable statement regarding the K-Ar and ^{40}Ar-^{39}Ar dating methods:

K-Ar and ^{40}Ar-^{39}Ar dating techniques have been applied to >100 rock and mineral samples from East Rudolf, but interpretation of the dates determined by these methods has not been straightforward. Geological and analytical factors have been postulated to explain the scatter of K-Ar and ^{40}Ar-^{39}Ar apparent ages obtained from volcanic sanidine-anorthoclase crystals separated from pumice cobbles in the tuffs.[23]

The authors did not imply that the radiometric dating workers were being dishonest. They did say that the interpretation of the dates involves hypothetical and philosophical assumptions that have a bearing on the results. (This, by the way, is exactly what creationists have been saying about all radiometric dating methods.) They also stated that their study was conducted because of the apparent conflict between the K-Ar and ^{40}Ar-^{39}Ar dating methods—something that was played down in previous studies.

Their conclusion was that the KBS Tuff has a date of 2.44 Myr. This was very close to the estimate of Fitch et al. published in the same issue of *Nature*. After describing their methodology they said:

> Using these techniques and a value for the ^{238}U spontaneous fission decay constant, λ of 6.85×10^{-17} yr^{-1} we have obtained ages on standard zircons which agree very closely with their independently known ages.[24]

This remarkable correlation of dates involving two independent dating techniques seemed to confirm all that the general public has been led to believe—that the dating methods can be trusted because independent methods give the same results.

However, in the 16 June 1977 issue of *Nature* appeared a letter from G. A. Wagner of the Max Planck Institute in West Germany. Wagner maintained that there is uncertainty as to the spontaneous fission constant of uranium 238, and that Hurford et al. should have used a different constant:

> . . . many fission-track specialists no longer use the 6.85×10^{-17} yr^{-1} value, but now use as the decay constant 8.46×10^{-17} yr^{-1}; there are good reasons for this preference. If this higher value for the decay constant is used, the fission-track age of the pumice in the KBS tuff recalculates to 1.98 Myr, which would lend support to the K-Ar age measured by Curtis *et al.*[25]

Hurford et al. defended their use of the uranium 238 constant by saying:

> When it is used in conjunction with the fission track glass standards of the U.S. National Bureau of Standards, we get the best agreement with the K-Ar ages of co-existing minerals and we use it for this reason.[26]

In other words, the true value of the spontaneous fission constant of uranium 238 is unknown. At least two values are currently in use. In matters of fission-track dating, one is thus free to use the value that gives him the answer he is looking for. One can make the age of the KBS Tuff agree with either Fitch or with Curtis, whatever one's pleasure might be. The difference in the two dates is almost half a million years in dealing with a date of only about two to two-and-a-half million. That hardly seems like precision dating.

Because they tended to confirm the older date for skull 1470, these two studies dating the KBS Tuff at 2.4 Myr obviously put more strain on the evolutionary establishment. Hurford et al. wrote: "Curtis has described the original 2.61 ± 0.26 Myr date for the KBS Tuff as being much questioned in private anthropological and paleontological circles."[27] Since anthropologists and paleontologists do not normally have technical expertise in the radiometric dating methods, they were not challenging the methodology or the assumptions of the dating methods. They were rejecting the older date solely because of its philosophical implications. The problem was the modern morphology of KNM-ER 1470 versus the demands of evolutionary theory.

A new study on the paleomagnetism of the Koobi Fora Formation was published in early 1977. It acknowledged that the previous paleomagnetic study had used ". . . the previously published age of 2.6 ± 0.26 Myr for the KBS Tuff as a fixed point. . . ."[28] The study cited additional paleomagnetic results as warranting a reevaluation of the magnetic stratigraphy.

The study gave two different interpretations of the data based upon the two different suggested ages of the KBS Tuff. It clearly revealed that dates arrived at by paleomagnetism are not independent confirmations of other dating results but are closely tied to the radiometric results they use as a starting point. Since the KBS Tuff is the top unit of the lower member of the Koobi Fora Formation, the following quotation reveals how different the results can be when different starting points are used in this dating game:

> In both interpretations the age of the upper member, which lies above the KBS Tuff, is between 1.2 and 1.8 Myr; however, the age of the top boundary of the lower members differs by 1 Myr.[29]

Around 1976, the name of the large lake on which Richard Leakey has his base of operations was changed by the Kenyan government from Lake Rudolf to Lake Turkana. This change has been a fruitful source of confusion, since fossils recovered east of the lake continue to carry the designation East Rudolf, whereas fossils recovered from west of the lake, representing more recent work, carry the designation West Turkana. Up to now we have consistently used the name Lake Rudolf. From now on, we will use the newer name Lake Turkana.

On 20 March 1980, two more dating studies on the KBS Tuff appeared in the pages of *Nature*. Remember that two earlier studies—one on fission-track dating of zircons and one on ^{40}Ar-^{39}Ar dating of orthoclase crystals—agreed closely that the age of the KBS Tuff was 2.4 Myr. They cited the close correlation of two independent dating methods as validating their accuracy. Now, two studies—one on fission-track dating of zircons and one on K-Ar dating of orthoclase crystals—agreed closely that the age of the KBS Tuff is 1.87 or 1.89 Myrs. They also cited the close correlation of two independent dating methods as validating their accuracy for the revised date. The new fission-track study was by A. J. W. Gleadow. The K-Ar study was by Ian Mc-Dougall, Robyn Maier, and P. Sutherland-Hawkes (all of the Australian National University, Canberra) and A. J. W. Gleadow. Then in late 1981, Ian McDougall published in *Nature* his ^{40}Ar-^{39}Ar study of the KBS Tuff, giving a date of 1.88 Myr. At that point, the ten-year controversy over the date of the KBS Tuff came to a close. Concordance on the more recent date had been achieved.

At first glance, it would seem to be a tremendous victory for evolution and the uniformitarian dating methods. We know that science often proceeds by trial and error and by controversy. The fact that an amazing correlation between the pig evidence and three different dating methods—fission track dating, K-Ar, and ^{40}Ar-^{39}Ar—had been achieved should be something to celebrate. The dating of the KBS Tuff was now a nonissue. Yet, there were factors that demand a closer look at the situation.

The Power of the Pigs

The dating of the KBS Tuff was not settled in 1980 and 1981 by the conformity of three different dating methods. The controversy was actually settled in 1975 by the pigs. Donald Johanson tells of attending the Bishop Conference on anthropology and geology in London. The dating of the KBS Tuff and its implications were major topics of conversation. Glynn Isaac, who accepted the older date, arrived with a "pig-proof helmet" to protect him against the pig men.

A major paper was presented by Basil Cooke (Dalhousie University, Halifax), who had studied the pig sequences at Omo (a fossil area in Southern Ethiopia on the Omo River, which flows into Lake Turkana from the north), at Hadar (the Ethiopian site where Johanson had worked), and at Olduvai Gorge (where Louis and Mary Leakey worked for many years). According to Cooke, the dating at Lake Turkana, based on the dating methods, was off by about 800,000 years. The pigs at Turkana told him so. He even wore a tie with the letters *MCP* woven into it. They stood for "male chauvinist pig," but Cooke claimed that they really stood for "*Mesochoerus* correlates properly." *Mesochoerus* was the species of fossil pig that was central to his argument. Johanson wrote of the 1975 conference: "Nearly everyone but the Lake Turkana team [Richard Leakey and his associates] went away convinced that the KBS tuff and the skull 1470 dates would have to be corrected."[30]

Astounding about the whole affair was that the anthropologists were rejecting the same objective, scientific data that they universally appeal to. At that time the radiometric evidence for the older date was very strong. There was internal consistency within the studies, and a high degree of conformity by five different dating techniques. The main thing the dates did not conform to was the concept of the evolution of pigs and of humans.

The evolution of the pigs is said to be the clear-cut answer to the dating problems at Koobi Fora as well as elsewhere in East Africa, but the evidence is less than impressive. In his phylogeny of the pigs (bushpig, forest hog, warthog, etc.) Basil Cooke presented family trees for three taxonomic groups.[31] Two of the groups have at their bases the phrase *hypothetical Sus-like ancestor*. The twenty species that make up these three groups are all shown in parallel lines connected only by dotted lines, indicating that there is no known relationship between any of the species. The parallel-line chart could just as well have been drawn by a creationist.

Most of the fossil-pig evidence consists of teeth. Several species are based on the skimpiest of evidence ("imperfectly known," "rare," "scarce") and the various relationships are largely judgment calls. Terms such as these appear:

is probably ancestral
seems to represent
suggest that . . . evolved independently
must at this stage also be giving rise to
probably ancestral to
suggests a derivation from
may have branched off
almost certainly branched off from
demand descent from a common ancestor

Cooke's article was written in response to one by White and Harris.[32] Cooke had three taxonomic groups while White and Harris had four. There are differences in the two taxonomies, but Cooke maintained that they were of no great moment. He then went on to explain why species that White and Harris had grouped together should be separate, and vice versa. This creation and annihilation of species by the whim of the taxonomist "due to our having different basic philosophies on the nature of species in paleontology"[33] reveals how plastic and subjective this science is. I am not minimizing the difficulty of the species concept in paleontology, nor am I debasing the attempt to sort things out. I am merely stating that the authority with which paleontologists make dogmatic statements about the evolution of the pigs is not warranted by the facts. As in every other area of paleontology, a great deal more humility would be appropriate.

The 1980 and 1981 studies on the date of the KBS Tuff contained so many criticisms of all of the earlier studies that they called into question the objectivity and validity of the dating methods themselves. Gleadow began the process:

K-Ar evidence of Curtis *et al.* suggesting that tuffs mapped as the KBS in Areas 105 and 131 were of slightly different age, has now been eliminated with the discovery of a systematic error in the lower (1.6 Myr) ages.[34]

After demonstrating the presence of contamination in all of his own samples, and the extreme difficulties in dating zircons in the 1-3 Myr time span, he continued:

It therefore seems highly likely that feldspars separated for K-Ar dating could also contain traces of much older basement feldspar. This supports the contention that older K-Ar and ^{40}Ar-^{39}Ar ages are the result of contamination. As discussed above, the fission track ages of Hurford *et al.* are thought to be too old for purely analytical reasons, in particular the mis-identification of a finite number of acicular inclusions or dislocations as tracks and possibly a biased choice of grains for counting.[35]

In the same issue of *Nature* was the report of a study by Ian Mc-Dougall et al. on the K-Ar dating of the KBS Tuff. They began by confessing that "Conventional K-Ar, ^{40}Ar/^{39}Ar and fission track dating of pumice clasts within this tuff have yielded a distressingly large range of ages."[36]

After explaining that Fitch and Miller actually reported results ranging from 0.52 to 2.64 Myr in one set of concentrates and ages from 8.43 to 17.5 Myr on another clast before settling on a 2.61 Myr date which they later revised to 2.42 Myr, they also explained how Curtis et al. arrived at their "concordant" ages:

> Disregarding four conventional K-Ar ages on feldspar from pumice clasts in the KBS Tuff in the range 2.01-6.9 Myr, thought to be caused by detrital contamination, Curtis *et al.* obtained concordant K-Ar ages on feldspar and glass from pumice clasts found in this horizon with mean value of 1.82 ± 0.04 Myr and 1.60 ± 0.05 Myr in two different areas, respectively. Subsequently, Drake *et al.* reported an error in potassium determinations on the samples previously dated by them that yielded the 1.6 Myr ages.[37]

McDougall et al. then stated how "remarkably concordant" their own dates were at 1.9 Myr after removing from consideration samples that gave ages of 4.11 and 7.46 Myr. They explained these anomalous ages as follows:

> We attribute these poorly reproducible ages to the presence of variable but small amounts of old detrital K-feldspar in the aliquants used in the argon extractions. *Careful petrographic examination of the mineral concentrate, however, did not lead to positive identification of detrital K-feldspar.* Nevertheless, there is no doubt that old detrital material was being brought into the East Turkana Basin during deposition of the sediments.[38]

With this clear victory of philosophy over observation, they then used the concordance of their results and agreement with the results of the study by Gleadow to give validity to their date for the KBS Tuff.

Since the fission-track dates and the K-Ar dates of the KBS Tuff had now been reconciled with the date demanded by the evolution of the pigs, the only remaining problem was the high ^{40}Ar-^{39}Ar dates that Fitch et al. had reported. McDougall solved this problem in 1981. He reported that in some of his work there was a greater scatter of data points than could be explained by experimental error, and that in the step-heating experiment it was necessary to exclude some data.

> Plateau and regression ages are derived using all data from each step heating experiment, as well as by excluding results from steps that give discordant ages. The criterion for exclusion of a datum was that the calculated age differed by more than twice its error (2σ) from that of the plateau.[39]

However, he maintained that the differences these matters made were small, and he expressed complete satisfaction in his date of 1.88 Myr for the KBS Tuff. He did not neglect to mention that this date was in excellent accord with the other recent dating studies of the KBS Tuff.

McDougall then issued one of the most stinging rebukes of a fellow scientist that I can remember seeing in the scientific literature. He referred to Fitch et al. and their older date for the KBS Tuff when he said:

> On the basis of the large scatter in the ages and the small proportion of ^{40}Ar* in the gas extracted from the anorthoclase concentrates, I suggest that the results are analytically less precise than given by these authors.

> I suggest that unrecognized analytical difficulties and larger than quoted errors must be invoked to explain these earlier ^{40}Ar/^{39}Ar results.[40]

McDougall was accusing Fitch et al. of invoking what is affectionately known in scientific circles as "the fudge factor" (deliberate falsification of data to achieve a desired result).

The study of the ten-year controversy in the dating of the KBS Tuff is tremendously revealing. Whereas the public is led to believe that these dating methods are highly objective and accurate, the scientific literature itself reveals that they are highly subjective. There is no question that rock samples are often manipulated to give the desired results. There is also no question that this manipulation is done in the utmost sincerity and with the noblest motives. But it is manipulation nonetheless. The "bad" material must be removed to allow the "good" material to be dated. But there is no way of knowing for sure which material is "good" and which is "bad."

The history of the dating of the KBS Tuff reveals that no matter how careful a scientist is in selecting his rock samples and in performing his laboratory work, if he gets the wrong date for his rocks he is open to the charge of using contaminated material and a defective methodology. The charges need not be proved. The fact that he got the wrong date is proof enough. The literature suggests that even if radiometric dating were valid in concept (which it is not), the practical matter of selecting rock samples that can be proven pure and uncontaminated requires an omniscience that is beyond the ability of mortal humans. The radioactive dating methods are a classic example of self-deception and circular reasoning. It is another of the myths of human evolution. Naeser et al. have said it well:

> The accuracy of any age can only be guessed at, in that we do not know the true age of any geologic sample. We can only strive for the best agreement with K-Ar and the other dating methods.[41]

I have no doubt that my evolutionist friends will protest that I have not been fair. "East Turkana," they will say, "is a most unique situation. It just isn't 'cricket' to take a unique situation with its many problems and imply that it is the norm." In this response, my friends are both right and wrong. There is no question that the geology of the Koobi Fora Formation, with the KBS Tuff, is exceedingly complex. However, Koobi Fora is far from the only fossil site that has a very complex geology. What is unique about Koobi Fora is something that so far has not been mentioned by anyone.

The radiometric date of 2.61 m.y.a. for the KBS Tuff was established before skull 1470 was discovered. It was supported by faunal correlation, paleomagnetism, and fission-track dating. Up until that time, the fossils and the artifacts that had been found in association with the KBS Tuff were more or less compatible with that older date. It is entirely possible that if skull 1470 had never been found, the KBS Tuff would still be dated at 2.61 m.y.a. We would continue to be told that it was a "secure date" based on the precision of radiometric dating and the "independent" confirmation of other dating techniques that acted as controls. It was the shocking discovery of the morphologically modern skull 1470, located well below the KBS Tuff, that precipitated the ten-year controversy.

What normally happens in a fossil discovery is that the fossils are discovered first. Then attempts are made to date the rock strata in which they are found. Under these conditions, a paleoanthropologist has a degree of control over the results. He is free to reject dates that do not fit the evolution scenario of the fossils. He is not even required to publish those "obviously anomalous" dates. The result is a very sanguine and misleading picture of the conformity of the human fossil record with the concept of human evolution. If, in many of these fossil sites the dates had been determined before the fossils had been discovered, evolutionists could not guarantee that the turbulent history of the dating of the KBS Tuff would not have been repeated many times.

The pigs won. In the ten-year controversy over the dating of one of the most important human fossils ever discovered, the pigs won. The pigs won over the elephants. The pigs won over K-Ar dating. The pigs won over ^{40}Ar-^{39}Ar dating. The pigs won over fission-track dating. They won over paleomagnetism. The pigs took it all. But in reality, it wasn't the pigs that won. It was evolution that won. In the dating game, evolution always wins.[42]

Endnotes

Chapter 1. The Family Gathering

1. Eric Delson, ed., *Ancestors: The Hard Evidence* (New York: Alan R. Liss, Inc., 1985), 1.
2. Delson, 1–2.
3. Bernard Wood, "A Gathering of Our Ancestors," *Nature* 309 (17 May 1984): 208.
4. "Old Bones Week," *Discover*, June 1984, 69.
5. Ian Tattersall and Niles Eldredge, "Fact, Theory, and Fantasy in Human Paleontology," *American Scientist* 65 (March–April 1977): 207.
6. Roger Lewin, *Bones of Contention* (New York: Simon and Schuster, 1987), 24.
7. Ellen Ruppel Shell, "Flesh and Bone," *Discover*, December 1991, 41.
8. Shell, 41–42.
9. Louis S. B. Leakey, *Adam's Ancestors* 4th ed. (New York: Harper and Row, Publishers, 1960), v–vi.
10. Becky A. Sigmon and Jerome E. Cybulski, eds., *Homo erectus: Papers in Honor of Davidson Black* (Toronto: University of Toronto Press, 1981), 5.
11. Roger Lewin, "Ancestors Worshipped," *Science* 224 (4 May 1984): 478.
12. Delson, 4.
13. Delson, 4–5.

Chapter 2. An Inexact Kind of Science

1. Carl Sagan, "Velikovsky's Challenge to Science," cassette tape 186–74, produced by the American Association for the Advancement of Science, Washington, D.C., 1974.
2. G. A. Kerkut, *The Implications of Evolution* (New York: The Macmillan Company, 1960), 7. Emphasis his.
3. Bernard Ramm, *The Christian View of Science and Scripture* (Grand Rapids: Wm. B. Eerdmans Publishing Company, 1954), 29–30. Emphasis his.
4. Richard Leakey, "The Search for Early Man," cassette tape interview produced by the American Association for the Advancement of Science, Washington, D.C., 1973.
5. David Pilbeam, "Rearranging Our Family Tree," *Human Nature* (June 1978): 40.
6. Pilbeam, 44. Parenthetical material added for clarity.
7. Pilbeam, 45.
8. Solly Zuckerman, *Beyond the Ivory Tower* (New York: Taplinger Publishing Company, 1971), 19.
9. Zuckerman, 64.
10. Andrew Hill, "The gift of Taungs," *Nature* 323 (18 September 1986): 209.
11. Norman Macbeth, *Darwin Retried: An Appeal to Reason* (Boston: Gambit Incorporated, 1971).
12. Phillip E. Johnson, *Darwin on Trial* (Washington, D. C.: Regnery Gateway, 1991).
13. Johnson, 81.
14. Johnson, 83. Bracketed material added for clarity.

Chapter 3. Dead Reckoning

1. Douglas J. Preston, "Four Million Years of Humanity," *Natural History* (April 1984): 12.
2. Boyce Rensberger, "Bones of Our Ancestors," *Science 84* (April 1984): 29. This publication has since been combined with the Time-Life publication *Discover*.
3. Richard E. F. Leakey, "The Search For Early Man," cassette tape interview, produced by American Association for the Advancement of Science, Washington, D.C., 1973.
4. Peter Stoler, "Puzzling Out Man's Ascent," *Time* (cover story) 7 November 1977, 77.
5. Vaughn M. Bryant, Jr., and Glenna Williams-Dean, "The Caprolites of Man," *Scientific American* (January 1975): 100.
6. John Reader, "Whatever Happened to Zinjanthropus?" *New Scientist* (26 March 1981): 802.
7. Constance Holden, "The Politics of Paleoanthropology," *Science* 213 (14 August 1981): 737.

Chapter 4. Monkey Business in the Family Tree

1. Carl Sagan, "Velikovsky's Challenge to Science," cassette tape 186–74, produced by the American Association for the Advancement of Science, Washington, D.C., 1974.
2. Vincent Sarich, Creation-Evolution debate, North Dakota State University, Fargo, April 28, 1979.
3. Since the revision of German orthography in 1901, the *h* in *Neanderthal* is omitted in the vernacular name; thus, *Neandertal*. However, according to the International Code for Zoological Nomenclature, the name of the species *Homo neanderthalensis* (King, 1864) must continue to be written with the letter *h* following the letter *t*, as well as the newer name *Homo sapiens neanderthalensis* (Campbell, 1964). Many writers still continue to use the older spelling for the vernacular.
4. Kenneth A. R. Kennedy, *Neanderthal Man* (Minneapolis: Burgess Publishing Company, 1975), 33. The quotation is from Boule, but Kennedy gives no reference.
5. Jerold Lowenstein, Theya Molleson, and Sherwood Washburn, "Piltdown Jaw Confirmed as Orang," *Nature* 299 (23 September 1982): 294.
6. Ronald Miller, *The Piltdown Men* (New York: St. Martin's Press, 1972).
7. Charles Blinderman, *The Piltdown Inquest* (Buffalo: Prometheus Books, 1986).
8. Stephen Jay Gould, "The Piltdown Conspiracy," *Natural History* (August 1980): 8–28.
9. John Winslow and Alfred Meyer, "The Perpetrator at Piltdown," *Science 83* (September 1983): 32–43.
10. Wilbur M. Smith, "In the Study," *Moody Monthly*, March 1954, 26.
11. The Piltdown hoax never ceases to surprise. After eighty-three years, the one man who up to now seemed to be above suspicion has been added to the list of suspects: none other than Sir Arthur Keith himself. See Frank Spencer, *Piltdown: A Scientific Forgery* (New York: Oxford University Press, 1990). There is still no smoking gun. Spencer's evidence, like all the rest, is entirely circumstantial. Spencer speculates that Keith and Dawson worked together. Dawson's motive was to gain entrance to the prestigious Royal Society. Keith's motive was to distract attention from the smaller-brained Java Man (*Pithecanthropus I*) that had been discovered about twenty years earlier, and to strengthen his own position that the brain of man evolved first, before bipedalism. No doubt, the list of suspects will continue to grow.

Chapter 5. Looks Isn't Everything

1. Vincent Sarich, Creation-Evolution debate, North Dakota State University, Fargo, April 28, 1979.
2. William W. Howells, "*Homo erectus* in human descent: ideas and problems," *Homo erectus: Papers in Honor of Davidson Black*, Becky A. Sigmon and Jerome S. Cybulski, eds. (Toronto: University of Toronto Press, 1981): 70–71. Bracketed material added.
3. F. B. Livingstone, "Gene flow in the Pleistocene" (abstract), *American Journal of Physical Anthropology* Supplement 12 (1991): 117.
4. T. C. Partridge, "Geomorphological Dating of Cave Openings at Makapansgat, Sterkfontein, Swartkrans, and Taung," *Nature* 246 (9 November 1973): 75–79.

5. Karl W. Butzer, "Paleoecology of South African Australopithecines: Taung Revisited," *Current Anthropology* 15:4 (December 1974): 382. Bracketed material added for clarity.
6. Butzer, 411.
7. Butzer, 404.
8. Phillip V. Tobias, "Implications of the New Age Estimates of the Early South African Hominids," *Nature* 246 (9 November 1973): 82.
9. Tobias, 82.
10. A. Walker, R. E. Leakey, J. M. Harris, and F. H. Brown, "2.5-Myr *Australopithecus boisei* from west of Lake Turkana, Kenya," *Nature* 322 (7 August 1986): 517–22.
11. Ian Tattersall, Eric Delson, and John Van Couvering, eds. *Encyclopedia of Human Evolution and Prehistory* (New York: Garland Publishing, 1988), 571. Bracketed material added for clarity.
12. Tattersall et al., 571.
13. Richard G. Klein, *The Human Career: Human Biological and Cultural Origins* (Chicago: University of Chicago Press, 1989), 113.
14. Klein, 113. Bracketed material added for clarity.
15. Bryan Patterson, Anna K. Behrensmeyer, and William D. Sill, "Geology and Fauna of a New Pliocene Locality in North-western Kenya," *Nature* 226 (6 June 1970): 918–21.
16. Bryan Patterson and W. W. Howells, "Hominid Humeral Fragment from Early Pleistocene of Northwestern Kenya," *Science* 156 (7 April 1967): 65. Bracketed material added for clarity. Originally the stratum was thought to be Pleistocene, but later it was determined to be of Pliocene Age. See endnote 15.
17. Patterson and Howells, 66.
18. Henry M. McHenry, "Fossils and the Mosaic Nature of Human Evolution," *Science* 190 (31 October 1975): 428.
19. David Pilbeam, *The Evolution of Man* (New York: Funk and Wagnalls, 1970), 151. The describer is W. W. Howells, not F. Clark Howell.
20. McHenry, 428.
21. Brigette Senut, "Humeral Outlines in Some Hominoid Primates and in Plio-Pleistocene Hominids," *American Journal of Physical Anthropology* 56 (1981): 275.
22. Oldest in the sense that it is the oldest fossil capable of a legitimate diagnosis. Another fossil found in the same general area, the Lothagam mandible fragment, KNM-ER 329, is dated at 5.5 m.y.a. A case could be made that this fossil also is truly human. However, the quality of the fossil is questionable, and there are legitimate diagnostic problems in dealing with mandible fragments. Humerus, mandible, and cranial fragments from the Chemeron Formation, Chesowanja, Kenya, and a cranial fragment from Belohdelie, Middle Awash, Ethiopia, are also considered to be of comparable age but are too fragmentary to diagnose.
23. Howells, 79–80. Emphasis added.

Chapter 6. With a Name Like Neandertal He's Got to Be Good

1. Technically, Neandertal fossils on Gibralter were discovered earlier, but their importance was not understood and they have remained in relative obscurity.
2. Donald Johanson and James Shreeve, *Lucy's Child* (New York: William Morrow and Company, Inc., 1989), 49.
3. Erik Trinkaus, "Hard Times Among the Neanderthals," *Natural History* 87:10 (December 1978): 58. See also R. L. Holloway, "The Neandertal Brain: What Was Primitive" (abstract), *American Journal of Physical Anthropology* Supplement 12 (1991): 94.
4. Trinkaus, 58.
5. Valerius Geist, "Neanderthal the Hunter," *Natural History* 90:1 (January 1981): 30.
6. Geist, 30.
7. Geist, 34.
8. Jared Diamond, "The Great Leap Forward," *Discover*, May 1989, 50–60.
9. Diamond, 50.
10. Diamond, 55.
11. "How Neanderthals Chilled Out," *Science News* (24 March 1990): 189.

12. Geist, 36.
13. B. Arensburg, A. M. Tillier, B. Vandermeersch, H. Duday, L. A. Schepartz, and Y. Rak, "A Middle Palaeolithic human hyoid bone," *Nature* 338 (27 April 1989): 759–60. See also B. Arensburg, L. A. Schepartz, A. M. Tillier, B. Vandermeersch, and Y. Rak, "A Reappraisal of the Anatomical Basis for Speech in Middle Palaeolithic Hominids," *American Journal of Physical Anthropology* 83:2 (October 1990): 137–46.
14. Christopher B. Stringer, "Fate of the Neanderthal," *Natural History* (December 1984): 12.
15. H. Valladas, J. L. Reyss, J. L. Joron, G. Valladas, O. Bar-Yosef, and B. Vandermeersch, "Thermoluminescence dating of Mousterian 'Proto-Cro-Magnon' remains from Israel and the origin of modern man," *Nature* 331 (18 February 1988): 614–16.
16. Y. Rak and B. Arensburg, "Kebara 2 Neanderthal Pelvis: First Look at a Complete Inlet," *American Journal of Physical Anthropology* 73 (1987): 227–31. See also Yoel Rak, "On the Differences between Two Pelvises of Mousterian Context from the Qafzeh and Kebara Caves, Israel," *American Journal of Physical Anthropology*. 81:3 (March 1990): 323–32.
17. Stephen Jay Gould, "A Novel Notion of Neanderthal," *Natural History* (June 1988): 20.
18. Chris Stringer, "The Dates of Eden," *Nature* 331 (18 February 1988): 565.
19. Chris R. Stringer and Rainer Grun, "Time for the last Neanderthals," *Nature* 351 (27 June 1991): 702.
20. Rebecca L. Cann, Mark Stoneking, and Allan C. Wilson, "Mitochondrial DNA and human evolution," *Nature* 325 (1 January 1987): 31–36.
21. Henry Gee, "Statistical cloud over African Eden," *Nature* 355 (13 February 1992): 583.
22. Marcia Barinaga, "'African Eve' Backers Beat a Retreat," *Science* 255 (7 February 1992): 687.
23. S. Blair Hedges, Sudhir Kumar, Koichiro Tamura, and Mark Stoneking, "Human Origins and Analysis of Mitochondrial DNA Sequences," *Science* 255 (7 February 1992): 737–739.
24. Allan C. Wilson and Rebecca L. Cann, "The Recent African Genesis of Humans," *Scientific American* (April 1992): 68.
25. Wilson and Cann, 68. Emphasis added.
26. Marcia Barinaga, "Choosing a Human Family Tree," *Science* 255 (7 February 1992): 687.
27. Wilson and Cann, 68. Bracketed material added.
28. Wilson and Cann, 72.
29. N. Mercier, H. Valladas, J.-L. Joron, J.-L. Reyss, F. Leveque, and B. Vandermeersch, "Thermoluminescence dating of the late Neanderthal remains from Saint-Césaire," *Nature* 351 (27 June 1991): 737–39.
30. Michael H. Day, *Guide to Fossil Man*, fourth edition (Chicago: University of Chicago Press, 1986), 128–29.
31. Day, 134.
32. Ranier Berger and W. F. Libby, *Radiocarbon Journal*, vol. 8 (1966): 480.
33. Richard G. Klein, *The Human Career: Human Biological and Cultural Origins* (Chicago: University of Chicago Press, 1989), 281–82. Emphasis added.
34. Klein, 279.
35. Geist, 34.
36. Klein, 283.
37. J. Lawrence Angel, "History and Development of Paleopathology," *American Journal of Physical Anthropology* 56:4 (December 1981): 512.
38. Francis Ivanhoe, "Was Virchow Right about Neandertal?" *Nature* 227 (8 August 1970): 577–79.
39. D. J. M. Wright, "Syphilis and Neanderthal Man," *Nature* 229 (5 February 1971): 409.

Chapter 7. Evolution's Illegitimate Children: Archaic *H.s.*

1. Kenneth A. R. Kennedy, *Neanderthal Man* (Minneapolis: Burgess Publishing Company, 1975), 55–56.
2. Carleton S. Coon, *The Origin of Races* (New York: Alfred A. Knopf, 1962), 630–32.

3. Günter Braüer, "A Craniological Approach to the Origin of Anatomically Modern *Homo sapiens* in Africa and Implications for the Appearance of Modern Europeans," *The Origins of Modern Humans: A World Survey of the Fossil Evidence*, eds. Fred H. Smith and Frank Spencer (New York: Alan R. Liss, Inc. 1984), 384, 387.
4. Richard G. Klein, *The Human Career: Human Biological and Cultural Origins* (Chicago: University of Chicago Press, 1989), 285–86, 288.
5. Braüer, 387.
6. S. M. T. Myster and Fred H. Smith, "The Taxonomic Dilemma of the Tangier Maxilla: A Metric and Nonmetric Assessment" (abstract), *American Journal of Physical Anthropology* 81:2 (February 1990): 273.
7. Klein, 226–27, 288–89.
8. Braüer, 359, 363.
9. Ian Tattersall, Eric Delson, and John Van Couvering, eds., *Encyclopedia of Human Evolution and Prehistory* (New York: Garland Publishing, 1988), 55.
10. Klein, 286–89.
11. Gail E. Kennedy, "The emergence of modern man," *Nature* 284 (6 March 1980): 11.
12. J. Radovcic and R. Caspari, "A new reconstruction of the Krapina D skull and a comparison with male western European Neandertals" (abstract), *American Journal of Physical Anthropology* 72:2 (February 1987): 244.
13. G. P. Rightmire, "Florisbad and Human Population Succession in Southern Africa," *American Journal of Physical Anthropology* 48:4 (May 1978): 475–86.
14. Klein, 235, 239–41.
15. R. Caspari and M. H. Wolpoff, "The morphological affinities of the Klasies River Mouth skeletal remains" (abstract), *American Journal of Physical Anthropology* 81:2 (February 1990): 203.
16. Tattersall, Delson, and Van Couvering, 55, 298.
17. M. H. Day, M. D. Leakey, and C. Magori, "A new hominid fossil skull (L. H. 18) from the Ngaloba Beds, Laetoli, northern Tanzania," *Nature* 284 (6 March 1980): 55–56.
18. Klein, 226–27, 288–89.
19. Klein, 235, 239.
20. Tattersall, Delson, and Van Couvering, 52.
21. Klein, 237–38.
22. Klein, 274, 288.
23. Klein, 226–27, 232.
24. Klein, 235, 243.
25. Klein, 226–27, 229–31.
26. Francesco Mallegni et al., "New European Fossil Hominid Material From an Acheulean Site Near Rome (Castel Di Guido)," *American Journal of Physical Anthropology* 62:3 (November 1983): 263–74.
27. Klein, 226–28.
28. Klein, 235, 242.
29. Klein, 191, 225–28.
30. Klein, 234–36.
31. Klein, 235, 243.
32. Klein, 226, 232–33.
33. Tattersall, Delson, and Van Couvering, 41.
34. Becky A. Sigmon and Jerome S. Cybulski, eds., *Homo erectus: Papers in Honor of Davidson Black* (Toronto: University of Toronto Press, 1981), 232.
35. Sigmon and Cybulski, 230.
36. Aris N. Poulianos, "Petralona Cave dating controversy," *Nature* 299 (16 September 1982): 280.
37. Klein, 235, 242.
38. William Howells, *Evolution of the Genus Homo* (Reading, Mass.: Addison-Wesley Publishing Company, 1973), 77.
39. Reinhart Kraatz, "Recent Research on Heidelberg Jaw," *Hominid Evolution*, ed. Phillip V. Tobias (New York: Alan R. Liss, Inc. 1985), 315.

40. David Pilbeam, "The Descent of Hominoids and Hominids," *Scientific American* (March 1984):93.
41. Ian Tattersall, "Recognizing hominid species in the late Pleistocene" (abstract), *American Journal of Physical Anthropology* 81:2 (February 1990): 306.
42. William Howells, "*Homo erectus*—Who, When, and Where: A Survey," *Yearbook of Physical Anthropology*, 23 (New York: Alan R. Liss, Inc., 1980): 15.
43. Klein, 224–25.
44. Klein, 225.
45. Klein, 226–31.
46. Arthur Smith Woodward, "A New Cave Man from Rhodesia, South Africa," *Nature* 108 (17 November 1921): 371.
47. Michael H. Day, *Guide to Fossil Man*, 4th ed. (Chicago: University of Chicago Press, 1986), 267.
48. Woodward, 371. Bracketed material added for clarity.
49. Giorgio Manzi, Loretana Salvadei, and Pietro Passarello, "The Casal de' Pazzi archaic parietal: comparative analysis of new fossil evidence from the late Middle Pleistocene of Rome," *Journal of Human Evolution* 19:8 (December 1990): 751–59.

Chapter 8. Java Man: The Rest of the Story

1. Bert Theunissen, *Eugene Dubois and the Ape-Man from Java* (Dordrecht: Kluwer Academic Publishers, 1989). Theunissen is on the staff of the Institute for the History of Science, University of Utrecht. The book, originally published in Dutch in 1985, was the author's Ph.D. dissertation. The English translation is an expanded version.
2. G. H. R. von Koenigswald, *Meeting Prehistoric Man*, Micheal Bullock, trans. (New York: Harper and Brothers, 1956), 38–39.
3. Theunissen, 49.
4. Theunissen, 38.
5. Alan Houghton Brodrick, *Early Man* (London: Hutchinson's Scientific and Technical Publications, 1948), 85.
6. Theunissen, 44, 68.
7. Brodrick, 85.
8. Theunissen, 121.
9. Theunissen, 122.

Chapter 9. Java Man: Keeping the Faith

1. For a definitive exposition of this popular eighteenth- and nineteenth- century concept, see Arthur O. Lovejoy, *The Great Chain of Being* (Cambridge: Harvard University Press, 1964).
2. See Ernst Mayr, "Evolution and God," *Nature* 248 (22 March 1974): 285–86.
3. Neal C. Gillespie, *Charles Darwin and the Problem of Creation* (Chicago: The University of Chicago Press, 1979), 137.
4. Bert Theunissen, *Eugene Dubois and the Ape-Man from Java* (Dordrecht: Kluwer Academic Publishers, 1989), 79–127.
5. Robert F. Heizer, ed., *Man's Discovery of His Past* (Englewood Cliffs, N.J.: Prentice Hall, Inc., 1962), 138.
6. Heizer, 135–36.
7. Theunissen, 68, 77.
8. Theunissen, 158.
9. Michael H. Day and T. I. Molleson, "The Trinil Femora," Michael H. Day, ed., *Human Evolution*, vol. XI, *Symposia of the Society for the Study of Human Biology* (London: Taylor and Francis, Ltd., 1973), 127–154. See also the comment by William W. Howells, *American Journal of Physical Anthropology* 81:1 (January 1990): 133–34.
10. G. H. R. von Koenigswald, *Meeting Prehistoric Man*, Micheal Bullock trans., (New York: Harper and Brothers, 1956), 34.
11. Richard G. Klein, *The Human Career: Human Biological and Cultural Origins* (Chicago: University of Chicago Press, 1989), 185.

12. von Koenigswald, 32.

13. Theunissen, 154–55.

Chapter 10. Wadjak Man: Not All Fossils Are Created Equal

1. Bert Theunissen, *Eugene Dubois and the Ape-Man from Java* (Dordrecht: Kluwer Academic Publishers, 1989), 41, 43.

2. G. H. R. von Koenigswald, *Meeting Prehistoric Man*, Micheal Bullock, trans. (New York: Harper and Brothers, 1956), 30.

3. Sir Arthur Keith, *The Antiquity of Man*, revised ed., 2 vols. (London: Williams and Norgate, Ltd., 1925) 2:439.

4. Carleton S. Coon, *The Origin of Races* (New York: Alfred A. Knopf, 1962), 399.

5. Theunissen, 176, and Keith, 2:440.

6. Theunissen, 44, 148.

7. Keith, 2:440–41.

8. Theunissen, 72.

9. C. Loring Brace, "Creationists and the Pithecanthropines," *Creation/Evolution* XIX (Winter 1986–1987): 16. *Creation/Evolution* is a publication of the American Humanist Association. Brace's article is the text of his presentation at the University of Michigan debate.

10. Michael H. Day, *Guide to Fossil Man*, first ed. (Cleveland: World Publishing Company, 1965), 247.

11. Theunissen, 147.

12. Day, 247.

13. Oakley, Campbell, and Molleson, ed., *Catalogue of Fossil Hominids*, Part III (London: Trustees of the British Museum, 1975), 115.

14. Dirk Albert Hooijer, "The Geological Age of Pithecanthropus, Meganthropus, and Gigantopithecus," *American Journal of Physical Anthropology*, vol. 9, N.S. No. 3 (September 1951): 275.

15. Hooijer, 275.

16. Hooijer, 278. Bracketed material added for clarity.

17. Hooijer, 278.

18. Kenneth P. Oakley, *Frameworks for Dating Fossil Man* (Chicago: Aldine Publishing Company, 1964), 7.

19. Sir Karl Popper, *The Logic of Scientific Discovery* (New York: Basic Books, 1959).

20. Werner Heisenberg, *Physics and Beyond*, Arnold J. Pomerans, trans. (New York: Harper and Row, Publishers, 1971), 63. Emphasis mine.

Chapter 11. The Selenka Expedition: A Second Opinion

1. M. Lenore Selenka and Max Blanckenhorn, *Die Pithecanthropus-Schichen auf Java* (Leipzig: W. Engelmann, 1911), 342

2. A. G. Tilney, "Pithecanthropus (Ape-Man): The Facts" (Stoke, England: Evolution Protest Movement, n.d.).

3. Sir Arthur Keith, "The Problem of *Pithecanthropus*," *Nature*, vol. 87 (13 July 1911): 49–50.

4. Alan Houghton Brodrick, *Man and His Ancestry*, revised ed. (Greenwich, Conn.: Fawcett Publications, 1964), 127.

5. Carleton S. Coon, *The Origin of Races* (New York: Alfred A. Knopf, 1973), 376.

6. George Grant MacCurdy, *Human Origins*, 2 vols. (New York: D. Appleton and Company, 1924), 1:316.

7. Bert Theunissen, *Eugene Dubois and the Ape-Man from Java*, (Dordrecht: Kluwer Academic Publishers, 1989), 127. Emphasis added.

8. G. H. R. von Koenigswald, *Meeting Prehistoric Man*, Micheal Bullock, trans. (New York: Harper and Brothers, 1956), 36.

9. Theunissen, 164.

Chapter 12. *Homo erectus*: A Man for All Seasons

1. Ian Tattersall, Eric Delson, and John Van Couvering, eds., *Encyclopedia of Human Evolution and Prehistory* (New York: Garland Publishing, 1988), 65–67.
2. L. Freedman and M. Lofgren, "The Cossack skull and a dihybrid origin of the Australian Aborigines," *Nature* 282 (15 November 1979): 298–300.
3. Tattersall et al., 65–67.
4. Kenneth P. Oakley, *Frameworks for Dating Fossil Man* (Chicago: Aldine Publishing Company, 1964), 171–73, 251–52, 314.
5. Kenneth P. Oakley, *Man the Tool-maker*, sixth ed. (Chicago: University of Chicago Press, 1976), 80. The dating of the Java Ngandong Solo people is highly controversial. I have dated them according to the associated artifacts as documented by Oakley. However, this association has been questioned. See Richard G. Klein, *The Human Career*, 206.
6. Tattersall et al., 53–54, and J. K. Austin, "Morphometric analysis of the Solo tibiae in relation to Upper Pleistocene hominid fossils" (abstract), *American Journal of Physical Anthropology* Supplement 12 (1991): 46.
7. A. G. Thorne and P. G. Macumber, "Discoveries of Late Pleistocene Man at Kow Swamp, Australia," *Nature* 238 (11 August 1972): 316–19.
8. Tattersall et al., 65.
9. Tattersall et al., 67.
10. Tattersall et al., 67.
11. Richard G. Klein, *The Human Career: Human Biological and Cultural Origins* (Chicago: University of Chicago Press, 1989), 396.
12. Tattersall et al., 67.
13. Eric Delson, ed., *Ancestors: The Hard Evidence* (New York: Alan R. Liss, Inc., 1985), 298.
14. Chris Stringer, "The dates of Eden," *Nature* 331 (18 February 1988): 565.
15. Günter Braüer, "A Craniological Approach to the Origin of Anatomically Modern *Homo sapiens* in Africa and Implications for the Appearance of Modern Europeans," *The Origins of Modern Humans*, Fred H. Smith and Frank Spencer, eds. (New York: Alan R. Liss, Inc., 1984), 347.
16. Michael H. Day, *Guide to Fossil Man*, first ed. (Cleveland: World Publishing Company, 1965), 114–15.
17. Delson, 299, 337.
18. Alison S. Brooks and Bernard Wood, "The Chinese side of the story," *Nature* 344 (22 March 1990): 288. The assignment of this fossil by Brooks and Wood to *Homo sapiens* is an arbitrary one based solely on the date. The fossil description better suits *Homo erectus*.
19. William W. Howells, "*Homo erectus*—Who, When, and Where: A Survey," *Yearbook of Physical Anthropology* 23 (New York: Alan R. Liss, Inc., 1980): 9.
20. Brooks and Wood, 288. Endnote 18 comment also applies to the assignment of this fossil.
21. Howells, 9.
22. Klein, 228.
23. Kenneth P. Oakley, Bernard Campbell, and Theya Molleson, eds., *Catalogue of Fossil Hominids*, part III (London: Trustees of the British Museum - Natural History, 1975), 135.
24. Klein, 123–24.
25. Brooks and Wood, 288.
26. Howells, 8.
27. Brooks and Wood, 288. Endnote 18 comment also applies to the assignment of this fossil.
28. Brooks and Wood, 288. Endnote 18 comment also applies to the assignment of this fossil.
29. Becky A. Sigmon and Jerome S. Cybulski, eds., *Homo erectus: Papers in Honor of Davidson Black* (Toronto: University of Toronto Press, 1981), 228.
30. Klein, 233.
31. Sigmon and Cybulski, 231.
32. Sigmon and Cybulski, 231.
33. Delson, 287.
34. Sigmon and Cybulski, 173, 231.
35. Klein, 189.

36. Brooks and Wood, 288. Endnote 18 comment also applies to the assignment of this fossil.
37. Klein, 247.
38. C. Loring Brace, Shao Xiang-qing, and Zhang Zhen-biao, "Prehistoric and Modern Tooth Size in China," *The Origins of Modern Humans*, Fred H. Smith and Frank Spencer, eds. (New York: Alan R. Liss, Inc., 1984), 485.
39. Brooks and Wood, 288.
40. Michael H. Day, *Guide to Fossil Man*, fourth ed. (Chicago: University of Chicago Press, 1986), 226, 230–31.
41. Brooks and Wood, 288. Endnote 18 comment also applies to the assignment of this fossil.
42. Tattersall et al., 133.
43. Sigmon and Cybulski, 231.
44. Klein, 189.
45. Oakley et al., *Catalogue of Fossil Hominids*, part III, 192.
46. Klein, 239.
47. Tattersall et al., 50–51.
48. Brooks and Wood, 288.
49. R. L. Ciochon, "Lang Trang Caves: A New Middle Pleistocene Hominid Site from Northern Vietnam" (abstract), *American Journal of Physical Anthropology* 81:2 (February 1990): 205. See also Russell Ciochon, John Olsen, and Jamie James, *Other Origins* (New York: Bantam Books, 1990), 128–149.
50. Klein, 120.
51. Day (1986), 177.
52. Oakley, *Frameworks for Dating Fossil Man*, 7.
53. Klein, 189, 191.
54. Tattersall et al., 580.
55. Klein, 189.
56. Brooks and Wood, 288.
57. Sigmon and Cybulski, 228.
58. Klein, 112.
59. G. Philip Rightmire, *The Evolution of Homo erectus* (Cambridge: Cambridge University Press, 1990), 12.
60. Klein, 120, 192.
61. Klein, 120, 192.
62. Oakley et al., *Catalogue of Fossil Hominids*, part I (Second ed., 1977), 8.
63. R. J. Clarke and F. Clark Howell, "Affinities of the Swartkrans 847 Hominid Cranium," *American Journal of Physical Anthropology* 37 (November 1972): 328.
64. Klein, 120.
65. Day (1986), 177.
66. Klein, 193. Chinese authorities date Yuanmou Man at 1.7 m.y.a. See J. Lanpo and H. Weiwen, *The Story of Peking Man* (Hong Kong: Oxford University Press, 1990), 79.
67. Brooks and Wood, 288.
68. Klein, 120.
69. Sigmon and Cybulski, 230.
70. Brooks and Wood, 288.
71. Klein, 120.
72. Day (1986), 177.
73. Klein, 120.
74. Day (1986), 178.
75. Klein, 191–92.
76. Klein, 120, 192.
77. Klein, 120.
78. Day (1986), 177.
79. D. Ninkovich and L. H. Burckle, "Absolute age of the base of the hominid-bearing beds in Eastern Java," *Nature* 275 (28 September 1978): 306–7.
80. Sigmon and Cybulski, 73, 89, 94.
81. Klein, 191–92.

82. Craig S. Feibel, Francis H. Brown, and Ian McDougall, "Stratigraphic Context of Fossil Hominids From the Omo Group Deposits: Northern Turkana Basin, Kenya and Ethiopia," *American Journal of Physical Anthropology* 78 (April 1989): 611.
83. Elwyn L. Simons, "Human Origins," *Science* 245 (22 September 1989): 1347.
84. Feibel et al., 613.
85. Michael H. Day, "Hominid Postcranial Material from Bed I, Olduvai Gorge," *Human Origins: Louis Leakey and the East African Evidence*, Glynn Isaac and Elizabeth McCown, eds. (Menlo Park, California: W. A. Benjamin, Inc., 1976), 369.
86. Klein, 120.
87. Day (1986), 177.
88. Feibel et al., 611.
89. Sigmon and Cybulski, 208.
90. Feibel et al., 613.
91. B. A. Wood, "Evidence on the locomotor pattern of *Homo* from early Pleistocene of Kenya," *Nature* 251 (13 September 1974): 135–36.
92. Feibel et al., 611.
93. B. Holly Smith, "Dental development in *Australopithecus* and early *Homo*," *Nature* 323 (25 September 1986): 329.
94. Feibel et al., 611.
95. Frank Brown, John Harris, Richard Leakey, and Alan Walker, "Early *Homo erectus* skeleton from west Lake Turkana, Kenya," *Nature* 316 (29 August 1985): 788–97.
96. Feibel et al., 611.
97. Rightmire, 99.
98. A. Walker, M. R. Zimmerman, and R. E. F. Leakey, "A possible case of hypervitaminosis A in *Homo erectus*," *Nature* 296 (18 March 1982): 248–50.
99. Feibel et al., 611.
100. Day (1986), 201, 212.
101. Feibel et al., 611.
102. Sigmon and Cybulski, 10, 79, 205–7.
103. Feibel et al., 611.
104. Smith, 329.
105. Randall L. Susman, "New Hominid Fossils From the Swartkrans Formation (1979–1986 Excavations): Postcranial Specimens," *American Journal of Physical Anthropology* 79 (August 1989): 451.
106. Frederick E. Grine, "New Hominid Fossils From the Swartkrans Formation (1979–1986 Excavations): Craniodental Specimens," *American Journal of Physical Anthropology* 79 (August 1989): 446.
107. R. L. Susman and F. E. Grine, "New *Paranthropus robustus* radius from Member 1, Swartkrans Formation" (abstract), *American Journal of Physical Anthropology* 78:2 (February 1989): 312.
108. Susman, 451, 470. Erik Trinkaus and Jeffrey C. Long, "Species Attribution of the Swartkrans Member 1 First Metacarpals: SK 84 and SKX 5020," *American Journal of Physical Anthropology* 83:4 (December 1990): 419–24.
109. Susman, 451.
110. Michael Day, "Pliocene Hominids," *Ancestors: The Hard Evidence*, Eric Delson, ed. (New York: Alan R. Liss, Inc., 1985), 92.
111. Rightmire, 116.
112. Teuku Jacob, "The absolute date of the Djetis beds at Modjokerto," *Antiquity* 46:182 (June 1972): 148.
113. Sigmon and Cybulski, 6.
114. Feibel et al., 613.
115. Rightmire, 108.
116. Feibel et al., 611.
117. Rightmire, 100.
118. Frank Brown et al., 791.
119. "Manlike fossils in China predate Peking Man," *San Diego Tribune*, 24 November 1988. For accuracy and authority, a newspaper source is a far cry from a scientific journal. How-

ever, at the time of writing, this is the only reference I have been able to locate on this particular fossil. It is difficult to get accurate information on the Chinese fossil discoveries.

120. Ann Gibbons, "Paleontology by Bulldozer," *Science* 247 (23 March 1990): 1408.
121. Bruce Bower, "Human Ancestors Make Evolutionary Change," *Science News* 127 (4 May 1985): 276.
122. Sigmon and Cybulski, 72, 153.
123. Klein, 183.
124. C. Loring Brace, "Creationists and the Pithecanthropines," *Creation/Evolution* 19 (Winter 1986–1987): 16.
125. Jared Diamond, "The Great Leap Forward," *Discover* May 1989, 52–53.
126. R. L. Susman, J. T. Stern, Jr., and M. D. Rose, "Morphology of KNM-ER 3228 and O. H. 28 innominates from East Africa" (abstract), *American Journal of Physical Anthropology* 60:3 (February 1983): 259. Emphasis added.
127. Tattersall et al., 67.
128. Delson, 298.
129. Day (1986), 177. Klein, 120.
130. Day (1986), 177. Klein, 120.
131. Bernard Wood, "Who is the 'real' *Homo habilis*?" *Nature* 327 (21 May 1987): 187.
132. Alun R. Hughes and Phillip V. Tobias, "A fossil skull probably of the genus *Homo* from Sterkfontein, Transvaal," *Nature* 265 (27 January 1977): 310–12.
133. Day (1986), 177–178.
134. R. L. Susman and J. T. Stern, Jr., "Functional Affinities of the *Homo habilis* postcranial remains from FLK, Olduvai Gorge" (abstract), *American Journal of Physical Anthropology* 57:2 (February 1982): 234–35.
135. R. C. Walter, P. C. Manega, R. L. Hay, R. E. Drake, and G. H. Curtis, "Laser-fusion ^{40}Ar/^{39}Ar dating of Bed 1, Olduvai Gorge, Tanzania," *Nature* 354 (14 November 1991): 145–149.
136. Donald C. Johanson et al., "New partial skeleton of *Homo habilis* from Olduvai Gorge, Tanzania," *Nature* 327 (21 May 1987): 205–9.
137. Noel T. Boaz and F. Clark Howell, "A Gracile Hominid Cranium from Upper Member G of the Shungura Formation, Ethiopia," *American Journal of Physical Anthropology* 46:1 (January 1977): 93–108.
138. Bennett Blumenberg, "Population Characteristics of Extinct Hominid Endocranial Volume," *American Journal of Physical Anthropology* 68:2 (October 1985): 270.
139. Day (1986), 177–78.
140. Walter et al., 145–49.
141. Feibel et al. 611., Day (1986) 214.
142. Donald C. Johanson and M. Taieb, "Plio-Pleistocene hominid discoveries in Hadar, Ethiopia," *Nature* 260 (25 March 1976): 297.
143. P. D. Gingerich and B. H. Smith, "Early *Homo* from Member l, Swartkrans" (abstract), *American Journal of Physical Anthropology* 72:2 (February 1987): 203–4.
144. Klein, 134.
145. Feibel et al., 611, 612.
146. Walter et al., 145.
147. Klein, 111.
148. See endnote 7.
149. See endnote 5.
150. Thorne and Macumber, 316.
151. A. G. Thorne, "Mungo and Kow Swamp: Morphological Variation in Pleistocene Australians," *Mankind* 8:2 (December 1971): 87.
152. Thorne and Macumber 316, 319.
153. Alan G. Thorne and Milford H. Wolpoff, "Regional Continuity in Australasian Pleistocene Hominid Evolution" *American Journal of Physical Anthropology* 55:3 (July 1981): 337–49.
154. Oakley et al., *Catalogue of Fossil Hominids*, part III, 200.
155. "Talgai Skull," *Science News* 93 (20 April 1968): 381.
156. Tattersall et al., 67.

157. L. Freedman and M. Lofgren, "Human Skeletal Remains from Cossack, Western Australia," *Journal of Human Evolution* 8 (1979): 285.
158. Ann Gibbons, "Jawing With Our Georgian Ancestors," *Science* 255 (24 January 1992): 401. This fossil was first announced in December 1991, and details are sketchy. It is said to be either 0.9 MY or 1.6 MY old.
159. Li Tianyuan and Dennis A. Etler, "New Middle Pleistocene hominid crania from Yunxian in China," *Nature* 357 (4 June 1992): 404–07.
160. Ann Gibbons, "An About-Face for Modern Human Origins," *Science* 256 (12 June 1992):1521.

Chapter 13. *Homo erectus*: All in the Family

1. Michael Day, "*Homo* turmoil," *Nature* 348 (20/27 December 1990): 688.
2. Day, 688.
3. Gabriel Ward Lasker, *Physical Anthropology* (New York: Holt, Rinehart and Winston, Inc., 1973) 284.
4. William S. Laughlin, "Eskimos and Aleuts: Their Origins and Evolution," *Science* 142 (8 November 1963): 644.
5. Milford H. Wolpoff, Wu Xin Zhi, and Alan G. Thorne, "Modern *Homo sapiens* Origins: A General Theory of Hominid Evolution Involving the Fossil Evidence From East Asia," *The Origins of Modern Humans*, Fred H. Smith and Frank Spencer, eds. (New York: Alan R. Liss, Inc., 1984), 465–66. Bracketed material added for clarity.
6. Franz Weidenreich, "The skull of *Sinanthropus pekinensis*," *Palaeontol Sinica* (n. s. D, No. 10, 1943) 246. Quoted by Wolpoff et al., 466.
7. Edmund White and Dale Brown, *The First Men* (New York: Time-Life Books, 1973), 14.
8. R. L. Susman, J. T. Stern, Jr., and M. D. Rose, "Morphology of KNM-ER 3228 and O. H. 28 innominates from East Africa" (abstract), *American Journal of Physical Anthropology* 60:3 (February 1983): 259.
9. Donald C. Johanson and Maitland A. Edey, *Lucy: The Beginnings of Humankind* (New York: Simon and Schuster, 1981), 144.
10. Stephen Molnar, *Races, Types, and Ethnic Groups* (Englewood Cliffs, N. J.: Prentice-Hall, Inc., 1975), 57.
11. Public lecture on the Neandertals by Erik Trinkaus, Colorado State University, Fort Collins, December 3, 1984.
12. C. Loring Brace, "Creationists and the Pithecanthropines," *Creation/Evolution* 19 (Winter 1986–1987), 23.
13. Harry L. Shapiro, *Peking Man* (New York: Simon and Schuster, 1974), 125.
14. Jared M. Diamond, "Extinctions, catastrophic and gradual," *Nature* 304 (4 August 1983): 397. Bracketed material added for clarity.
15. Becky A. Sigmon and Jerome S. Cybulski, eds., *Homo erectus: Papers in Honor of Davidson Black* (Toronto: University of Toronto Press, 1981), 227.
16. Chris B. Stringer and Rainer Grun, "Time for the last Neanderthals," *Nature* 351 (27 June 1991): 701.
17. M. D. Leakey, "Primitive Artifacts from Kanapoi Valley," *Nature* 5062 (5 November 1966): 581.
18. Lawrence H. Robbins, "Archeology in the Turkana District, Kenya," *Science* 176 (28 April 1972): 360.
19. Eileen M. O'Brien, "What Was the Acheulean Hand Ax?" *Natural History* (July 1984): 20–23.
20. O'Brien, 23.

Chapter 14. Back to the Future

1. James B. Conant, *Science and Common Sense* (New Haven: Yale University Press, 1951), 259–60. Emphasis mine.
2. Margaret Schabas, "The Idea of the Normal" (Book Review), *Science* 251 (15 March 1991): 1373.

3. Eugenie C. Scott, "'Creation Science' and Philosophy of Science: Reflections" (abstract), *American Journal of Physical Anthropology* 75:2 (February 1988): 269.
4. Michael J. Oard, *An Ice Age Caused by the Genesis Flood* (San Diego: Institute for Creation Research, 1990).
5. Oard, 116.
6. Oard, 95.
7. "Rudolf Virchow," *Encyclopedia Americana*, 1963 edition.
8. See L. Harrison Matthews, "Introduction," *The Origin of Species*, by Charles Darwin (London: J. M. Dent & Sons, Ltd., 1971), xi.
9. D. R. Brothwell, *Digging Up Bones*, 3rd ed., revised (Ithaca: Cornell University Press, 1981), 163.
10. Francis Ivanhoe, "Was Virchow Right about Neandertal?" *Nature* 227 (8 August 1970): 578. Emphasis mine.
11. Jeffrey Laitman, "Australia," *Encyclopedia of Human Evolution and Prehistory*, eds. Ian Tattersall, Eric Delson, and John Van Couvering (New York: Garland Publishing, 1988), 67.
12. Richard G. Klein, *The Human Career: Human Biological and Cultural Origins* (Chicago: The University of Chicago Press, 1989), 396. Bracketed material added for clarity.
13. A correspondent, "Late Pleistocene Man at Kow Swamp," *Nature* 238 (11 August 1972): 308. Editorials in *Nature* often accompany significant articles. Until recently, those editorials, although written by authorities in their particular fields, were unsigned and listed as being from "a correspondent."
14. A. G. Thorne and P. G. Macumber, "Discoveries of Late Pleistocene Man at Kow Swamp, Australia," *Nature* 238 (11 August 1972): 319.
15. L. Freedman and M. Lofgren, "The Cossack skull and a dihybrid origin of the Australian Aborigines," *Nature* 282 (15 November 1979): 299.
16. Laitman, 67.
17. Klein, 396.
18. Kenneth P. Oakley, Bernard Campbell, and Theya Molleson, *Catalogue of Fossil Hominids*, part III (London: Trustees of the British Museum—Natural History, 1975), 199 (plate 5).
19. Chris Stringer, "Homo Sapiens," *Encyclopedia of Human Evolution and Prehistory*, eds. Ian Tattersall, Eric Delson, and John Van Couvering (New York: Garland Publishing, 1988), 274.
20. Phillip J. Habgood, "The Origin of the Australian Aborigines: An Alternative Approach and View," *Hominid Evolution: Past, Present and Future*, ed. Phillip V. Tobias (New York: Alan R. Liss, Inc., 1985), 375.
21. Eric Delson, *Ancestors: The Hard Evidence* (New York: Alan R. Liss, Inc., 1985), 298.
22. L. Freedman and M. Lofgren, "Human Skeletal remains from Cossack, Western Australia," *Journal of Human Evolution* 8 (1979): 295.
23. Brothwell, 49.

Chapter 15. *Homo habilis*: The Little Man Who Isn't There

1. Louis S. B. Leakey, "Finding the World's Earliest Man," *National Geographic*, September 1960: 421. Emphasis mine.
2. Louis S. B. Leakey, Phillip V. Tobias, and John R. Napier, "A New Species of the Genus *Homo* from Olduvai Gorge," *Nature* 202 (4 April 1964): 7–9.
3. Marvin L. Lubenow, "Reversals in the Fossil Record: The Latest Problem in Stratigraphy and Evolutionary Phylogeny," *Creation Research Society Quarterly* 13 (March 1977): 185–90.
4. Richard G. Klein, *The Human Career: Human Biological and Cultural Origins* (Chicago: The University of Chicago Press, 1989), 161.
5. Richard E. Leakey, "Skull 1470," *National Geographic*, June 1973, 819.
6. *Science News* 102 (18 November 1972): 324.
7. Richard E. F. Leakey, "Evidence for an Advanced Plio-Pleistocene Hominid from East Rudolf, Kenya," *Nature* 242 (13 April 1973): 450.
8. *AnthroQuest*: The Leakey Foundation News, No. 43 (Spring 1991): 13.

9. Roger Lewin, *Bones of Contention: Controversies in the Search for Human Origins*, (New York: Simon and Schuster, 1987), 160. Emphasis mine. Bracketed material added for clarity.

10. Donald C. Johanson, Fidelis T. Masao, Gerald G. Eck, Tim D. White, Robert C. Walter, William H. Kimbel, Berhane Asfaw, Paul Manega, Prosper Ndessokia, and Gen Suwa, "New partial skeleton of *Homo habilis* from Olduvai Gorge, Tanzania," *Nature* 327 (21 May 1987): 205–9.

11. Dean Falk, "Cerebral Cortices of East African Early Hominids," *Science* 221 (9 September 1983): 1073.

12. Johanson et al., 209.

13. Klein, 155, 156, 158 (2), and 182.

14. Charles E. Oxnard, "The place of the australopithecines in human evolution: grounds for doubt?" *Nature* 258 (4 December 1975): 389.

15. Matt Cartmill, David Pilbeam, and Glynn Isaac, "One Hundred Years of Paleoanthropology" *American Scientist* 74 (July–August 1986): 419.

16. R. A. Foley, "Evolutionary History of the 'Robust' Australopithecines" (Book Review) *American Journal of Physical Anthropology* 82:1 (May 1990): 113.

17. Oxnard, 389–95. See also Charles Oxnard, *The Order of Man* (New Haven: Yale University Press, 1984).

Chapter 16. Fossil Failure on a Grand Scale

1. Richard G. Klein, *The Human Career: Human Biological and Cultural Origins* (Chicago: University of Chicago Press, 1989), 396.

2. Klein, 300.

3. Klein, 396.

4. Ian Tattersall, Eric Delson, and John Van Couvering, eds., *Encyclopedia of Human Evolution and Prehistory* (New York: Garland Publishing, 1988), 56, 87.

5. Alison S. Brooks and Bernard Wood, "The Chinese side of the story," *Nature* 344 (22 March 1990): 288.

6. Klein, 304, 327.

7. N. Minugh-Purvis and J. Radovcic, "Krapina A: Neandertal or Not?" (abstract), *American Journal of Physical Anthropology* Supplement 12 (1991): 132.

8. Klein 226–27.

9. See chapter 10.

10. C. B. Stringer, R. Grun, H. P. Schwarcz, and P. Goldberg, "ESR dates for the hominid burial site of Es Skhūl in Israel," *Nature* 338 (27 April 1989): 756–58.

11. Rainer Grun, Peter B. Beaumont, and Christopher B. Stringer, "ESR dating evidence for early modern humans at Border Cave in South Africa," *Nature* 344 (5 April 1990): 537–39.

12. G. Philip Rightmire and H. J. Deacon, "Comparative studies of Late Pleistocene human remains from Klasies River Mouth, South Africa," *Journal of Human Evolution* 20:2 (February 1991): 131–56.

13. M. J. Mehlman, "Early *Homo sapiens* in northern Tanzania: some evidence from Lake Eyasi basin" (abstract), *American Journal of Physical Anthropology* 75:2 (February 1988): 249.

14. Klein, 286–91.

15. Klein, 241.

16. C. B. Stringer, J. J. Hublin, and B. Vandermeersch, "The Origin of Anatomically Modern Humans in Western Europe," *The Origins of Modern Humans*, Fred H. Smith and Frank Spencer, eds. (New York: Alan R. Liss, Inc., 1984), 58–59.

17. See chapter 7.

18. See chapters 8, 9, 10, and 11.

19. See chapter 11.

20. Anna K. Behrensmeyer and Leo F. Laporte, "Footprints of a Pleistocene hominid in northern Kenya," *Nature* 289 (15 January 1981): 167–69.

21. Brigette Senut, "Humeral Outlines in Some Hominoid Primates and in Plio-pleistocene Hominids," *American Journal of Physical Anthropology* 56:3 (November 1981): 275–83.
22. Michael Day, "Hominid Postcranial Material from Bed I, Olduvai Gorge," *Human Origins: Louis Leakey and the East African Evidence*, Glynn Ll. Isaac and Elizabeth R. McCown, eds. (Menlo Park, Calif.: W. A. Benjamin, Inc., 1976), 364-65.
23. B. A. Wood, "Evidence on the locomotor pattern of *Homo* from early Pleistocene of Kenya," *Nature* 251 (13 September 1974): 135-36.
24. Craig S. Feibel, Francis H. Brown, and Ian McDougall, "Stratigraphic Context of Fossil Hominids From the Omo Group Deposits: Northern Turkana Basin, Kenya and Ethiopia," *American Journal of Physical Anthropology* 78 (April 1989): 613.
25. See chapter 15 and the appendix.
26. Henry M. McHenry, "Fossils and the Mosaic Nature of Human Evolution," *Science* 190 (31 October 1975): 427.
27. Feibel, Brown, and McDougall, 613.
28. Alan Walker, "The Koobi Fora Hominids and their bearing on the origins of the genus *Homo*," *Homo erectus: Papers in Honor of Davidson Black*, Becky A. Sigmon and Jerome S. Cybulski, eds. (Toronto: University of Toronto Press, 1981), 203.
29. Feibel, Brown, and McDougall, 611.
30. Louis S. B. Leakey, "Adventures in the Search for Man," *National Geographic*, January 1963, 147.
31. Sonia Cole, *Leakey's Luck* (New York: Harcourt, Brace, Jovanovich, 1975), 248.
32. Russell H. Tuttle, "The Pitted Pattern of Laetoli Feet," *Natural History*, March 1990: 60-65. See also chapter 16.
33. See chapter 5.
34. R. C. Walter, P. C. Manega, R. L. Hay, R. E. Drake, and G. H. Curtis, "Laser-fusion ^{40}Ar/^{39}Ar dating of Bed I, Olduvai Gorge, Tanzania," *Nature* 354 (14 November 1991): 145-49.
35. Mary D. Leakey, "Primitive Artifacts from Kanapoi Valley," *Nature* 212 (5 November 1966): 581.
36. Mary D. Leakey, "Footprints in the Ashes of Time," *National Geographic*, April 1979, 446.
37. Tuttle, *Natural History,* 64.
38. R. H. Tuttle and D. M. Webb, "The Pattern of Little Feet" (abstract), *American Journal of Physical Anthropology* 78:2 (February 1989): 316.
39. R. H. Tuttle and D. M. Webb, "Did *Australopithecus afarensis* make the Laetoli G footprint trails?" (abstract), *American Journal of Physical Anthropology* 1991 Supplement: 175.
40. Tuttle, *Natural History,* 64.
41. Russell H. Tuttle, "Primate Origins and Evolution" (Book Review), *American Journal of Physical Anthropology* 85:2 (June 1991): 244.
42. Tuttle, *Natural History,* 64.
43. William Howells, "*Homo erectus* in human descent: ideas and problems," *Homo erectus: Papers in Honor of Davidson Black*, Becky A. Sigmon and Jerome S. Cybulski, eds. (Toronto: University of Toronto Press, 1981), 79–80. Emphasis added.
44. Tattersall et al., 67.
45. Tattersall et al., 65–67.
46. Rainer Grun, Christopher B. Stringer, and Henry Schwarcz, "ERS dating of teeth from Garrod's Tabun cave collection," *Journal of Human Evolution* 20:3 (March 1991): 231.
47. Kenneth A. R. Kennedy, *Neanderthal Man* (Minneapolis: Burgess Publishing Company, 1975), 45–46.
48. Tattersall et al., 55.
49. R. Caspari and M. H. Wolpoff, "The morphological affinities of the Klasies River Mouth skeletal remains" (abstract), *American Journal of Physical Anthropology* 81:2 (February 1990): 203.
50. Minugh-Purvis and Radovcic, 132.
51. J. Radovcic and R. Caspari, "A new construction of the Krapina D skull and a comparison with male western European Neandertals" (abstract), *American Journal of Physical Anthropology* 72:2 (February 1987): 244.

52. R. S. Corruccini, "The forgotten Skhul crania and the 'neopresapiens' theory" (abstract), *American Journal of Physical Anthropology* 81:2 (February 1990): 209.
53. William Howells, *Evolution of the Genus Homo* (Reading, Mass.: Addison-Wesley Publishing Company, 1973), 121–22.
54. Mehlman, 249.
55. Kennedy, 54–55.
56. Klein, 238–239.
57. See chapters 8, 9, 10, and 11.
58. See chapters 8, 9, 10, and 11.
59. Behrensmeyer and Laporte, 167–69.
60. Feibel, Brown, and McDougall, 611, 613.
61. Wood, 135–6.
62. Feibel, Brown, and McDougall, 611, 613.
63. See chapter 15 and the appendix.
64. Feibel, Brown, and McDougall, 613.
65. Mark Ridley, "Who Doubts Evolution?" *New Scientist* 25 (June 1981): 831.
66. Ridley, 830.
67. Ridley, 831. Emphasis added.
68. Charles Darwin, *The Origin of Species* (Everyman's Library Edition, London: J. M. Dent and Sons, Ltd., 1967), 292–3.
69. David Pilbeam, "Rearranging Our Family Tree," *Human Nature* (June 1978): 44.
70. David Pilbeam, "Patterns of Hominoid Evolution," *Ancestors: The Hard Evidence*, Eric Delson, ed. (New York: Alan R. Liss, Inc., 1985), 53.
71. Mary Leakey, *Disclosing the Past* (Garden City, New York: Doubleday and Company, 1984), 214.
72. J. S. Jones and S. Rouhani, "How Small Was the Bottleneck?" *Nature* 319 (6 February 1986): 449.
73. Robert Martin, "Man Is Not An Onion," *New Scientist* 4 (August 1977): 285.

Chapter 17. Remember Baby Fae?

1. "Ethics questions linger after Baby Fae's death," *Coloradoan* (Fort Collins, Colo.), November 17, 1984.
2. "Interview," *U. S. News and World Report*, November 16, 1984, 58.
3. Deborah Erickson, "Blood Feud," *Scientific American* (June 1990):17.
4. William L. O'Neill, "An Environmentalist Lineage" (Book Review), *Science* 244 (19 May 1989):835.
5. Erickson, 17.
6. Quoted by O'Neill, 834.
7. Charles S. Nicoll and Sharon M. Russell, "Animal Rights Literature," *Science* 244 (26 May 1989): 903.
8. Bartell Nyberg, *Denver Post*, December 6, 1987.
9. Erickson, 17.
10. Francis Darwin, ed., *The Life and Letters of Charles Darwin*, 2 vols. (New York: D. Appleton and Co., 1919).
11. Ernst Mayr, "Evolution and God" (Book Review), *Nature* 248 (22 March 1974): 285.
12. Neal C. Gillespie, *Charles Darwin and the Problem of Creation* (Chicago: University of Chicago Press, 1979), xi.
13. Mayr, 285.
14. Thomas Huxley, "On the Reception of 'the Origin of Species'," *The Life and Letters of Charles Darwin*, Francis Darwin, ed. (New York: D. Appleton & Co., 1919) 1:556. Emphasis mine.
15. William Paton, "Is the reductionist beyond belief?" (Book Review), *Nature* 324 (11 December 1986): 522.
16. Donald K. McKim, "Doctrine of Creation," *Evangelical Dictionary of Theology*, Walter A. Elwell, ed. (Grand Rapids: Baker Book House, 1984), 281.

17. David L. Hull, "The God of the Galapagos" (Book Review), *Nature* 352 (8 August 1991): 486.
18. John T. Robinson, Creation/Evolution debate, University of Wisconsin, Madison, February 10, 1978.
19. George Gaylord Simpson, *Life of the Past* (New Haven: Yale University Press, 1953), 155. Bracketed material added for clarity.
20. James Rachels, *Created from Animals: The Moral Implications of Darwinism* (New York: Oxford University Press, 1990), 197–98.
21. Rachels, 199–223.
22. Rachels, 194.

Chapter 18. Is the Big Bang a Big Bust?

1. Douglas LeBlanc, "Worship at the Relativist Shrine" (Book Review), *Christianity Today* 35:14 (25 November 1991), 35. Bracketed material added for clarity.
2. Sir Arthur Eddington, *The Expanding Universe* (Ann Arbor: University of Michigan Press, 1958), 17. Emphasis his.
3. Sir Bernard Lovell, lecture at Schoolcraft College, Livonia, Michigan, October 12, 1971. Recorded by the author. Emphasis added. Bracketed material added for clarity.
4. Robert Wagoner, lecture on cosmology, annual meeting of the American Association for the Advancement of Science, Boston, Massachusetts, February 18–24, 1976. AAAS cassette tapes 76T, No. 240B, Washington, D.C., 1976. Bracketed material added for clarity.
5. Ronald F. Youngblood, *The Book of Genesis*, 2nd ed. (Grand Rapids: Baker Book House, 1991), 42.
6. Raphael G. Kazmann, "It's about time: 4.5 billion years," *Geotimes*, September 1978, 18.
7. Kazmann, 18. Emphasis and bracketed material added. A careful study of the recording of the session confirms that Kazmann has quoted John Eddy accurately.
8. Douglas J. Futuyma, *Science on Trial* (New York: Pantheon Books, 1983).
9. Roger K. Ulrich, "Solar Neutrinos and Variations in the Solar Luminosity," *Science* 190 (14 November 1975): 619.
10. J. W. Bieber, D. Seckel, Todor Stanev, and G. Steigman, "Variation of the solar neutrino flux with the Sun's activity," *Nature* 348 (29 November 1990): 407.
11. Bieber et al., 407. Emphasis added.
12. Jessie L. Greenstein, "Stellar astronomy," *1972 Britannica Yearbook of Science and the Future* (Chicago: William Benton, Publishers, 1971), 190. Emphasis added.
13. Lovell, Schoolcraft College lecture.
14. Faye Flam, "In Search of a New Cosmic Blueprint," *Science* 254 (22 November 1991): 1106.
15. William K. Hartmann, *Moons and Planets: An Introduction to Planetary Science* (Belmont, California: Wadsworth Publishing Company, Inc., 1973), 6–7. Bracketed material added for clarity. Dingle was referring primarily to a unique form of creationism proposed one hundred years ago by Philip Gosse. I have omitted the portion of the quotation that would apply specifically to Gosse's view, and have quoted only that which would apply to any literal view of Genesis. By the use of this quotation, I do not wish to imply that either Dingle or Hartmann accept the Genesis cosmology.

Chapter 19. Genesis: The Footnotes of Moses

1. See John Korgan, "Free Radical: a word (or two) about linguist Noam Chomsky," *Scientific American* (May 1990):40–44.
2. P.J. Wiseman, *Ancient Records and the Structure of Genesis*, ed. Donald J. Wiseman (Nashville: Thomas Nelson Publisher, 1985).
3. R. K. Harrison, *Introduction to the Old Testament* (Grand Rapids: Wm.B.Eerdmans Publishing Co., 1969), 542–53.
4. Derek Kidner, *Genesis* (Downers Grove, Illinois: InterVarsity Press, 1967), 80.
5. Kidner, 24.
6. Some recent evangelical writers question that the *TOLeDOT* phrases are colophons. However, their objections are not weighty and can be adequately explained. See Kidner

23–24; Allen P. Ross, *Creation and Blessing* (Grand Rapids: Wm. B. Eerdmans Publishing Company, 1988), 69–74; and Duane Garrett, *Rethinking Genesis* (Grand Rapids: Baker Book House, 1991), 94–96.

7. There is some question as to just when Hebrew came into common use as a language. There is abundant reason to believe that it was in common use in Moses' day.
8. For examples of colophons in *Enuma Elish*, a Babylonian creation account, see Alexander Heidel, *The Babylonian Genesis*, 2nd ed.(Chicago: University of Chicago Press, 1951), 25–45.

Chapter 20. Adam and the Evangelical

1. Bernard Ramm, *The Christian View of Science and Scripture* (Grand Rapids: Wm. B. Eerdmans Publishing Company, 1954).
2. See Marvin L. Lubenow, "Progressive Creationism," a paper presented to the Evangelical Theological Society (Midwest Section), Trinity Evangelical Divinity School, March 1975.
3. Ramm, 215.
4. Terence E. Fretheim, "Were the Days of Creation Twenty-Four Hours Long? Yes," *The Genesis Debate*, ed. Ronald F. Youngblood (Grand Rapids: Baker Book House, 1990), 19–20.
5. Ramm, 229.
6. Ramm, 222. Emphasis his.
7. Ramm, 321–322. Bracketed material added for clarity.
8. Ramm, 215.
9. Ramm, 315.
10. Ramm, 272.
11. Carl F. H. Henry, *God, Revelation and Authority*, 6 vols. (Waco, Texas: Word Books, 1983) 6:205. Emphasis added.
12. Robert C. Newman and Herman J. Eckelmann, Jr., *Genesis One and the Origin of the Earth* (Downers Grove, Ill.: InterVarsity Press, 1977), 105–123.
13. William Henry Green, "Primeval Chronology," quoted in Newman and Eckelmann, 123. Emphasis added.
14. Ramm, 325.
15. Newman and Eckelmann, 60. Bracketed material added for clarity.
16. Davis A. Young, *Creation and the Flood* (Grand Rapids: Baker Book House, 1977).
17. Young, 114.
18. Young, 127, 131, 132.
19. Richard K. Bambach, "Responses to Creationism" (Book Reviews), *Science* 220 (20 May 1983): 853.
20. Young, 152.
21. Young 153, 155.
22. Pattle P. T. Pun, *Evolution: Nature and Scripture in Conflict?* (Grand Rapids: Zondervan Publishing House, 1982).
23. Pun, 260.
24. Pun, 266.
25. Cassette recording of radio interview with Pattle Pun, "Open Line" broadcast on October 11, 1989, Moody Broadcasting Network, Chicago.
26. James O. Buswell III, "Creationist Views on Human Origin," *Christianity Today*, 8 August 1975, 5.
27. Buswell, 4.
28. Cassette recording of a lecture by Davis A. Young at the science symposium, "The Harmonization of Scripture and Science," Wheaton College, Wheaton, Illinois, March 23, 1990.
29. Ronald F. Youngblood, *The Book of Genesis: An Introductory Commentary*, 2nd ed. (Grand Rapids: Baker Book House, 1991), 48.
30. Youngblood, 47.

31. See Clifford Wilson and Donald McKeon, *The Language Gap* (Grand Rapids: Zondervan Publishing House, 1984), and Lauren J. Harris, "Uniquely Human" (Book Review), *Science* 254 (11 October 1991): 313–314.
32. Gleason L. Archer, Jr., *A Survey of Old Testament Introduction* (Chicago: Moody Press, 1964), 187.
33. Archer, 188.
34. Archer, 189.
35. John R. W. Stott, *Understanding the Bible* (Glendale, Calif: Regal Books, 1972), 63.
36. See John Calvin, *Institutes of the Christian Religion*, 2 vols., trans. Ford Lewis Battles, ed. John T. McNeill (Philadelphia: The Westminster Press, 1960) 1:188–189.
37. See John H. Gerstner, *The Rational Biblical Theology of Jonathan Edwards*, 3 vols. (Orlando, Florida: Ligonier Ministries, 1991) 1:592.
38. Derek Kidner, *Genesis* (Downers Grove, Illinois: InterVarsity Press, 1967), 48.
39. Kidner, 28.
40. Kidner, 29.
41. Kidner, 30.
42. John H. Walton, "Genesis 1–3: A New Look at Human Origins," a paper presented at the 42nd annual meeting, The Evangelical Theological Society, New Orleans, November 15, 1990.
43. Gordon R. Lewis and Bruce A. Demarest, *Integrative Theology*, 3 vols. (Grand Rapids: Zondervan Publishing House, 1990) 2:46.
44. Gale E. Christianson, *In the Presence of the Creator* (New York: The Free Press, 1984), 41. Emphasis his.
45. See Stephen Jay Gould, "Fall in the House of Ussher," *Natural History* (November 1991): 16.

Appendix

1. F. J. Fitch and J. A. Miller, "Radioisotopic Age Determinations of Lake Rudolf Artifact Site," *Nature* 226 (18 April 1970): 226–28.
2. Fitch and Miller, 226.
3. Fitch and Miller, 226.
4. Fitch and Miller, 226.
5. Fitch and Miller, 228.
6. *Detroit Free Press*, November 10, 1972.
7. R. E. F. Leakey, "Evidence for an Advanced Plio-Pleistocene Hominid from East Rudolf, Kenya," *Nature* 242 (13 April 1973): 447.
8. Vincent J. Maglio, "Vertebrate Faunas and Chronology of Hominid-bearing Sediments East of Lake Rudolf, Kenya," *Nature* 239 (13 October 1972): 379–85.
9. A. Brock and G. Ll. Isaac, "Paleomagnetic stratigraphy and chronology of hominid-bearing sediments east of Lake Rudolf, Kenya," *Nature* 247 (8 February 1974): 344–48.
10. Brock and Isaac, 347.
11. Brock and Isaac, 344.
12. Brock and Isaac, 347–48.
13. Brock and Isaac, 346.
14. Anthony J. Hurford, "Fission track dating of a vitric tuff from East Rudolf, North Kenya," *Nature* 249 (17 May 1974): 236.
15. Frank J. Fitch, Ian C. Findlater, Ronald T. Watkins, and J. A. Miller, "Dating of a rock succession containing fossil hominids at East Rudolf, Kenya," *Nature* 251 (20 September 1974): 214.
16. Richard E. Leakey, "Skull 1470," *National Geographic*, June 1973: 819.
17. G. H. Curtis, Drake, T. Cerling and Hampel, "Age of KBS Tuff in Koobi Fora Formation, East Rudolf, Kenya," *Nature* 258 (4 December 1975): 395.
18. Curtis et al., 398.
19. Curtis et al., 396.
20. F. J. Fitch, P. J. Hooker, and J. A. Miller, "^{40}Ar/^{39}Ar dating of the KBS Tuff in Koobi Fora Formation, East Rudolf, Kenya," *Nature* 263 (28 October 1976): 741.

21. Fitch, Hooker, and Miller, 742.
22. Fitch, Hooker, and Miller, 740.
23. Anthony J. Hurford, A. J. W. Gleadow, and C. W. Naeser, "Fission-track dating of pumice from the KBS Tuff, East Rudolf, Kenya," *Nature* 263 (28 October 1976): 738–39.
24. Hurford, Gleadow, and Naeser, 739.
25. G. A. Wagner, letter, *Nature* 267 (16 June 1977): 649.
26. Naeser, Hurford, and Gleadow, letter, *Nature* 267 (16 June 1977): 649.
27. Hurford, Gleadow, and Naeser, 739.
28. J. W. Hillhouse, J. W. M. Ndombi, A. Cox, and A. Brock, "Additional results on palaeomagnetic stratigraphy of the Koobi Fora Formation, east of Lake Turkana (Lake Rudolf), Kenya," *Nature* 265 (3 February 1977): 411.
29. Hillhouse et al., 414.
30. Donald C. Johanson and Maitland A. Edey, *Lucy: The Beginnings of Humankind* (New York: Simon and Schuster, 1981), 240. Bracketed material added for clarity.
31. H. B. S. Cooke, "Suid Evolution and Correlation of African Hominid Localities: An Alternative Taxonomy," *Science* 201 (4 August 1978): 460–63.
32. T. D. White and J. M. Harris, "Suid Evolution and Correlation of African Hominid Localities," *Science* 198 (7 October 1977): 13–21.
33. Cooke, 460.
34. A. J. W. Gleadow, "Fission track age of the KBS Tuff and associated hominid remains in northern Kenya," *Nature* 284 (20 March 1980): 229.
35. Gleadow, 230.
36. Ian McDougall, Robyn Maier, P. Sutherland-Hawkes, and A. J. W. Gleadow, "K-Ar age estimate for the KBS Tuff, East Turkana, Kenya," *Nature* 284 (20 March 1980): 230–31.
37. McDougall et al., 231.
38. McDougall et al., 232. Emphasis added.
39. Ian McDougall, "$^{40}Ar/^{39}Ar$ age spectra from the KBS Tuff, Koobi Fora Formation," *Nature* 294 (12 November 1981): 123.
40. McDougall, 124.
41. Naeser, Hurford, and Gleadow, letter, *Nature* 267 (16 June 1977): 649.
42. Chapters 9 and 10 of Roger Lewin's *Bones of Contention* contain his account of the ten-year history of dating the KBS Tuff. Since my account was written independently of his, it would be an enlightening experience to read his account also. By omitting many of the details that I have included, he is able to make the affair a graphic victory for the dating methods. Accounts like his explain why many people continue to put almost unlimited faith in the dating methods.

Index of Persons

288

Index of Fossils

Index of Topics

LIFE SPACE INTERVENTION

THE AMERICAN UNIVERSITY
WASHINGTON, D.C.

TO: Eric – For all the good times,
work times, and caring times
we spent together –

Nicholas J. Long

4400 Massachusetts Avenue, N.W., Washington, D.C. 20016
return postage guaranteed

We dedicate this book to Norman Wood and Jody Long, who shared our Life Space during this project by being supportive, nurturing, loving, and just plain fun.

LIFE SPACE INTERVENTION

*Talking with
Children and Youth
in Crisis*

Mary M. Wood

Nicholas J. Long

pro·ed

8700 Shoal Creek Boulevard
Austin, Texas 78758

Printed in the United States of America

Library of Congress Cataloging-in-Publication Data

Wood, Mary M.
   Life space intervention : talking with children and youth in
crisis / Mary M. Wood, Nicholas J. Long.
     p.   cm.
   Includes bibliographical references.
   ISBN 0-89079-245-3
   1. Personnel service in education. 2. Crisis intervention
(Psychiatry) 3. Oral communication. I. Long, Nicholas James,
1929–   . II. Title.
LB1027.5.W59 1990
371.4'6—dc20   90-32277
                                        CIP

pro·ed

8700 Shoal Creek Boulevard
Austin, Texas 78758

10 9 8 7 6 5 4 3 2 1   90 91 92 93 94

Contents

PART TWO: ANATOMY OF LIFE SPACE INTERVENTION

PART THREE: APPLYING LSI TO PARTICULAR TYPES OF PROBLEMS

Example of LSI Used to Build Values
That Strengthen Self-Control: "I
Don't Want Them to Think I'm
Weird or Anything"
What We Hope You Have Learned
from This Chapter

Example of an Abbreviated
     LSI: "Saga of the Yellow Towel"
Summary
What We Hope You Have Learned
     from This Chapter

Preface

Life Space Intervention (LSI) is about talking with children and youth who are in crisis. It is a process that can be used in almost any situation or location because it requires no props or equipment, only a skilled and understanding adult. Yet obtaining that skill is not easy, and the difficulties of helping today's troubled students are enormous. Skilled verbal strategies are essential requirements for adults in helping roles. Every crisis *requires talk*! An adult's skills in using verbal strategies will directly influence both the immediate solution to a crisis and the long-term effect on the student. Crisis handled well can make positive, long-lasting changes; crisis managed ineptly will contribute to a devastating cycle of alienation, hostility, and aggression.

The decay and dysfunction of the family and the shocking social problems in communities have created a new level of deviancy and disturbance never before seen by educators and other adults who work with children and youth. We are involved daily with students who come to school struggling with painful reality—problems such as alcoholism, drugs, suicide, gang warfare, rape, physical and psychological abuse, crime, parental neglect and abandonment, brutality as entertainment, and poverty. Violence at home and in the neighborhood is a common experience for many students today.

William Raspberry, a feature writer of the *Washington Post*, wrote a commentary about the frequency of violence in the lives of New York City youth. He quoted from his interview with Deborah Meier, principal of Central Park East Secondary School, about students overwhelmed, suspicious, and numbed to violence with three brutal rapes, a fire, and a murder in a housing project near her school.

"The thing that is so hard for outsiders to understand," she said, "is the utter routineness of violence in the lives of so many of the children."

"Violence is normal in the lives of today's adolescent," she reported. "Even worse, it is glamorous and appealing. In advisory meetings, where people are frank and open, the boys will acknowledge that their ideal of manliness exudes violence . . . to be a man is to sneer in the face of the weak. To let your guard down is an invitation to danger or cruel jests, at the very least. Weakness [is] equated with sissiness. To be a thoughtful person is to invite a rep for being a homosexual."

"Middle-class kids often see this conforming cruelty as a temporary necessity of adolescence, whereas working class and poor kids seem more prone to the view that this is the way the world is." (Raspberry, 1989, p. 4a)

What Meier has described is just the tip of the iceberg when we take an in-depth look, nationwide, at the stresses our students face in schools, at home, and in their neighborhoods. These problems will not disappear by themselves. Teachers, counselors, administrators, social workers, juvenile correction workers, and special educators are demanding more sophisticated training in crisis intervention skills to help these troubled students. Daily we have witnessed this critical need for advanced skills in crisis intervention.

This book is our response. We believe in the power and effectiveness of LSI as a crisis intervention strategy with *long-term benefits* in the lives and personalities of students who participate. More than 50 combined years of work with disturbed children and youth lead us to the conviction that it is important and possible to teach effective LSI skills to others. LSI is not easy to learn, since it involves adults in the complex, often irrational, defensive, disorganized, and, at times, fantasy world of troubled students. Crisis often occurs in chaotic or destructive social settings, always involves others, and taps emotions of the student, the group, and the adult. Once an adult learns to see a crisis through the eyes of the student, greater empathy, support, realistic problem solving, and behavioral self-control can occur. When an effective LSI is done, a crisis situation that could otherwise end up as a destructive and deprecating experience for the student instead becomes an instructional and insightful experience. This is what LSI is all about.

The "talking" strategies we describe are based on in-depth clinical interviewing skills developed from Fritz Redl's (1959a) concept about Life Space Interviewing. Redl described the LSI process as "a mediating role between the child and what life holds for him" (1966, p. 40). The intent is to convey the adult as mediator among the stress, the student's behavior, the reactions of others, and the private world of feelings that students are sometimes unable to handle without help. This remains an accurate way to describe the expanded uses of LSI—as a mediating process. The substance of the original LSI process has not been changed, only amplified. For a full description of the development of LSI, we recommend the classic works by Redl and Wineman, *Children Who Hate* (Redl & Wineman, 1951), *Controls from Within* (Redl & Wineman, 1952), and *When We Deal with Children* (Redl, 1966). A monograph edited by Ruth Newman and Marjorie Keith (1964), *The School-Centered Life Space Interview*, offers rich reading about early applications of LSI in school settings. We include a history of the extensive field validation of LSI in Appendix A.

William Morse (1981) describes LSI as "a living, action process" with a direct connection to a student's past experience:

> One can and should know all of the personal and situational antecedents possible. But it [LSI] has a life of its own that is not constricted by case histories. It is a slice of life action. Rather than past oriented, it is future oriented about resolutions. One begins to develop the structure of the youngster's self-concept from the current behavior—the emotional state, distortions, attributions, expectations, values, and hopes for the future. Of course, one includes that part of the past which has present currency. But one is free of the dominance that results from looking backwards. If one knows children and knows disturbance, one comes to each situation with a vast [information base] . . . and it becomes easier to associate the relevant past with the present. (p. 70)

In this book, the term *interview* has been changed to *intervention* to emphasize that crisis *always evokes verbal intervention*. The quality of an adult's verbal intervention is the key to success or failure in obtaining a therapeutic outcome of a crisis. When first used outside of clinical settings, the term *interview* was sometimes misinterpreted to mean questioning a student to extract information about an incident and resulting behavior. To some, the term seemed to suggest interrogating students in hopes of getting "confessions" of rule violations or admissions of wrongdoing. The term also was misinterpreted to be something done one time in response to a specific problem. None of these interpretations is accurate. Such limited approaches do not produce positive, lasting behavior change. As an alternative, we have substituted the term *intervention* to emphasize the dynamic nature of the interactions between adult and student at the time of crisis. Talk happens; it is a form of intervention, and it can be therapeutic if skillfully done.

In this book, we hope we have captured the excitement and complexity of LSI and have conveyed the significance of the contributions made by Fritz Redl, David Wineman, William Morse, and Ruth Newman. We also want to thank our colleagues and friends who have encouraged us to bring LSI forward into the 21st century.

PREPARING TO DEAL WITH STRESS AND CRISIS 1

Because you have chosen to read this book, we assume that you are working in some capacity with children and youth. And we also know that to work with them in today's complex world is the toughest of challenges. There are plenty of problems! How to guide young people through stressful experiences; how to help them make sense out of the ineptness and meanness they see around them; how to make the most of themselves; how to help them view situations in which they have messed up: These are our concerns. Adults who can do these things have something significant to contribute. We focus on Life Space Intervention (LSI) in this book because it is a way to accomplish these goals. We use LSI every day in our work with troubled children and youth because we believe we have responsibility to teach them to deal constructively with stress and crisis. This book is our dialogue with you about how we do it.

In Part One we introduce LSI as a verbal intervention strategy designed especially to use in crisis. We describe the Conflict Cycle as a way of understanding the elements that create crisis; outline a procedure for identifying feelings and anxieties that often are expressed in students' inappropriate behavior; and discuss what you need to understand about yourself and your student as you begin every LSI.

"I hate school!"
"I hate reading!"
"I hate Teddy!"
"I hate you!"

CRISIS IS A TIME FOR LEARNING

When students tell you things like this, the next move is yours. You are a part of their daily life experience, their "life space." All who work with children and youth are confronted frequently with such incidents that may require adult intervention. Sometimes incidents may seem trivial or insignificant on the surface; other times, they are of such magnitude as to be overwhelming, with no apparent solution. In any event, what is needed is a way to help students assume responsibility for changing their own dissocial behavior in ways that produce constructive and long-lasting results.

Students seldom assume responsibility for changing their own behavior (as opposed to relying on outside authority and control for behavior change) until they are psychologically empowered to make choices about their behavioral alternatives and are ready to accept the consequences of these choices. But how do they acquire this empowerment to regulate their own behavior? Self-regulation emerges from *understanding* of people and events in their environment, *motivation* to change unpleasant conditions, and *trust* in adults. These three essential dimensions of self-regulation can be described in operational terms this way:

UNDERSTANDING AN EVENT

- Acknowledgment of the part personal behavior and feelings contribute;

- Awareness of reactions of others;

- Social perception about the sequence of consequences that follow; and

- Recognition of alternatives that can modify a chain of events.

MOTIVATION TO CHANGE

- Desire to improve existing conditions;
- Belief that change for the better is possible;
- Sufficient self-esteem to believe that improvement is deserved; and
- Confidence to try something different.

TRUST IN ADULTS

- Confidence in adults' respect for students' feelings;
- Conviction that adults value students;
- Belief that adults recognize students' attributes;
- Belief that adults use authority and power wisely;
- Confidence that adults can solve problems in satisfactory ways; and
- Willingness to accept adult guidance.

Adults who work with students know that when these conditions are present students have a working process for problem solving and are regulating their own behavior. These qualities exist because of many successful interactions between student and adults, around both ordinary and extraordinary events—some pleasant and some painful. Talking together, adults and students build these conditions.

But most troubled students do not have these qualities. Their perceptions of events are often shortsighted or distorted. They frequently fail to understand that their behavior upsets others. Their feelings take over and flood their rational minds. They make matters worse by behaving in impulsive, defensive, or destructive ways. Most students also are bombarded with daily reality full of frustrations and disappointments that warrant supportive adult intervention.

Guiding children and youth in this endeavor is an awesome responsibility. Their reactions to stress and its by-product, crisis, are among the most difficult challenges confronting teachers, child care providers, youth workers, and parents. This is the topic of the entire book: talking with students in crisis. The goal is self-regulated, value-based behavior. Here we offer a way of talking with students to achieve this goal. It is a process that teaches students to solve a crisis by understanding the behavioral reactions and feelings of those involved, including themselves. Talking creates conditions where motivation to change exists and builds a student's trust in adults sufficiently to accept guidance. The more skilled we are in talking with students, the more effective we will be in helping them learn to manage their own problems. With understanding and sensitivity to students' feelings and the circumstances surrounding an incident, we can provide the emotional support they need while we teach them to problem-solve more effectively and with more satisfying results. If we are effective in this task, students will develop beyond the need to rely on adult authority for behavior control. They will have learned to manage their own behavior, control their impulses and feelings, live by rules and values, make constructive decisions, and deal with others in positive ways.

LSI: A PROCESS FOR TALKING WITH STUDENTS IN CRISIS

Life Space Intervention (LSI) is a therapeutic, verbal strategy for intervention with students in crisis. It is conducted at the time the crisis occurs. The process uses students' reactions to stressful incidents to (a) change behavior, (b) enhance self-esteem, (c) reduce anxiety, and (d) expand understanding and insight into their own (and others') behavior and feelings. LSI can be used with children and youth of any age in situations where reaction to stress is a concern, and with students who are unable to control or manage their own behavior appropriately.

LSI focuses on crisis that occurs when an incident escalates into conflict between a student and others. Because such crisis involves a student's immediate life experience (the "life space"), it is an optimal time for learning. Students are intensely involved in situations that hold personal significance or have disrupted their sense of well-being. Adults who work with students in crisis need to understand the conflict from the student's point of view, while also promoting the student's active choice and personal responsibility for behavior. It has been our experience that too many students in crisis defend themselves against their feelings by denying them, displacing them on others, blaming others, regressing, or rationalizing. One of the most important steps in helping troubled students become more realistic and responsible for their behavior is to help them understand the feelings that drive their behavior.

It is helpful for students to see the connections between their feelings and their behavior. Students who show emotional and social maturity can say yes to their feelings but no to improper behavior to express the feelings. This basic concept of supporting the feelings but not the behavior frequently is lost as adults struggle to guide and teach students in crisis. For LSI this is a fundamental concept, as is commitment to promote self-awareness and behavioral change. Unlike psychotherapy, LSI is not an open and permissive process; yet unlike strict behavioral programs, LSI does not deny the power of personal feelings.

In a comprehensive behavior management program, the priority goal is to teach students prosocial behaviors that enable them to function appropriately. This does not negate the need for adults to handle crisis skillfully. A fire department offers an analogy for LSI as a strategy for fighting the fires of crisis. While a fire department states its first priority to be the prevention of fire, this priority does not prevent it from having sophisticated fire-fighting skills. Similarly, LSI is a strategy that is used to put out emotional fires that flare up and ignite into disruptive and destructive behavior. This is the primary purpose of LSI—crisis intervention. It should be used as the strategy of choice only after strategies such as positive talk, proximity control, bonus points, appeal to rules and values, and stated behavioral consequences have been tried. In this continuum, LSI is compatible with behavior modification and social learning theory; it is part of a continuum of behavior management strategies.

Occasionally students may resist any attempt through LSI to alter personal views of a problem or their behavior. In such situations, the adult has responsibility to reflect reality and consequences. When inappropriate behavior continues, the adult must set up a specific intervention plan based on rules and behavioral contingencies. In these situations, LSI moves from interpersonal acceptance of feelings to planned strategies for change with clear reality consequences and no excuses accepted for making poor behavioral choices.

Originally formulated as the "Life Space Interview" more than three decades ago by Fritz Redl (1959a, 1966; Redl & Wineman, 1952), LSI was initially limited in use to those practitioners who worked in clinical settings with children and youth who had extremely serious social, emotional, or behavioral problems. As discussed in the preface, the term *interview* has been changed to more accurately convey a broader application for those who do not work in clinical settings. When first applied in more natural environments, the interview was sometimes misinterpreted to mean questioning a student to extract information about the incident and resulting behavior. To others, *interview* seemed to suggest interrogating students in hopes of getting confessions of rule violations or admissions of wrongdoing. The term also was misinterpreted to be something done once in response to a specific problem. None of these interpretations is true. Such limited approaches do not produce positive, lasting changes.

As an alternative, we have substituted the term *intervention* to denote "mediation for problem solving." The intent is to convey the role of adult as mediator among the stress, the student's actions, the reactions of others, and the private world of feelings that students seem to be unable to handle without help. This idea of a mediator was used originally by Redl to describe the intent of the LSI as "a mediating role between the child and what life holds for him" (Redl, 1966, p. 40).

This remains an accurate way to describe the expanded uses of LSI. The substance of the original LSI process has not been changed. For a full description of the original process in developing LSI, we recommend Redl's classic work on the subject, *When We Deal with Children* (1966), and the original publication, *Controls from Within* by Redl and Wineman (1952). We also have included a brief history of LSI in Appendix A.

LSI currently is used in educational programs, correctional facilities, and alternative living homes—wherever the paramount concern is to teach troubled children and youth better ways to cope with the social, emotional, and behavioral crises of life. LSI is effective in these many different programs because it offers a way to provide emotional support on the spot. It provides a process for understanding behavior and feelings, and it outlines a procedure for adults to teach children and youth better ways to cope with stress, to change behavior, and to resolve conflict.

WHAT HAPPENS DURING LSI

As LSI begins, there is an exploration of the student's understanding of the event. LSI then expands to feelings that evoked the behaviors and the reactions of others to those behaviors. As the incident is clarified and expanded, the central issue is formed. During the process, the focus often shifts from the incident to a deeper, more serious, underlying concern not easily or directly expressed by the student. At this point, the process takes a turn toward problem solving and away from problem exploration. Together, adult and student explore ways to ameliorate both the immediate incident and associated long-term problems. Behavioral alternatives are selected to resolve the present crisis and to achieve better outcomes when stress occurs again in the future.

Figure 1.1 summarizes the six steps of LSI. We illustrate them in a V shape. Going down the V, Steps 1 through 3 focus on the problem. This brings you to the bottom of the V in the illustration. Steps 4 through 6 are concerned with the solution. These last three steps put you on the upward side of the V as you and the student focus on outcomes, consider solutions, and form a plan of action. This visual format seems to help some people recall the LSI framework during crisis. (Remember: Three steps for the problem, and 3 steps for the solution.)

This current version of LSI is presented in detail, step-by-step, in Part Two of this book. Here we simply summarize the objectives and content of each step in the sequence. Keep in mind the distinct purpose and content of each step. When linked together, these steps build a communication bridge to a fuller understanding of the student's distress, the resulting crisis, and alternative solutions that require behaviors in keeping with the student's ability to cope successfully.

Figure 1.1 Summary of Steps in Life Space Intervention

THE INCIDENT

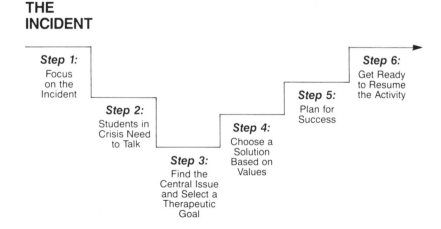

Step 1: Focus on the Incident

PURPOSE: To convey support and understanding of the student's stress and to start the student talking about the incident.

CONTENT: The incident itself—the event that actually brought about the need for LSI—is identified.

Step 2: Students in Crisis Need to Talk

PURPOSE: To talk in sufficient detail to clarify and expand understanding about the reality components of the incident, and to decrease student's emotional intensity while increasing reliance on rational words and ideas.

CONTENT: A sequence of events, a time line, is established to obtain details of the student's view of the incident, the associated stress, and personal involvement.

Step 3: Find the Central Issue and Select a Therapeutic Goal

PURPOSE: To explore the student's perception of the incident and associated feelings and anxieties until you have sufficient understanding to concisely state the central issue and decide what the therapeutic goal should be.

CONTENT: Determine the extent to which the student's behavior is driven by feelings and anxiety, the depth and spread of this conflict, the amount of rational control the student can exercise over these emotions, and what the long-term and short-term outcomes should be for the student as a result of this specific LSI.

Step 4: Choose a Solution Based on Values

PURPOSE: To select a solution that the student values as beneficial and claims with a sense of genuine ownership. If a student is not able to do this, the adult chooses a solution that establishes group values and reality consequences that will work in the student's behalf.

CONTENT: The solution is selected from several alternatives, representing the student's own changing insights and beliefs about what constitutes a satisfactory solution, considering subsequent consequences. When a student denies responsibility or cannot choose, the adult structures the solution for the student around group values and social norms that are within the student's capacity to use successfully.

Step 5: Plan for Success

PURPOSE: To rehearse what will happen and anticipate reactions and feelings (of self and others) when the chosen solution is actually put into action.

CONTENT: Selected behaviors are specifically practiced as rehearsal for reacting and problem solving successfully when the student faces the consequences of the original incident and when a similar problem may occur in the future.

Step 6: Get Ready to Resume Activity

PURPOSE: To plan for the student's transition back into the group's ongoing activity, and to close down private topics or feelings that may have surfaced during the talk.

CONTENT: The adult shifts the focus to help the student anticipate how to manage reentry into the peer group. If there is to be a short-term consequence to the original incident, the student is prepared as that goes into action. This final step is essential also for closing down emotions and reducing the intensity of the relationship that may have occurred during the LSI between student and adult.

PREREQUISITE SKILLS NEEDED BY STUDENTS

To know if LSI is an appropriate strategy for a particular student, it is necessary to assess the student's readiness skills for successful participation. These skills include five general cognitive and communication processes: (a) attention span for listening and retaining what has been said, (b) minimal verbal skills to use language spontaneously and with sequential thought, (c) sufficient comprehension to understand the meanings of words, (d) mental reasoning to understand the essence of the incident and the problem it produced, and (e) trust in the adult. These basic prerequisites are detailed in Table 1.1.

We use this list to do an informal assessment of each student *before* there is a crisis. In this way, we can screen those for whom LSI is not recommended, and we make a notation in these students' individual educational or treatment plans to that effect. When we identify students who have some, but not all, of these prerequisite skills, we note this also and recommend that an abbreviated LSI form be used. This abbreviated form of LSI, for young children and those who are developmentally delayed, is described in Chapter 12. When a student is severely incapacitated by drugs or psychosis, the abbreviated LSI form also may be useful because it focuses on teaching needed communication and cognitive skills which often are impaired in such students. As these skills gradually develop or improve, these students will be able to participate in a full LSI.

OUTCOMES OF EFFECTIVE LSIs

Both short-term and long-term gains can be expected from an effective LSI. The specific program objectives established before the crisis should guide the problem-solving steps of LSI. Every program has its own goals and procedure for developing individual students' objectives. Overall, long-range program goals and specific objectives for small, incremental gains reflect individual students' needs and circumstances. Incorporating these program objectives and goals into LSI is essential.

In addition to specific individual program objectives and goals, every LSI must have a responsible and satisfactory resolution of the incident. Any number of other short-term outcomes may occur also. Here are several examples:

Table 1.1 Prerequisite Skills Needed by Students

- Some degree of awareness of self, events, and other people.
- Ability to attend to verbal, interpersonal stimuli (the adult speaking).
- Sufficient self-regulation of body and body functions to sustain attention.
- Sufficient understanding (receptive vocabulary) to comprehend the words used by the adult.
- Ability to comprehend the stream of content connecting the adult's words.
- Sufficient memory to recall a simple sequence of events.
- Ability to produce words or signs sufficiently complex to represent the crisis event.
- Sufficient trust that the adult really cares to cooperate, seek, or respond positively to the adult.
- Willingness and ability to share minimal information with the adult.
- Ability to describe simple characteristics of self and others.
- Ability to describe personal experiences, even if in a simple form.
- Ability to give simple reasons why events occur.

TOMMY

For Tommy, one immediate outcome of LSI was his first attempt to use words to solve a problem instead of having a temper tantrum and using his fists to lash out at anyone standing in his way.

Short-term outcome: Tommy tells another student, "I don't like what you're doing. I want you to stop."

JOAN

For Joan, LSI created new awareness that her verbal attacks on friends created enemies instead of the effect she wanted: to have friends admire her and seek her out.

Short-term outcome: Joan's plan of action is to find something to say to compliment her friend, for example, "I think she's pretty smart. I'll tell her I want to study spelling with her."

RICK   For Rick, LSI provided enough emotional sup-
              port to assume responsibility for his own role in
              vandalizing school property.

Short-term outcome: Rick plans a way to work off most of the repairs: "I'm facing the music."

Gradually, over a series of effective LSIs, long-term goals will be evident. A student should gain increased understanding of relationships among behavior, feelings, and the reactions of others. But understanding is not enough. Students also must gain more successful problem-solving behavior for coping with stress, emotions, and crisis. As these two long-term outcomes become evident, there will be a third outcome: emergence of greatly improved self-esteem. Here is an example:

TONY   Tony was skilled at walking into a group and set-
              ting someone off. He always managed to get
              another student in trouble within a matter of
              minutes. After a series of LSIs, Tony began to
              understand that he got a great sense of power
              from doing this, but it was always ending with
              Tony in trouble. He and his teacher began talk-
              ing about other ways he could behave that would
              bring better results and still give him a sense of
              power and pride. Tony tried out a few of these
              behavioral alternatives, rather reluctantly at first.
              Gradually, he saw for himself that other students
              were beginning look to him for leadership rather
              than in fear. They also sought his friendship and
              advice, calling him "Doc, the Answer Man." This
              clear change in the reactions of others grad-
              ually changed Tony's view of himself from
              "destroyer" to "magnet."

An effective LSI is a complex interpersonal process built on relationship and trust. It constructs bridges of understanding between a student's obvious behavior, words, and body language and underlying feelings, concerns, and anxieties. Positive change in self-esteem is the end result, in the broadest terms.

ATTRIBUTES OF THOSE WHO CONDUCT SUCCESSFUL LSIs

LSI is conducted by someone who is seen by the student as being in the student's natural environment—parents, teachers, principals, counselors, child care workers, and youth workers. For adults, the LSI process is an adventure in interpersonal relationships—a double struggle to keep ourselves in control (of ourselves) while conveying objectivity, control, and concern about the crisis. "Dispassionate compassion" has been used to describe what is needed from adults using the LSI. It requires us to know and be sensitive to the emotions and private side of the student. Adults doing LSI are bombarded by an enormous amount of emotion (our own and the student's). It is essential to monitor and control our own reactions, feelings, and behavior. It is not a process in which we lack control over our faces, body language, or what we are feeling. Nor is it a process in which adults (who have the answers?) freely give advice about how to handle problems in a better way.

Adults who do successful LSIs seem to hold and convey some fundamental beliefs about students' needs for protection, gratification, relationship, and responsibility—basic immunizations against excessive stress and anxiety.

Protection by adults provides emotional support while enhancing a student's capacity to face stress constructively. Protection gives students confidence that they will not be exposed to embarrassment, ridicule, or failure. When adults provide protection, students are not asked to do what they cannot accomplish successfully. Protection gives assurance that adults will not violate a student's private space—that secret sore spots will not be violated. Protection also assures that adults will maintain an environment where students will not be hurt by others. Protection is a fundamental premise in every LSI.

Gratification is a necessity of life. Somehow, every student must have hope that things will get better and experiences will bring pleasure. Without hope and pleasure life loses meaning and there is little, if any, energy available for learning or for emotional and social growth. A student's storehouse of memories must contain, on balance, a greater amount of positive than negative emotional memories. Effective LSI expands this memory bank and uses a student's need for gratification as motivation for change. It holds out the promise of better things to come . . . through change!

Relationship is connection. Adults and students "connect" for all sorts of reasons, all having to do with their own unique needs. Adults have motivation to guide and teach students, seeing themselves as helpers and advocates of students or as nurturers and care providers. Students connect for protection and gratification; responding to the experience, power, wisdom, knowledge, or skills of adults. They listen to adults when connections have been made. They model adult behavior and attitudes, identifying with an admired adult as someone they want to be like. Adults often see this as guidance that students seek. Possibly true; but even more, an adult's effectiveness with students is in large part built on the quality of the connection—conveyed attitudes of affirmation, respect, care, enhancement, and just plain liking. Effective LSI is built on these dimensions of relationship.

Responsibility is the basis for solving problems and coping successfully with stress. Yet few students in crisis show a sense of responsibility. It requires value-directed behavior, in which a student recognizes the need to self-regulate to assure that what happens is what the student values. Students' capacity for responsibility evolves through the development of a coherent sequence of thoughts and feelings about their obligations to themselves and others. This developmental sequence begins with responsibility to oneself (for basic needs), and gradually expands into values about law and order, fairness and justice, and concern (and obligation to do something) about the needs of others. Effective management of stress, and solutions to crisis by students, always draws on some value. Whatever the extent of a student's sense of responsibility, it is a necessary theme to be introduced, in some form, in every LSI. In the most basic form, it will be responsibility for looking after one's own interests. When adults keep in mind that ultimately a student must assume responsibility, it also helps to restrain the natural impulse to take over and tell or direct a student about how to handle a problem or solve a crisis.

These four basic psychological concepts are conveyed by adults who use LSI successfully. These messages contribute to a climate that can counter the assault of excessive stress, frustrations, and disappointments experienced daily by our students. These are the antidotes to students' feelings of unworthiness and helplessness to control what happens to them. Throughout the book you will see these four attributes applied.

In the next three chapters, we review the basic background we believe to be essential for effective LSI. Then, in Part Two, we go through each LSI step in detail. We describe the specific purposes and objectives for the steps, outline the dynamics of each step in charts, discuss adults' roles and strategies, and give examples of how (and when) to end a step and bridge into the next one. We also include a chapter on self-monitoring and fine-tuning your own proficiency in using LSI.

In Part Three, we provide material for those who have mastered the basic LSI material in the first two parts of the book. We expand applications for those who have used the basic LSI procedures and now want to work with increasingly difficult emotional and behavioral problems. We also provide an adaptation for using LSI with young children and those who are developmentally delayed.

EXAMPLE OF LSI

Before going ahead into the next chapters, study the following LSI example, "The Cap and Gown Crisis." We selected this real-life LSI as a basic illustration; the crisis is neither so complex that it might obscure the LSI process nor so simple that it might mislead you into concluding that LSI is a simple procedure. The crisis is typical of daily life for many students of many ages. This example takes the reader through the six LSI steps in a straightforward sequence. We note when each new step begins, and we also note key decision points in each step.

As you study this LSI example, become familiar with the basic sequence. Also learn the names of the steps. By doing this, you should find it easy to recall the general purpose of each. This structure will help you to keep even the most difficult and complex LSI on track and moving in a constructive direction.

After you are acquainted with the sequence of steps and the purpose of each, read this LSI again looking for the places where the teacher has conveyed those essential concepts of protection, gratification, relationship, and responsibility. You will find some of the examples are explicit; others are implied in the student's body language or conveyed by the supportive remarks the teacher makes to affirm the student throughout the LSI.

We will come back to this example again, several times, to illustrate the background concepts presented in the next three chapters. In this way, we hope to show how effective LSI skills can be learned step-by-step. It is not necessary to know how to apply everything in this book before you begin using LSI. But, gradually, as you use LSI, you will find yourself increasingly insightful and skilled in talking with students in crisis.

"The Cap and Gown Crisis"

The setting is Towne High School, an ordinary high school in a district that allows corporal punishment, administered at the discretion of the principal and with written approval by a parent. (We do not want to debate here the efficacy of this discipline procedure; it is sufficient to acknowledge that it still exists in some school systems and represents a reality for some students to deal with.)

The teacher using LSI has been this student's homeroom teacher for his first year in Towne High. It has been a tough year for teachers and students alike. The English teacher left after 2 weeks, and a series of substitutes and replacements filled in during the year. The principal, Mr. Martin, has spent the last 10 years at Towne High and likes the students. But he is a disciplinarian and has little tolerance for students who behave inappropriately.

Our student (S) has been in trouble so frequently that he is known as one of the troublemakers. His homeroom teacher (T) has counseled him frequently and believes that he is making genuine progress. The last month has produced a real turnaround in his behavior and in his academic work. They are now in the last week of the school year.

STEP 1: FOCUS ON THE INCIDENT	**T :** Mr. Martin tells me you're in big trouble.
	S: (Angrily.) Get out of my face!
	T: It seems like something has really gone wrong with your day.
	S: Shut up! It's none of your business.
	T: Your business is my business, especially when I can see you're in trouble.
	S: (Steps toward teacher angrily.) Get out of my way! I'm leaving this stinking place!

T: Walking away from it won't help.

S: (No response; looks out the window, picks at fingers.)

T: What happened?

S: (Looks briefly at teacher.) Well, that dumb substitute was in my math class again. She's so stupid she has to have a security guard.

T: Why was the security guard there?

S: I don't know. Maybe she's scared of us. (Nervous laugh.) Maybe because she had to go . . . to the office.

T: So, the security guard was there, and the teacher left?

S: Yeah. She told the security guard that if anybody gets out of line to put their name down on paper. I got up to get my brother's cap and gown for graduation because he couldn't come to school today, and she told me to sit down. But I left and got it anyway, and when I came back she wrote my name down. (Big sigh.)

T: You were trying to do something for your brother and got into trouble.

S: (Nods in agreement and looks at teacher.)

STEP 2:
STUDENTS
IN CRISIS
NEED TO
TALK

T: Getting your brother's cap and gown was important. He would have been really mad at you if you had messed up with his cap and gown.

S: It was important, so he could graduate! He'd beat up on me if I lost it! (Kicks table leg and picks at fingers again.)

T: Graduation is an important step in your brother's life; you had a big responsibility.

S: (Nods again and leans body toward teacher.)

T: What happened after you got the cap and gown?

S: The security guard said she told me to sit down and be quiet. But I didn't realize she told me.

T: Maybe you didn't want to hear her.

S: No! That's not it! (Kicks table leg.)

T: Did she know that you had a responsibility to get your brother's cap and gown for him?

S: I don't know. I'm not mad or nothin'.

T: I can see that you are in control.

S: I know why she put my name down and everything—for not saying where I was going. She did say to sit down and be quiet, even though I didn't hear it. I just got up to get the cap and gown. It probably made her mad and she wrote my name down.

T: You're exactly right; teachers expect students to follow directions. Did anything else happen?

S: She took down someone else's name before mine and they gave her a false name, so she had to mess around with that for a while and was getting pretty mad.

T: Do you suppose she was kind of tuned in to what was going on?

S: I don't know. I think she thought we were pretty wild. People started talking and playing around and she got all upset. (Kicks table leg.) So she sent me to the office.

T: Did she think you were responsible for the problem?

S: No. She seemed like she was calm. She just picked me out to send to the office. But I wasn't really doing anything. It was the others in the back of the room.

T: There's been trouble before from that group in the back of the room. I'm sorry you got yourself involved, just as you were learning to see things differently. Your new behavior was beginning to pay off for you.

S: Yeah. I haven't been in trouble for a month.

T: That certainly must make you feel good!! But what made it all fall apart today?

S: I don't know. Maybe she was having a bad day.

T: So it wasn't necessarily just you that she was upset with?

S: No. It was the others, too.

T: So other things were going on in the classroom, not just what you were doing. But you were taken to the office!

**STEP 3:
FIND THE
CENTRAL
ISSUE AND
SELECT
A THERA-
PEUTIC
GOAL**

T: Now, Mr. Martin says you're in big trouble because you failed to follow the rules. And you believe that you had a responsibility to get your brother's cap and gown.

Teacher begins to shape the central issue as conflict between emerging independence and following rules.

S: That's it! It doesn't seem fair to punish me when I had to get his cap and gown.

T: When you went to the office, what happened there?

S: The security guard told Mr. Martin what happened and left me sitting in the office, and a few minutes later a student came from that class to get the security guard again because a student went out the window, or somethin'.

T: What went through your mind at that point?

S: I was thinkin' what they were gonna do because I didn't think what I had done was bad enough to get me punished. Mr. Martin told the security guard that I did need to keep up with that thing, that if I lost something it would be pretty hard to get another one.

T: You're talking about the cap and gown?

S: Yeah. I didn't want my brother to miss out on graduation.

T: That's excellent, to be a responsible brother. Were you concerned about what might happen to you because you were in trouble again?

> Teacher probes for student's insight.

S: In a way I was and in a way I wasn't. I was concerned because I was afraid the security guard would tell that I had done something before that, or something like that. I thought I was going to get sent home.

T: That concerned you. You were worried about being sent home.

S: Yeah, in a way. (Looks out the window.)

T: And what happened when Mr. Martin talked with you?

S: He told me I had used up my last chance but I could choose my punishment.

T: He was remembering what he told you the last time you were in trouble?

S: Yeah. He remembered the trouble I used to get into.

T: So, what did he mean, about choosing your own punishment?

S: He said I could get suspended, or transfer to alternative school, or take a paddling.

T: He was leaving the choice to you? Those are tough choices. He must believe you are showing more responsibility.

Teacher probes for student's motivation to change.

S: Yeah. (Looks at floor, shuffling nervously.)

T: Let's review this. It seems like you had a choice to make in the classroom that didn't work out very well for you.

S: That's about it. Mr. Martin says I have to choose my punishment. And I don't think I should be punished for getting my brother's cap and gown.

T: It isn't easy to face punishment when you thought you had done something responsible.

S: (Looks at the floor and kicks table leg.)

T: This is your first year here at Towne High. Has it been difficult adjusting to a different school?

S: I thought it would at first, because my brother went to Central High and then transferred here. I thought I would get to go there because that's where all my friends was. I didn't think I'd like this school, but I like it.

T: But you've had a few problems here. Are they different from the problems you had at the Middle School?

S: Some of them.

T: What problems were different? Different teachers? Or what?

S: What they're teaching me here is a lot different from what I had last year. Some of my classes are easier, but a couple of my classes are harder.

T: It seems as if teachers expect more of older students. I've heard some good things about you from your new English teacher, Mr. Smith. Your work has been excellent. But that particular class has a few students in there that try to keep things stirred up.

S: There's a problem because they'll start up in the back of the room and the whole room gets going. I told Mr. Smith I'd like to be put in another room. I told the counselor and everybody else. Because once they start cuttin' up, I start cuttin' up with them.

T: That's a mature way to handle it. Have they told you why they can't put you in another class?

S: Not really. I talked to Mr. Smith and all. At first I was supposed to be in another class. But I'm not sure what happened about that.

T: Is it your most difficult class?

S: Naw. The work's easy.

T: I mean, about your behavior.

S: That's the only class that I have trouble; maybe Ms. Dundford's too. I used to sit in the back of the room in the beginning of the year, where most of the noise would start up. Then I moved to the front and would try to ignore it, but it didn't work.

T: It's hard to not notice and get involved, especially sixth period when everybody seems to be thinking about getting out. How is your school work in that class?

S: Mr. Smith had a talk with me, and since then I've pretty much been keeping up with my work.

T: He's taking a personal interest in you. He believes you can really do well. Do you think he has helped you?

S: Yeah, because ever since the last time he talked to me about not doing my work I've been doing it.

T: That's great! Mr. Smith seems to expect responsibility from students, to control their own behavior.

S: Yeah, I think if the students that always talk would sit there and take notes, the teacher wouldn't be as mad at the class all the time.

T: So you think those students are not taking enough responsibility upon themselves for their own behavior, just relying on the teacher. What about you? What can you do to make things better for yourself, and the class?

S: I'm not sure, because I've tried everything!

T: It sounds like you're kind of frustrated.

S: Yeah. (Looks directly at teacher.)

T: So it seems you've tried everything. But I've been noticing a change for the better. Things in general seem to be going well for you.

S: I guess so. I used to get into trouble all the time.

T: Maybe this is why you are so upset about what happened today. The fact is, you've stayed out of trouble for a month. When you meet with Mr. Martin this afternoon, you have another chance to show him how much progress you have made.

S: He won't remember.

T: Could you tell him specific ways you are more mature than you used to be?

Teacher identifies therapeutic goal—to teach new skills for independent crisis resolution.

S: I guess so. (Doubtfully.)

T: Sometimes just plain, calm talk can communicate ideas. I'm impressed with the way you can look back and understand what happened and why it happened, and then tell it clearly without getting angry.

S: (Straightens up and sits taller in the chair.)

STEP 4: CHOOSE A SOLUTION BASED ON VALUES

T: Do you think Mr. Martin is going to follow through with the choices he gave you?

S: I'm pretty sure he will.

T: If you choose suspension, what will happen when you go home?

| Teacher starts student's review of alternative solutions. |

S: The first thing is my mom would ask me why and I would tell her and she would come up to talk to them about it.

T: Will she be real concerned?

S: Yeah. (Looks at floor again.)

S: What will happen then?

S: If she comes to school and they tell her I've been acting up the whole day, or somethin' like that, she'd probably put me on restriction for a while.

T: Your mom sounds like she cares about what's going on with you at school and that you do what you're supposed to do.

S: (No response.)

T: Is this a problem you want to handle on your own instead?

S: Yeah, I guess. I don't think I should be punished for just getting that cap and gown.

T: You thought you were being responsible by getting the cap and gown like you were told to?

S: Yes! I had to get them.

T: And you also knew that you were breaking a rule when you did it?

S: Yes, I guess so.

T: So, you were facing a double problem. If you looked after what your brother needed at this time, you would have to leave class and in turn break a rule—not leaving the classroom without permission.

> Teacher recognizes student's values include personal responsibility. Student cares about living up to expectations of others and is not motivated primarily by fear of punishment.

S: That's right.

T: Do you think Mr. Martin knows this?

S: Probably not.

T: If he asks you which choice you've made, what will you tell him?

S: I'm not sure. My dad, my stepdad, he would say to go ahead, get the paddling.

T: Just to get it over with?

S: I don't know. I'll probably go to alternative school.

T: If you choose alternative school, what does that mean?

S: I've never been there; I don't know.

T: You think it might be a better choice than paddling?

S: Yes, because if I come here in the morning, or whenever he does the paddling, that'll spoil my whole day. I may say something smart to a teacher and get in trouble again.

T: You're exactly right! It would be real difficult to come in and face something like that. So you think the alternative school is a better choice?

S: Yes, if I have to go; but I don't particularly want to go.

T: Could you tell Mr. Martin exactly how you should have handled the cap and gown problem this morning?

S: What I should have done was walk into the classroom, told the security guard where I was going, and ask if I could go get the cap and gown, and come back, and sit down quietly.

T: That's a calm and mature way to let him know you understand the rules.

STEP 5: PLAN FOR SUCCESS

T: Mr. Martin is expecting you in his office this afternoon. When you walk in, what do you think he will say to you?

S: He'll ask me if I made my decision.

T: What will you say?

S: "Mr. Martin, I don't want paddling. I guess, alternative school; unless I can stay here."

T: What will he say?

S: I don't know. Maybe he'll paddle me anyway. Or ship me off to that other place. Or call my parents.

T: Is there anything else you could say that might change his mind?

S: I could say, "Mr. Martin, I know what I should have done; like asking permission before I leave to get the cap and gown."

T: You're right! You understand that you are responsible for keeping yourself out of trouble. I hope you can work it out with Mr. Martin, to show him you have a new, more mature sense about how to follow rules.

S: (Nods in agreement.)

T: But what if Mr. Martin follows through and you have to make a choice?

S: I guess I'll just have to take the punishment.

T: You're showing maturity when you can deal with hard choices.

STEP 6: GET READY TO RESUME THE ACTIVITY

T: It's time to meet Mr. Martin. He said he would see you at 3:30.

S: Maybe he forgot. (Nervous laugh.)

T: (Smiles back.) Just your luck, he never forgets anything!

S: So he'll remember what I used to do.

T: Unless you can use some different behavior to change his mind. Remember how clear and calm you were when you told me about a different way to handle it?

S: Okay, let's have at it!

T: Now, tell me one more time, how are you going to answer Mr. Martin?

S: I'll tell him, I really know I should have followed the security guard's directions and explained how important it was to pick up my brother's cap and gown. I'll tell him I know I should have gotten permission to go.

Student describes behavior necessary for reentry.

T: That's the mature way to handle problems—communicate. When you talk to people about what's on your mind, events don't get out of hand. Everybody wins!

S: I hope he believes me.

T: And if he doesn't?

S: I guess I'll have to choose a punishment.

T: Do you think you can handle it now?

S: I think so.

T: I see so much maturity in your approach to this that I believe you can face whatever comes.

> **Teacher affirms and empowers student to face the new stress realistically and with self-esteem.**

T: I'll be down in my room doing some paperwork. When you finish talking with Mr. Martin, I want to see you.

S: (Mumbles.) Okay.

T: Mr. Smith needs to see you also. He has the assignment you missed this afternoon. He said you can do it at home if you can't stay here to finish it today.

S: I don't feel like doing any work.

T: I can understand that! So, let's talk about how you feel, after your meeting with Mr. Martin. Then we can go down together to see Mr. Smith. Whatever the outcome, we'll work it out together.

With this LSI, in addition to supporting the student to take personal responsibility for resolving the immediate crisis, progress has been made toward long-term goals: Student has used rational control for directing behavior, shown insight into how others react to his behavior, suggested new ways to behave, and conveyed attitudes of responsibility and self-esteem.

WHAT WE HOPE YOU HAVE LEARNED FROM THIS CHAPTER

- Why crisis is a good time for learning new behaviors.

- Three aspects of self-regulation of behavior.

- Four purposes in using LSI when students are in crisis.

- The name and purpose of each step in the LSI sequence.

- Five general prerequisite skills needed by a student to participate in LSI.

- One overall, short-term outcome of every effective LSI.

- Three overall, long-term outcomes of effective LSIs.

- Four fundamental needs of students that adults can provide in LSI to protect them from stress and anxiety.

Stress arouses feelings.
Feelings trigger behavior.
Behavior incites others.
Others increase stress.
And around it goes!

Chapter 2

One way of looking at crisis is to see it as the product of a student's stress, kept alive by the reactions of others. When a student's feelings are aroused by stress, the student will behave in ways that buffer against the painful feelings. This behavior usually is viewed as negative by others (adults and peers), causing them to react negatively to the student. This reaction from others causes additional stress for the student. We call this the Conflict Cycle. It is a way of looking at crisis by analyzing the interactions among a student's feelings, behavior, and the reactions of others in the environment. If this cycle, produced by these actions and reactions, is not broken it will inevitably explode into crisis.

This series of events could also be called the Crisis Cycle or the Stress Cycle, depending upon whether you emphasize the end point (crisis), or the beginning point (stress). But we prefer to call it a Conflict Cycle because this term includes the idea of conflict between two opposing forces: needs within the student clashing against the expectations of others. Healthy adjustment results when these two opposing forces are minimized or resolved; maladjustment results when these two opposing forces continue at odds.

To break the cycle, we must first be able to recognize that it is happening. Then we can analyze the ingredients and plan ways to intervene. Figure 2.1 illustrates the major elements in the Conflict Cycle, beginning with an event that causes stress and producing an incident. Unless this first cycle is interrupted and redirected, the Conflict Cycle spirals. The nature of the stress changes, different feelings may be unleashed, new behavioral defenses are called into play, and reactions from others expand in intensity. Invariably, the result is crisis, a critical moment demanding action. Something must be done!

A crisis generally takes shape from an unresolved incident. It has an intensity and urgency that come from the emotions it arouses in the student and those who are involved. It is not unusual for a crisis to develop from a minor incident that has spiraled. One cycle, uninterrupted, leads to another, and yet another; each expanding into increasingly complex, private depths of stored feelings and evoking behavior increasingly driven by emotional rather than rational processes. Figure 2.2

Figure 2.1   The Student's Conflict Cycle

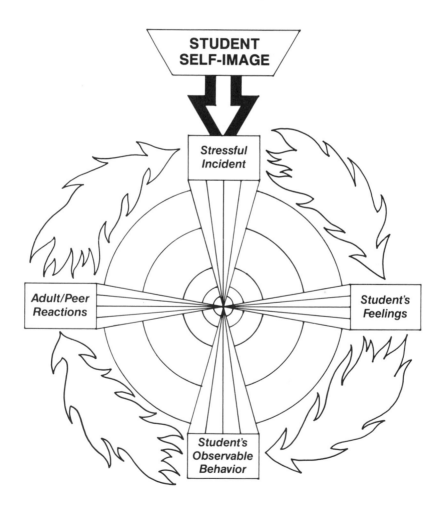

Figure 2.2 Unbroken, the Conflict Cycle Spirals into Crisis

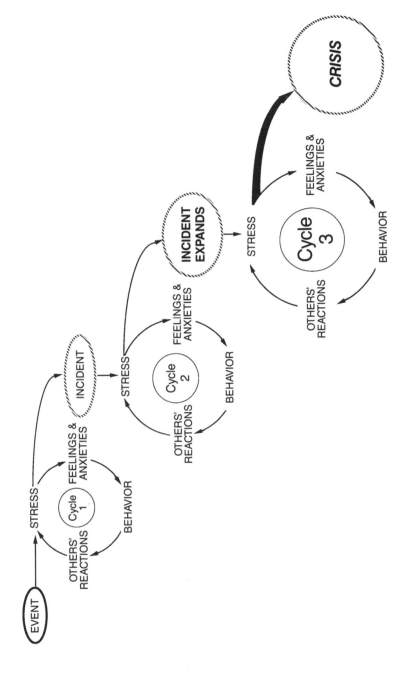

illustrates this spiraling phenomenon. Uninterrupted, it starts from an ordinary event, expands into an incident, and then becomes a personal crisis for the student.

INGREDIENTS OF THE CONFLICT CYCLE

Every crisis can be analyzed into a series of Conflict Cycles, each of which provides four different points for intervening in a crisis:

- Modifying the stress

- Alleviating the student's distressed feelings

- Changing the student's behavior

- Changing the behavior of others

Below we discuss each of these four major dimensions of the Conflict Cycle. When adults understand these critical aspects of crisis, they focus their LSI strategies with greater accuracy and effectiveness.

The First Ingredient: Stress

The sources of stress for students are enormous: reality stresses created by society, profound psychological stresses occurring in the fabric of contemporary family life, and developmental stresses embedded in the normal process of growing up. As a nation, we are failing to provide basic social and psychological conditions where our young can develop in security. This situation puts children and youth at high risk for educational, personal, and social failure (Coles, 1967). Unless we expand our emphasis to teach students ways to cope with stress and crisis successfully, the outlook for children, youth, and society is bleak.

A national epidemic of stresses is impacting on our children and youth. Consider the record. We are a society with one of the highest infant mortality rates among the industrialized nations of the world; the highest murder rate; and a steadily increasing rate of fetal alcohol damage, fetal cocaine addiction, fetal AIDS and HIV, child abuse, child crime, adolescent suicide, child and teen pregnancy, and substance abuse among children and youth.

The majority of the nation's children and youth are living at or near the poverty level (Children's Defense Fund, 1986). Estimates are made that 25% of the nation's children are now born out of wedlock, a similar proportion will be on welfare at some time during their lives, and 42% will be raised by a single parent. About 60% of children in school have working mothers, and a fourth of the mothers who are not working report that they would do so if adequate child care was available. This dramatic increase in working mothers has resulted in a national crisis in child care. We also are failing to provide psychological security. Our anxious, adult-oriented society has unrestrainedly indulged its own preoccupations without attempt to protect, limit, or shield our children and youth. We bombard them with alienation, violence, and grief—in popular music, television, advertising, news, movies, books, fashions, entertainment, pornography, and art. We are, in short, a society that seldom stops to consider that our children and youth may not be equipped mentally or emotionally to deal with this bombardment of adult experience (Coles, 1986). Nor are we teaching them ways to successfully cope with these stresses.

Such concerns are widely held by child development specialists. They are in agreement that children without a close relationship to an important adult, without awareness of being loved and protected (or who perceive themselves as unloved or unworthy) suffer intense emotional pain. The resulting anxiety is debilitating. It compounds other problems of general development. Without alleviation, this sense of abandonment can persist into adolescence and adult life, permeating self-esteem, heightening sensitivity to stress, and compounding the problems of making and keeping satisfying relationships with others.

A number of research studies provide support for the need to be concerned about the social and emotional development of our nation's children—their behavior, the quality of family relationships, and their future status as adults. One example is the Berkeley Guidance Study of four generations used by Elder, Liker, and Cross (1984) to study economic and psychological stresses on families. They examined interrelationships among family members, generational characteristics, irritability, hostile and inconsistent patterns of parenting, and interactional patterns among parents and children. The results are not surprising: Relationships within families significantly affect children. Hostile and inconsistent patterns of parenting tend to lead to negative behavior patterns in children, especially aggressive and antisocial behavior.

Similarly, the New York Longitudinal Study (Thomas & Chess, 1984) followed a group from the first year of life to age

26. They found that psychopathology in adults had antecedents of disturbed behavior in early childhood. No disturbed adults were found in the study who were not disturbed as children. Thomas and Chess also found that the best single predictor of adult mental status was the presence (or absence) of parental conflict during the preschool years.

In reflecting on the impact of contemporary society on children, youth, and adults, Jerome Kagan (1986, pp. 44–45) sees two major sources of chronic stress. His first stress theme is a concern for being loved: "All members of our society feel relatively deprived." He suggests that this may be a result of the failure of status, accomplishment, and power to gratify; with the one remaining source of gratification being a close relationship with another person. Yet such relationships seem to be increasingly difficult to establish or maintain. Kagan's second stress theme is uncertainty about one's value and worth—the inability to be in control of events that lead to attainment of goals. Failure to achieve gratifications that provide reassurance about worthiness contributes to this "modern mood of anxiety."

The result is a generation of stressed children and youth, living in a psychologically complex, unpredictable, generally nonsupportive, anxious, adult-oriented society. It is a problem so widespread that it has reached epidemic proportions. It touches every classroom and family in America, every income level, every race, sex, personality, and intellectual level.

Some say schools reflect society. Probably true, for schools today generally are failing to alleviate stress. The social and psychological conditions in schools leave considerable doubt about our national priority and capability to provide quality of life and educational excellence for all children and youth. Schools make unique demands on each student, setting expectations to conform, achieve, share, perform, produce, and bend to the will of others. Endless scenarios exist that illustrate the stress promoted in schools. Here are a few examples of stress-producing situations that occur daily in classrooms:

- Personal put-down

- Failure to do something correctly

- Failure with friends

- Ridicule and derision

- Scapegoating

- Being left out

- Infringement on rights

- Failure of justice

- Deprivation of a valued object or opportunity

- Confusion

- Boredom

- Confinement

- Not understanding what is expected

- Threats of harm

- Expectations beyond capacity

Sometimes the consequences are obvious: embarrassment, restrictions, failure, punishment, deprivation, or denial of opportunity. But often the consequences of failing the expectations of others are subtle and psychological: estrangement, rejection, alienation, disapproval, or a judgment of being unworthy. When the expectations of others and a student's personal needs and feelings collide, the tension becomes so great that stress occurs. This stress sets the Conflict Cycle in motion. An incident may seem innocuous in itself and hardly worth the conflict that results. But keep in mind, the Conflict Cycle builds on itself. Seemingly minor incidents often escalate into major events—sometimes to traumatic proportions. The original stress becomes masked in a floodgate of actions and reactions between the student and others.

We must address these stress themes if we are to assist children and youth in the difficult process of adjustment and achievement in today's complex society. Earls (1986) suggests two major streams of effort to help children and youth survive the intense stresses they experience. First, teach parents, teachers, and other primary care providers to be major sources of emotional support, buffering and protecting students from excessive stress and explaining the meaning of daily events in ways that are understandable and reassuring. Second, Earls suggests strengthening each student's innate or learned capacity to cope successfully with adversity, stressful events, and daily pressures. LSI is a procedure that includes both priorities: providing emotional support and teaching behaviors for coping successfully. By using LSI, adults can teach students how to manage stress more constructively, regulate their own behavior, and break the Conflict Cycle.

The Second Ingredient: Feelings and Anxieties

There is an obviousness on the surface of every event that classifies it as "public," with witnesses and players. While everyone may not agree on what happened or why, they have a common topic on which to focus. The event has occurred in a common psychological and physical space. It is a public reality. It is this observable aspect of an event that provides the point for using LSI.

Each incident also has a private reality for each individual involved. This is the emotional memory bank where thoughts, feelings, and anxieties are stored. This private perspective is the energizing drive behind public words and behavior, seldom expressed in an open or direct way. Private realities are among the most powerful forces that drive individual behavior, forming a large portion of personality and specific behavioral responses. It is here, in the personal storehouse of thoughts, feelings, and anxieties, that memories of previous experiences are filed away with emotional notations. They lie in wait for expression and can be triggered by stress from even the smallest, insignificant event. This is no minor force! Understanding this private dimension of a student is central to an effective LSI.

The Conflict Cycle builds on the stress created around these public and private realities. At the beginning of this chapter, we mention the great number of pervasive stresses permeating contemporary life. Our students are caught up in life-styles that are hurried, unpredictable, and bombarded with stress. No age group is exempt. Relationships and personal, emotional needs often seem to be ignored in the push for things, action, and performance. Then, when clashes between private, unmet emotional needs and the expectations set by others occur and are unresolved, stress develops that demands relief.

Some of the feelings and anxieties stored in a student's emotional memory bank result from daily frustrations and disappointments. Other emotional memories are of painful or assaultive individual life experiences. These are the psychosocial stressors discussed previously—memories of unhappy and hurtful associations in the past. Protecting oneself from reexperiencing such feelings is a major motivation. People will go to great length to avoid these feelings.

Other feelings and anxieties occur in response to the demands made on a student in the course of normal development. Called "developmental stressors," these are standard demands for performance made on most children and youth at various ages. Here is a basic illustration: Our expectations for performance from infants are uncomplicated. We expect them to smile back at us, cease crying when we comfort them, and swallow their food when fed. If they fail to respond, we react. It may be with concern, frustration, disgust, anger, or a sense of ineptness or helplessness. The reaction is cycled back to the infant in the form of "doing something." If the situation is ameliorated and the infant gives us the desired response, the rewards for both the infant and adult are significant. Feelings of well-being permeate the emotional memory banks of both. In contrast, if the infant does not give back the desired behavior, the result is stress for both, and the emotional memory banks are crowded with feelings of unhappiness and discomfort.

Project that same interaction further, as development continues. Expectations for the toddler's performance have increased enormously. Bowel and bladder control are important; talk is essential, as are obedience to simple commands and self-feeding. The first preschool experience requires additional sitting and sharing behaviors. The primary grades demand reading, writing, and arithmetic. Upper elementary school demands appropriate group behavior governed by rules and concepts of fairness. In high school, students face complex demands for employment, vocation, and intellectual expansion. Simultaneously, they are expected to somehow stay out of trouble while thrown into a sea of adult social ills.

Such age-related expectations for performance are developmental stressors—experienced in some form by all children and youth during the normal stages of their social and emotional development. If the expected performance is not forthcoming, adults react. When adults' reactions are supportive and oriented toward helping students achieve success in overcoming these developmentally incurred problems, crises may be averted. When students achieve developmental expectations, we say they are "well adjusted." If they fail to perform as others expect, we talk about "problem" students and "poor adjustment." For these students, developmental stressors, like reality stressors, have produced painful feelings and defensive behaviors. These memories seldom disappear. They are simply stored in emotional memory banks, where they are protected by defensive behavior. Because feelings and anxieties play a central role as antecedents for crisis and are essential elements in breaking the Conflict Cycle with LSI, we continue this topic in greater depth in the next chapter.

The Third Ingredient: Behavior

Observed behavior is the visible sign of self-protection in action. In a Conflict Cycle, an event always triggers feelings that, in turn, result in defensive behavior. We call these behaviors defensive because they are attempts by students to protect or insulate themselves from the feelings and anxieties evoked by the stress of an event. We believe that almost all behavior problems are defensive maneuvers by students to cope with stress, whether resulting from developmental stresses or the stresses of contemporary life.

Clinicians refer to this idea of psychological defenses when they use the term *defense mechanisms*. By identifying a student's behavior as a particular type of defense, you will know (a) that the student is sufficiently anxious about the event to need psychological protection (defense against painful feelings and anxieties) and (b) the general defense strategies the student is using.

At every age, defensive behaviors enable students to cope with stress. They go to great lengths to protect themselves from the emotional pain of unmitigated stress, whether it arises from normal developmental stressors or from assaultive life experiences. When successful, defensive behaviors enable a student to cope comfortably. But if they fail to produce the needed relief, new behaviors will be substituted. If these also fail to produce relief, students' behavior becomes increasingly dominated by excessively defensive maneuvers, perceptions are distorted, and functioning is impaired. Chronic defensiveness can permeate students' personality development and influence the way they learn (or fail to learn) and how they respond to adults, peers, and daily events in the classroom.

Defensive behaviors can be loosely grouped into three categories: denial, escape, and substitution. Table 2.1 defines each of the major defensive behaviors.

Denial.   Behaviors in the first group are those in which the student actively denies ownership of the problem:

"Nothing happened." (denial)

"I wouldn't hurt anybody." (repression)

"I thought the test was tomorrow, so I made a dentist appointment for today." (rationalization)

Table 2.1 Definitions of Major Defensive Behaviors

Denial

1. *Denial*—A defense mechanism in which the individual protects from unpleasant aspects of reality by ignoring or refusing to perceive them, remaining unaware of facts that could create one side of a conflict.

2. *Repression*—A defense mechanism in which the individual's painful or dangerous thoughts and desires are excluded from consciousness without awareness of what is happening.

3. *Rationalization*—A defense mechanism in which logical, socially approved reasons to justify past, present, or proposed behavior are used. It helps the individual to justify actions and beliefs, and it aids in softening disappointment connected with unattainable goals.

4. *Projection*—A defense mechanism in which the individual transfers blame for shortcomings, mistakes, and misdeeds to others or attributes to others his or her own unacceptable impulses, thoughts, and desires.

Escape

5. *Defense through the opposite (reaction formation)*—A defense mechanism in which dangerous desires and impulses are prevented from entering consciousness or from being carried out in action by the fostering of opposed types of behavior and attitudes.

6. *Withdrawal*—A defense mechanism involving emotional, intellectual, or physical retreat from a situation.

7. *Intellectualization*—A defense mechanism in which the affective charge in a hurtful situation is cut off or incompatible attitudes are separated into logic-tight compartments.

8. *Regression*—A defense mechanism in which the individual retreats to an earlier developmental level involving less mature responses and a lower level of aspiration.

Substitution

9. *Displacement*—A defense mechanism in which there is a shift of emotion, symbolic meaning, or fantasy from a person or object toward which it was originally directed to another person or object.

10. *Compensation*—A defense mechanism in which the individual covers up weakness or feeling of inadequacy by emphasizing a desirable trait or makes up for frustration in one area by overgratification in another.

11. *Sublimation*—A defense mechanism that involves the acceptance of a socially approved substitute goal for a drive whose normal channel of expression or normal goal is blocked.

12. *Identification*—A defense mechanism in which the individual identifies with some person or institution, usually of an illustrious nature.

"I'm trying to play by the rules, and he keeps breaking them." (projection)

Escape. In the second group of defensive behaviors, the student tries to escape from facing a problem by mentally or physically withdrawing:

> A student spends much class time staring out the window or drawing doodles around the edge of his papers. (daydreaming/fantasy)
>
> The student simply gets up and walks out of class when things don't go the way she wants them to go. (withdrawal)
>
> Whenever the 12-year-old student cannot get his way, he goes into a temper tantrum. (regression)

Substitution. In the third group of defensive behaviors, the problem is modified or distorted by the student so that it is somehow changed into something more manageable by substitution of a more acceptable behavior that still provides control of painful or dangerous feelings.

> A science student is angry at his teacher, so he takes apart his 6-week project, "It's no good, it won't work!" (displacement)
>
> A student who is without friends and generally ignored by the other students carries snacks in her purse. (compensation)
>
> An adolescent handles his aggressive impulses by playing the drums in the high school band. (sublimation)
>
> Just before final exams, a student becomes sick and has to leave school. (psychosomatic reaction)
>
> A student imitates the mannerisms, expressions, and styles of a current television star. (identification)

The Fourth Ingredient: Others' Reactions

In the Conflict Cycle, a student's defensive, negative behaviors typically evoke hostile or defensive reactions and widespread alienation from peers and adults. These reactions produce greater stress on the student, and the Conflict Cycle goes into a second round, again tapping into feelings and anxieties that require more defensive behavior, producing more negative reactions from others. Unless it is interrupted, the cycle will continue to expand in intensity and spread.

For an adult involved in a Conflict Cycle, the situation is pivotal. The student's behavior encourages adult counterbehaviors. Student comments, such as "No!" "Make me!" or "So what?" beg for an aggressive response. Such students, who defy authority or refuse to participate, are detrimental to constructive group activity. What complicates our adult position is that it seems justified and natural to react in a negative way and put such students in their place. After all, adults are responsible for the maintenance of law and order!

An amazing concept of interpersonal relationships is how students under stress can create their own feelings of anger, frustration, helplessness, or insecurity in adults—to the point where adults will behave in counteraggressive, impulsive, or rejecting ways. As a result, the Conflict Cycle is fueled. Defensive adult reactions create new stress for the student. The student now has to deal with the adult's rejection or anger in addition to the original stress. For an adult to respond with *any* counteraggressive behavior is self-defeating. If you act out the feelings you have and "do what comes naturally," your behavior will perpetuate the cycle by mirroring the student's own aggression or inability to control behavior. As a result, the statement, "Aggression elicits aggression," becomes true, and the student becomes the one determining the adult behavior. The more "involved" you become in struggling with a student, the more likely it is that you will be the one who ends up in crisis. Even if the student loses the battle but succeeds in getting you to escalate counteraggression by expressing open dislike, hostility, or rejection, the student wins the emotional war by demonstrating the self-fulfilling prophecy that adults are hostile and cannot be trusted.

What do adults need to do to stop this destructive cycle? They must understand the cycle, see clearly how they can be seduced by their own feelings and reaction, and not fall into the trap! An experienced adult sees this potential for student aggression to evoke counteraggression from the adult. It requires understanding and self-control to disconnect from the struggle, putting aside personal emotions that rise in reaction to a student's behavior. The expression "dispassionate compassion" is a useful way to think about such emotional reactions during the Conflict Cycle.

Reject the natural instinct to win out over the student; it is not necessary.

With careful self-monitoring, you can stop your own potentially destructive involvement in the Conflict Cycle. By recognizing the existence of your own counteraggressive feelings, you have choice and control over your own behavior; you do not allow yourself to be pulled into the Conflict Cycle. In Part Two of this book, we describe the changing role of the adult during each step of LSI. This should be of help in keeping yourself focused on the task, conveying to students that you are a responsible, dependable adult who understands what they are feeling, respects them, and offers ways to help bring about solutions to their problems.

SUMMARY OF THE CONFLICT CYCLE

Students seldom learn successful behavior for handling stress on their own. Adults must carefully and methodically set about to break the Conflict Cycle by teaching students to regulate their own behavior in ways that bring more satisfying results. Without changes in their behavioral responses to stress and the underlying feelings and anxieties, students continue to behave in a less than satisfactory manner to protect themselves from stress.

Students' negative behaviors can evoke hostile or defensive reactions and widespread alienation from peers and adults. Such feelings spiral into further alienation with more inept attempts to handle problems. Conflict can become contagious to the group, causing further breakdown in students' emotional and social competence. In such an environment, learning can be brought to a standstill. But crisis also represents a potential turning point and opportunity for new learning. A crisis can be one of the best times to teach students how to handle crisis constructively, and LSI provides the process for doing this.

EXAMPLE OF THE CONFLICT CYCLE

It is useful to develop the habit of analyzing each incident into the four components of the Conflict Cycle: stress, feelings and anxieties, behavior, and reactions of others. If you do this as you work through the first three steps of LSI with a student, it helps keep the central points clear. From time to time you may find yourself digressing from the real issue; or a student may unload a lot of feelings or thoughts that may cloud the issue. This is not a problem if you have those four components sorted out in your own mind.

To illustrate how we use the Conflict Cycle to analyze an incident, let's go back to "The Cap and Gown Crisis" (example at the end of Chapter 1). You may recall, the homeroom teacher (T) is using LSI with a student (S) who is on the brink of serious punishment by the principal for a rule violation. As they talk, the student describes the events leading up to the crisis point when he is sent to the principal. We see the original stress spiraling into three Conflict Cycles before it is broken by the teacher using LSI (Figure 2.3). By stripping away all but the essential information about events that lead to the LSI, this analysis reveals major differences between adults and student in perceptions of events. It also shows that the student has verbal ability to communicate but failed to see that communication was the appropriate behavior to use when the incident began.

The original stress that started Cycle 1 was a fairly simple problem that could have been handled by talk about the student's need to leave (getting the cap and gown). The student was expected to understand rules and obey; he also was expected to know when to appeal rules and how to do it. Obviously, he failed in this expectation, and no adult intervened at this point to avoid a spiraling cycle.

Cycle 2 reveals the student's view that adults are unreasonable, and the adults' view that the student never follows rules (student's history coloring perspective?). As this cycle ends, the student begins to see himself in a no-win position, and his feelings of anger and frustration intensify. Again, adult intervention increases rather than breaks the Conflict Cycle.

Cycle 3 illustrates the student's intensified attempts to protect himself. Two types of feelings seem to be prevalent by the time the LSI starts: distress over the impending punishments, and frustration and anger that adults do not recognize his responsibility to get the cap and gown. There is another, minor theme of being helpless against adults who do not understand.

As the teacher goes through this LSI, she touches these themes in every step. Sometimes she addresses them by reflecting what she sees in the student's responses or body language. Other times she defuses the intensity by countering with affirmations and praise of the student's maturity and the sense of responsibility he shows in handling the crisis. At no time is there

Figure 2.3   Illustration of the Conflict Cycle in "The Cap and Gown Crisis"

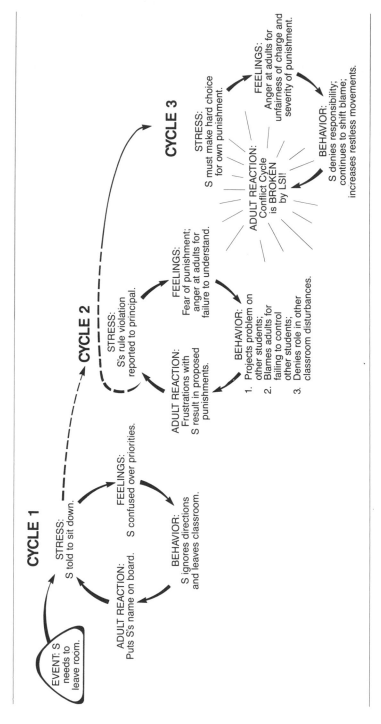

a sense that the teacher blames the student. (Whether or not the student is to blame is moot.) What is important is that the teacher can see the crisis as a sequence of stress, feelings, behavior, and reactions of others.

WHAT WE HOPE YOU HAVE LEARNED FROM THIS CHAPTER

- The two opposing needs that produce the Conflict Cycle.

- How an event can spiral into an incident and then into a crisis.

- How the idea of a Conflict Cycle is used in LSI.

- Four different points for intervening to break the Conflict Cycle.

- Two general sources of stress for students that can set off a Conflict Cycle.

- Examples of ordinary, stress-producing situations in classrooms.

- Examples of typical developmental stresses for each age group.

- Why students use defensive behaviors.

- Three categories of defensive behaviors.

- An example of each of the major defensive behaviors.

- Why understanding the Conflict Cycle is basic to doing effective LSI.

"You think you know
what I said.
But what you don't
know is,
what I said is not
what I meant."

Chapter 3

Students in crisis seldom see connections among what they feel, how they behave, and how others respond. Their responses to stress tend to be behaviors fueled by feelings, perpetuating conflict and crisis. Not only are most students unable to recognize feelings, they are not able to talk about them. But unless they are able to do this, it is difficult, if not impossible, for a student to make lasting changes from behavior driven by feelings to behavior regulated by rational processes.

Some adults are equally inept in this process. An adult can be of little help to a student until the adult recognizes behavior that is expressing feelings and knows how to convey this understanding to a student. An adult cannot teach a student something that the adult does not understand. Learning to do this is a major part of learning to use LSI effectively. It is a three-step process for adults. First, learn to recognize feelings and anxieties expressed in students' behavior. Next, learn to convey this understanding to the student. In LSI, adults must be the first to make connections between behavior and feelings when their students cannot do so themselves. Finally, a skilled adult teaches students to do this themselves during LSI, recognizing the part feelings play in their own behavior and that of others.

In this chapter we focus on the foundation needed by adults for teaching students to recognize feelings conveyed through their own behavior and that of others. First, we outline a framework for identifying students' developmental anxieties that are a natural part of personality development for all children and youth. This material provides a background framework for identifying general types of deep concerns of students of all ages. Second, we

Abandonment

"No one cares!"

Abandonment is the fundamental anxiety, arising from fears of physical or psychological deprivation, abuse, or annihilation of the self. It occurs as a normal developmental crisis in infants and very young children under age 2. The need to have basic physical and psychological nurturing predominates behavior. Unresolved, anxiety about abandonment permeates personality development and all subsequent relationships. In older children, adolescents, and adults, this anxiety is expressed in attempts to satisfy primitive needs at any cost—by eating, hoarding, stealing, pursuit of sexual gratifications, and superficial emotional attachments.

Developmental anxiety about abandonment can be resolved with consistent care; expressed, unconditional affection; and an environment that provides consistency, pleasure, comfort, care, and security.

Inadequacy

"I can't do anything right."
"You won't like it."

In normal development, the anxiety of inadequacy typically becomes the emotional crisis of preschool children as they become increasingly aware of the expectations of the important people around them. The dynamics of this anxiety include self-doubt and fear of the unknown, the need to avoid failure by denying association, and avoiding experiences with uncertain outcomes. Imagination runs rampant, and students see most events as a clash between right and wrong, good and bad. Punishment is always imminent from powerful adults and other external sources (magical forces, nature, rules, authority). There is no control or escape. When anxiety about inadequacy continues in school-age students, they seek to avoid blame, criticism, punishment, or failure at any cost. The driving motivation is to cover up mistakes (by blaming others, denying responsibility, lying, or by avoiding failure or being put in a position of looking ''bad'').

Anxiety about inadequacy can be overcome as students learn skills to be successful, build self-confidence, and receive adult approval as individuals who can do things successfully. They learn that problems have solutions (not just punishment), and they begin to see themselves with some power to control events around them.

Guilt

"I'm no good; I'm a loser."

Guilt is a more complex form of inadequacy in which the source of judgment about a student's unworthiness changes from others to self. Typically, this is a normal developmental crisis for students during the first 3 years in school (about ages 6 to 9 years). Guilt signals the presence of a basic value system. The expectations and rules of others are accepted as one's own. The student then serves as judge and jury to the self, determining that personal inadequacy (failure) deserves punishment.

Unresolved, guilt arrests a student at a developmental level where self-denigration predominates. Failure to meet the standards of others is absorbed as excruciating evidence of unworthiness. Older students with this anxiety often allow themselves to become scapegoats, to be exploited by adults and peers. Their behavior is outrageous. They do unacceptable acts to prove their own unworthiness and to be punished as atonement. Or they may violate rules so clumsily that it is evident that they want to be caught.

Another defensive strategy seen in students with guilt is the massive cover-up. Clinicians sometimes call this passive aggression. These students are so aware of the expected standards of behavior and so in need of approval from others that they cannot afford to put themselves in the position of being rejected. Yet they are so troubled by their own personal failings and so angry at others for their miserable situation that they express the anger in devious ways to avoid blame. Every adult has worked with such students, often calling them manipulators.

Developmental guilt normally resolves as the student expands successfully in independent experiences with friends and adults. Adult sanction of this new independence is essential. Students gradually develop alternative standards to judge their own

If each developmental anxiety is successfully resolved, a student has mastered significant milestones of behavior in social and emotional development. Equipped with age-appropriate behavior, a student is prepared to face new stresses produced by the next stage of development. If unresolved, a developmental anxiety at one stage contributes to the stresses of the next stage of development, compounding the student's problems and impacting negatively on the student's capacity to develop further.

Central to each LSI is the need to recognize which developmental anxiety students are protecting, so that you can teach them to recognize connections among their own feelings, inappropriate behavior, and resulting reactions of others. Every LSI provides abundant opportunity to teach new behaviors and alleviate developmental anxieties. We show these connections to students by decoding their behavior during crisis.

DECODING BEHAVIOR

Decoding is connecting what students are doing and saying to what they are feeling. Decoding has three purposes: (a) to teach students to recognize specific feelings that drive their inappropriate behavior, (b) to build students' confidence that they do not have to be victims of their own bad feelings because there are alternative behaviors that bring better feelings, and (c) to convey that talking about feelings and anxieties may not be as awful as anticipated. Unless decoding is done, feelings continue to fester, finding outlets in all sorts of inappropriate behavior.

Decoding begins with objective observation of students' behavior and careful listening to what they say. Accurate decoding also requires sensitivity to body language, to what is implied and left unsaid. From this information, connections are made for the student between specific behavior and associated feelings. To fail to decode is to ignore the most potent force driving the behavior of the students you work with.

To decode, you must have some understanding of a student's concern. We suggest using the general framework of developmental anxieties, described in the first part of this chapter, to guide you in identifying these concerns. Sometimes, as you begin LSI, you may find a student's concern elusive, or your understanding of the student's anxieties incomplete. As you talk, question, reflect, and think about the student's responses, construct hypotheses about the nature of the student's concerns. Form an initial hypothesis from the framework of the developmental anxieties, beginning with the anxiety that is naturally associated with the same age group as your student. You can expect that particular anxiety to be present as a normal part of the life experience of most students of the same age, even without stress or crisis. Then, refine your idea about the student's concerns by observing and listening carefully to what the student says and does during the first three steps of the LSI. You may see behavior and hear remarks that lead you to consider the presence of other developmental anxieties. Also, gather information about feelings and behaviors that are specific to that student and the incident under discussion.

The accuracy of your decoding will be evident in the student's responses. Typically a student will react when you decode accurately. The reaction may be vigorous denial or passive quietness. In either case, the decoded message has touched the student's concern. When decoding misses its mark, students generally do not change their level of affect. They may continue the same behavior, appear to be genuinely disinterested, or simply go on to another subject without an emotional change.

As students learn that talking about feelings and connecting them to behavior is not invasive, they learn to do this themselves. They begin to decode for you, explaining what they did and the feelings that were a part of their behavior. It is essential to keep in mind that when you decode, or students learn to decode their own behavior and that of others, it is not the end of the LSI. Students must also learn to change their behavior in ways that bring about better feelings and more satisfying results. For this reason we use different types of decoding during the first three steps of LSI while exploring the problem, the issues, and the feelings involved. Then, during the last three LSI steps, as we work with students to plan and implement solutions, we prefer to use less decoding and refocus instead on new, more appropriate behaviors and the more pleasant results (better feelings) a student can expect from a change.

The First Level of Decoding: Acknowledging

We begin decoding by simply acknowledging that feelings are embedded in a student's words and behavior. Sometimes this strategy is called reflection (reflecting a student's behavior or expressed feelings back). This basic level of decoding does not attempt to interpret complex responses or characterize the feelings in any particular way. It requires no particular understanding about the type of developmental anxiety involved or other deep concerns. It is particularly useful when you are just beginning a series of LSIs with a student, when you know very little about a student or the incident, or when you need to convey acceptance and support while disapproving of the behavior.

The intent in acknowledging feelings is to communicate your awareness that a student is feeling unhappy (awful, miserable, terrible, sad, etc.). When acknowledging feelings, you are not asking the student to describe feelings or motivations. Neither are you condoning the inappropriate behavior the student used to express the feelings. You are not making interpretations about why or how. You are not even asking a student to tell about the feeling. The feeling exists; you recognize how unpleasant or painful it is for the student. That is enough at the beginning. By simply acknowledging the presence of feelings, you are not intruding too rapidly into a student's private space. This respect for a student's feelings can help establish your credibility and can lead to trust. It is essential to acknowledge feelings during LSI, especially during Step 1.

Here are examples of how we use this type of decoding to acknowledge feelings in ways that need no response on the student's part:

SALLY — When asked to stop her project, Sally blurts out, "I hate you!"

Teacher decodes: "You're telling me clearly how you feel."

CALVIN — Calvin curls up in a ball on the floor.

Teacher decodes: "This isn't much fun for you."

JEFF — Jeff is pulled away from a student he has been pounding on the playground.

Teacher decodes: "Teachers have to control you until you can control yourself."

Decoding behavior by simply acknowledging feelings before an incident occurs often will defuse or diminish the intensity of feelings brought to school as "excess baggage." Here are observations made by two teachers that illustrate the potential for crisis from nonspecific sources:

> "He came to school today with a chip on his shoulder." [Student is angry about something and will use any opportunity to unload the anger.]

> "She's an emotional basket case." [Student is so emotionally overloaded that feelings spill over into anything she does.]

Each student is conveying emotionally driven behavior. Unless there is some intervention, you can predict accurately that there will be a crisis. Even the most innocuous event can cause behavior to spiral out of control. The incident may not be an issue, but merely used by the student as an outlet for feelings. Decoding these feelings by simply acknowledging them can avoid misconnecting them to events that are not really issues.

The Second Level of Decoding: Surface Interpretation

Connecting freely expressed, undisguised feelings to specific behavior is the next level of complexity in decoding. The strategy we use is explaining the emotional meaning expressed in the behavior. Sometimes this strategy has been called surface interpretation. For this type of decoding, the behavior is so obviously related to feelings that decoding can be done in a straightforward way. It is easy to recognize the expressed feeling and respond to it. For example, there is a fairly direct connection between the action of a student throwing a book across the room and an associated feeling. To accurately decode that behavior for the student you would need to have information about the sequence of antecedent events and know enough about the student's history, developmental anxieties, or self-image to understand how the event could be stressful. Depending upon such information, the teacher might decode the book-throwing behavior in one of several ways:

> "You thought the work was too difficult, but throwing the book won't make the problem go away."

> "It made you angry when they didn't wait for you to finish, and the book was handy."

"Throwing a book is one way to say you don't want to do the work, but there are better ways to get the message across."

Here are several other examples of decoding that carry clearly conveyed messages explaining feelings communicated by students' behavior.

TOM — Tom is not chosen to play on his friends' team, so he walks off the playground without saying anything to anyone.

Teacher decodes: "When friends hurt your feelings, it's natural to want to get away from them."

ANN — Ann loses control and flies into a rage because her friend will not share paints.

Teacher decodes: "You do things for friends and then when they don't do things for you, you feel betrayed. It makes you angry."

KATHY — Kathy refuses to go with the group to the library: "Nobody tells me what to do. I'll damn well do as I please!"

Teacher decodes: "People are always telling you what to do. It makes a person feel like nothing when they have to do things they don't want to do."

Sometimes adults are afraid to make such statements because they fear that students will interpret the decoded message as tacit approval of inappropriate behavior. This may be true, unless the adult continues to emphasize how feelings cause behavior that results in reactions from others that are not always in the student's best interest. During the first three steps of LSI, decoding with such follow-up discussions occurs frequently. The adult guides (and sometimes leads) a student from acknowledgment of feelings, through interpretation of surface feelings associated with the behavior, to the unpleasant reactions of others. Such discussions are essential before you and a student converge on a solution during the last three steps of the LSI, to break the Conflict Cycle.

The Third Level of Decoding: Secondary Interpretation

Vigorous denial by a student in reaction to decoding usually requires a third, difficult, type of decoding we call secondary interpretation. It is difficult because it requires shifting focus from the student's original behavior to the student's denial. It also requires decoding a second time, connecting feelings to the denial. Second decodings of denials are reformulations of the original decoding, put in different ways that are more palatable and supportive but just as accurate. As in the first type of decoding, a student is not asked to describe or verify the feelings involved. It is the adult who acknowledges the presence of feelings in the observed behavior. To decode a student's denial, you need a solid understanding of which anxiety is at the root of the behavior and the types of defensive behaviors the student is using.

When students are first exposed to decoding, it is not unusual for them to react with strong denial of what you say. It often is uncomfortable and hard for them to believe that you can understand and talk about the feelings behind their behavior. They usually will deny the accuracy of the decoded message. Vigorous denial may indicate several things:

1. The student does not trust you (or adults in general) sufficiently to let you into the private domain of personal feelings.

2. The student may be asking for greater psychological "space" (distance) because the message is about feelings that have been avoided because they are too painful to face.

3. You may have the wrong explanation of the feelings and concerns behind the behavior.

The first two situations call for some form of secondary interpretation. The student's reactions indicate that you have struck an accurate (and sensitive) chord. They also tell you that the student does not want to hear the message, even if it is true. In these situations, secondary decoding often reiterates the original message in a less threatening and more supportive way. Here are examples of decoding with a second interpretation that focuses on the denial:

CARLA   Carla stalks sullenly into the room:
                      "I ain't doin' nothin' today."

Teacher decodes: "You've got something else on your mind today that makes school work seem pretty unimportant."

Carla: "Shut up and mind your own business. There's nothin' wrong with me. I just ain't doin' nothin' today!"

Teacher decodes again: "Some problems are so big that it seems no one can do anything about them."

MARK

Mark throws his pencil on the floor: "I can't do this stupid work!"

Teacher decodes: "This work seems really hard. Sometimes it's embarrassing to ask a teacher for help."

Mark: "I don't need help. Everybody hates me, but I don't care!"

Teacher decodes again: "It makes people feel awful when they can't do something and they think no one wants to help them."

TONY

After a fight in the bathroom, Tony shouts at the teacher, "You're picking on the wrong person!"

Teacher decodes: "It's hard to face up to a problem when you have hurt someone."

Tony: "No! You've got it all wrong! That's not what happened. I was just standing there."

Teacher decodes again: "Sometimes it seems like the best thing to do is put the blame on others."

You probably noticed that this type of decoding requires no specific answer from the students. The intent is to get the message out in the open, to be built upon as the students can tolerate the insights. When a student responds to this second decoding with less emotion, less denial, no further response, or changes the subject, you know that the message has connected.

The Fourth Level of Decoding: Disguised Behavior

The most difficult type of decoding is interpretation of behavior when students make statements or behave in ways that are exactly the opposite of their real feelings. Decoding opposite reactions intrudes into students' strong, protective defenses and brings out thoughts and feelings too painful or unacceptable to be expressed openly by them.

When students form opposite reactions, what they do and say are attempts to keep you away from their emotional "sore spot"—those stored anxieties and feelings. Their verbal and nonverbal behavior may have little or no specific significance in itself but alert you to the presence of defense mechanisms, telling you that the students are vigorously behaving in ways to protect themselves from feelings and ideas that they want to avoid facing. The most secure way to do this is to behave in exactly the opposite way!

To decode this type of behavior accurately, you must be able to recognize the behavior as an opposite reaction and then directly interpret this for the student. Accurate decoding in these situations requires considerable understanding of a student's hidden feelings, which anxiety is being protected, and what defenses are being used to hide it.

Here are examples of decoding inappropriate behavior by interpreting the opposite feelings. The third example illustrates several rounds of repeated decoding with a sophisticated, manipulative student.

JACK   Jack refuses to leave the table and join the other students. When the teacher calls him he ignores her. She goes over to him and puts her hand on the back of his chair. "Get away from me," he snarls.

Teacher decodes: "You don't believe teachers can really help (care, understand). So you want to keep everyone away."

SALLY   Sally refuses to play a group game: "They're stupid and they don't play fair. I'd rather read."

Teacher decodes: "You think maybe they don't like you, so you'll reject the whole situation."

SAM   Sam refused to go with the other students to get a drink of water. When they returned, he burst into tears and cried out, "I didn't get my drink of water!"

Teacher decodes: "You think your rights have been violated, but you chose not to go."

Sam: "You're not fair to me! You're jumping down my throat!"

Teacher decodes again: "I can see how upset you are. Let's talk about what exactly happened."

Sam: (Grimacing.) "Nothing happened."

Teacher: "Maybe you didn't make a good choice and got left out."

Sam: (Looks at floor.)

This form of decoding is particularly difficult for adults because it requires us to make interpretations and be blatant in bringing out hidden feelings. We often are somewhat unsure of the accuracy of our interpretations of a student's behavior. As people who are sensitive to the feelings of students and know the value of a calm environment, exposing hidden, unpleasant feelings also has potential to reinforce unpleasant feelings, which disrupt and cause greater outbursts from students.

SUMMARY OF DEVELOPMENTAL ANXIETIES AND DECODING BEHAVIOR

In doing effective LSI, it is essential to have an understanding of the feelings associated with the inappropriate behavior a student has used in response to stress. Without this understanding, it is not possible to teach a student to self-regulate behavior and the emotions that drive the behavior. We find out about feelings by objective observation of behavior and body language, questioning for information about the sequence of events, assessing the level of emotional intensity in the student's behavior, and careful listening to what a student says (and avoids saying). We put this information into a framework of five sequential anxieties which occur as a natural part of social and emotional development: abandonment, inadequacy, guilt, conflict, or identity. When you understand which anxiety a student is protecting, the student's inappropriate behavior can be understood as defensive strategies for protection from further psychological pain.

This understanding is communicated to a student in a process called decoding. Used extensively during the first three steps of LSI, decoding is the means for making connections for the student between behavior and feelings. Depending upon the student's behavior, there are four different types of decoding. The least complicated form of decoding requires only acknowledgment that feelings are present. More complicated forms of decoding involve interpretation of surface feelings conveyed through behavior, and secondary interpretation when a student denies the first decoding. The most difficult type of decoding requires interpretation of feelings that are masked by a student's opposite behavior.

By decoding, an adult conveys support and understanding of what a student is feeling during crisis. Decoding that is accurate and supportive enhances the trust between student and adult. The student sees the adult as understanding, and a reliable source for assistance in a difficult time. The student also begins to learn about the connections between behavior and reactions from others. And by changing behavior, the student learns that outcomes can be changed for the better.

EXAMPLE OF DEVELOPMENTAL ANXIETIES

To be effective with LSI, decoding is an essential skill. It requires practice and knowledge about students' feelings and anxieties. To illustrate how developmental anxieties are used along with decoding, let's go back to "The Cap and Gown Crisis" example at the end of Chapter 1. We always begin our search for the predominating anxiety by considering the student's actual age. We look for evidence of how that anxiety is being expressed. This gives a key to what feelings and anxiety would normally be present as a process of social and emotional development. Then we observe the defensive behaviors actually used in LSI to see which defenses are being used and in which specific situations. From these observations, we consider other anxieties that may also be present and perhaps may be a central issue for the current crisis. With this information, we decode it for the student in response to the student's specific, observed behavior. Let's try this process out with the high school student (S) participating in LSI with his homeroom teacher (T) from Chapter 1:

AGE OF STUDENT: 14 years

DEVELOPMENTAL ANXIETY FOR THIS AGE GROUP: Identity ("What sort of person am I?")

HOW STUDENT CONVEYED THIS CONCERN: Student attempts to convey himself as an independent person of "responsibility," defending his choice to get his brother's cap and gown rather than follow the directions of adults.

DEFENSIVE BEHAVIORS OBSERVED: Denial that he did anything wrong; projection of responsibility onto peers; displacement of frustration and anger toward adults who appeared to be unfair to him, by seeing them as incompetent to handle other, difficult students; displacement of feelings by increased movement and restlessness; and occasional psychological withdrawal by looking away from teacher or out the window.

ANOTHER DEVELOPMENTAL ANXIETY: Conflict between the desire to be independent (making a choice that ran against adults' rules and expectations) and wanting approval from adults for being responsible.

DECODING

Acknowledging Feelings

> **S:** . . . and when I came back she wrote my name down. (Big sigh.)
>
> **T:** You were trying to do something for your brother and got into trouble.
>
> **S:** . . . It was important, so he could graduate!
>
> **T:** That's an important step in his life, and you felt like you had a big responsibility.

Surface Interpretation of Feelings

> **S:** . . . I thought I was going to get sent home.
>
> **T:** That concerned you. You were worried about being sent home.
>
> **S:** . . . I've tried everything!
>
> **T:** It sounds like you're kind of frustrated.
>
> **S:** . . . I used to get into trouble all the time.
>
> **T:** Maybe this is why you are so upset about what happened today . . .

Secondary Interpretation of Denial of Feelings

> **S:** . . . I didn't realize she told me.

T: Maybe you didn't want to hear her.

S: (Kicks table leg.) No! That's not it!

T: Sometimes adults give students directions that seem to mess up what a student wants to do, and then there is lack of communication.

Interpretation of Behavior Opposite of Feelings

S: . . . And I don't think I should be punished for getting my brother's cap and gown.

T: It isn't easy to face punishment when you thought you had done something responsible.

In going through "The Cap and Gown Crisis" line by line, you probably noticed that this student was in rational control all the time. He was attempting to solve the crisis that he had gotten himself into but lacked skills to do so. An underlying current of feelings was also conveyed by his remarks and body language. Without LSI, the crisis would have played out into an expanding Conflict Cycle with the student feeling increasingly negative about himself and others.

To verify the accuracy of the decoding, go back through the LSI again to locate the lines where decoding was used. Look at the student's responses immediately after the teacher decodes. The student reacts nonverbally in each instance. Sometimes the reaction is movement toward the teacher; other times the reaction is heightened restlessness; or, when the decoding is highly supportive, the student continues to expand on the decoded message. These are all responses that verify the accuracy of the teacher's decoding. You can see it culminating during Step 3 when the teacher first shapes the issue as conflict between independence and following rules. Then, the decoding is further verified by the student's body language at the end of Step 3 when the teacher has determined that the student is ready to resolve this crisis by facing the principal with independent responsibility.

WHAT WE HOPE YOU HAVE LEARNED FROM THIS CHAPTER

- Why it is important to recognize students' feelings when using LSI.

- A general definition of developmental anxieties.

- Five types of developmental anxiety and the general causes of each.

- The typical age range when each developmental anxiety emerges as a part of normal personality development.

- Characteristic behavior of students with each of the developmental anxieties.

- How adults can combat each developmental anxiety.

- A general definition of "decoding behavior."

- Why decoding is essential for doing effective LSI.

- How you can tell if you are decoding accurately.

- Four levels of decoding and the purpose of each.

- The LSI steps in which decoding is most frequently used.

- How observations of defensive behavior, understanding of developmental anxieties, and decoding are used together in effective LSIs.

"Who's in charge here?"
"Who's going to handle
   this problem?"
"How do I know it's
   going to be all right
   for me?"

Chapter 4

To your students, you are the representative adult and authority, their surrogate for adults in general. As such, you will be the "stand-in" and recipient of a student's behavior, feelings, and attitudes toward other adults. Students bring a history of experience with other adults into every classroom and crisis. Your own behavior and intentions may be the best, your relationship with students amiable, but there is no avoiding the student's history with other adults as you and a student interact. The student has learned to respond to adults in certain ways from past experience; this behavior is what you will receive from a student in crisis. Three questions from this history arise repeatedly during LSI: Who is the authority? Whose responsibility is it to come up with a solution to the crisis? What assurance does the student have that personal needs will be considered?

These are the concerns of every student in crisis—worries about emotional security and what adults will do. The first question reflects the need of all students to be reassured that a higher authority (adult power) will not let events get out of control. All children and youth want to believe that anarchy will not reign and life can be experienced with a minimum of distress. They view adults as the ultimate authority for this reassurance. The second question conveys students' uncertainty about the extent to which they can rely on adults to handle problems. The third question conveys the doubts students have that their own needs will be considered as a crisis is resolved. This concern triggers students' tendencies to rely on established, defensive behaviors to protect themselves and meet their own needs. Tapping into such needs provides motivation for independent action, but this independence can produce unacceptable rather than constructive behavior.

During LSI, such pervasive concerns are always present. Unresolved, they preoccupy a student and dominate the LSI. The specific crisis will be eclipsed and crisis resolution will not occur. To use LSI effectively, an adult must recognize the specific form these concerns are taking and deal with them. It is essential to recognize that the student's behavior is probably not directed against you as an individual but against you as the representative adult. It also is essential to understand a student's emotional attitudes and history of relationships with adults. We suggest that you actually try to answer each question for yourself at the beginning of every LSI. Then, you must be able to convey answers to the student that will be most beneficial to the student's progress. Because the meanings and feelings behind these questions are complex, we use this entire chapter to provide a background. Then in Parts Two and Three you will find these ideas put into practice throughout LSI. Let's take the questions one by one.

"WHO'S IN CHARGE HERE?" (Authority, Adult Roles, and Influence

Authority is a term with many meanings: those in command, the source of support for a statement or defense for an action, power to influence thought or behavior, independence or freedom granted, and expertness. We use the term here to include all of these meanings because the adult using LSI effectively must be able to convey these many forms of authority, as needed. What we *do not* mean in using the term is blind obedience, which defines a different approach—authoritarian.

Adult authority stands among the most dynamic forces in the lives of all students. Whether accepting adults or rejecting them, students generally acknowledge the authority and power of adults. They also have strong social and emotional needs for adults, even if they reject adults. Students recognize that adults possess special knowledge and skill; have power to order others around, make rules, decide what is right and wrong, judge, and reward or punish. Adults also solve problems, dispense care, and affirm students. Adults are the source of approval and disapproval. Students tend to see themselves in the mirror of the views adults hold about them.

At every age, students are influenced by their needs for certain behavior from significant adults. The way adults have responded in the past to those needs contributes to forming the behavior students use with adults in the present. Students' histories are a series of fulfilled and unfulfilled expectations. When students' expectations about adults have been met, they are left with feelings of well-being; adults can be counted on to handle problems, encourage and praise, provide, and do what is needed (right, just, kind, helpful, caring, bountiful). Unmet expectations leave students with a range of unsettled feelings and confusion about adults.

Changing Roles of Adults

Specific types of behavior from adults toward students convey different types of authority. In the normal process of development, the type of adult authority needed by children and youth changes as they mature. These changing roles of adults in the normal developmental process are well known. When used deliberately with students at particular stages of development, specific adult roles can promote students' social and emotional development. Table 4.1 contains a synopsis highlighting typical adult roles based on normal social and emotional development of students in various age groups. By determining where a student is in this developmental progression, you can select the general type of adult authority and role needed to facilitate a student's particular stages of development. It is not too difficult to keep a particular role in mind as you begin LSI with a student. When you do this, you will find the gap narrowing between adult expectations and student needs. This seems to reduce stress, and the tendency diminishes for a student to displace anger onto you as a stand-in for other adults who have failed to fulfill expectations in the past.

Adult Strategies That Influence Student Behavior

Once you have determined the type of adult authority and role you should convey in a particular LSI, you have a range of intervention strategies you can choose from to convey a particular role effectively. Because there are so many strategies to choose among, we find it helpful to group strategies according to style of influence and fit them to an individual student's view of adults. Adult styles of influence, when used selectively, cause students to do spontaneously what they might not ordinarily do on their own. When an adult is able to assist students in spontaneously changing behavior, a major gain has been made toward independent regulation of behavior.

The most frequently used strategies for adult influence can be grouped into four general forms: motivation, friendship, skills to share, and fear of consequences. Here is a brief summary of each of these styles of influence.

Motivation. Motivation is the preferred form of influence, in which the adult is able to influence a student's behavior while

Table 4.1 The Changing Role of Adults

ADULT ROLE	APPROXIMATE STUDENT AGE	TYPICAL SOCIAL-EMOTIONAL NEEDS
Satisfier of basic needs	Birth to age 2	Babies need a particular type of adult behavior —nurture and care. To be effective, adults must provide security and a sense of comfort, conveying that the world is a pleasant place to be.
Teacher of basic social and learning skills, motivator, director of behavior	Ages 2 to 6	Preschool children need success and mastery in their first steps toward independence.
Facilitator of group process, upholder of law and order, expert teacher	Ages 6 to 9	Primary grade students need new learning to expand mental and physical achievement and group skills.
Group leader, advocate of individuals, social role model	Ages 10 to 12	Upper elementary students need guidance as they shift from external to internal regulation of behavior and form new attachments to peers.
Counselor, advisor, confidant	Adolescence	Youth need models for values and new roles for independence, identity, new relationships, and responsibilities.

Note. Adapted from *Developmental Therapy in the Classroom*, by M. M. Wood, 1986, Austin, TX: PRO-ED. Reprinted with permission of the publisher.

not directly controlling it. This style of influence emphasizes the student as the central player and works through motivation to foster spontaneous behavior control and participation. The emphasis is on using indirect, motivating strategies that enable students to participate spontaneously and self-direct in acceptable ways. Indirect strategies and highly motivating materials provide intrinsic feedback for success, catching students' interest and leading them to successful outcomes without the appearance of adult influence.

Peer pressure, mobilized by an adult, is another example of influence through motivation. When students participate in an activity spontaneously, or without resistance, they have been motivated by some intrinsic aspect of the activity that seldom is overtly associated with adult control.

Strategies that influence students through motivation are highly effective with students who are struggling with the developmental anxiety of conflict between the need to be independent and the need to be cared for by adults (see Chapter 3). Students who think they need to be independent of adult control often relax their defenses when they begin to experience success on their own with materials and activities and then receive abundant recognition from adults and peers for the accomplishment. Strategies that motivate also are effective with students who distrust adults intensely and those who have developed passive-aggressive strategies. Passive-aggressive students have highly developed skills in manipulating without confronting peers or adults directly. They are experts in manipulating for social power. When you counter with effective motivation, it allows them to participate and experience success in an endeavor without the problem of confrontation. At some point, however, adults working with such students will need to switch from this indirect form of influence as the student becomes less defensive and a greater relationship between adult and student develops. Direct verbal confrontation usually is necessary at some point with these students to directly face the impact of their unacceptable behavior (see Chapter 8 for more about how this is done during LSI).

Friendship. Friendship is a form of adult influence built from intervention strategies that rely on the adult's personal characteristics and relationship with the student to influence the student's behavior (see Chapter 11 for more about introducing concepts of friendship to students during LSI).

The goal in using strategies that influence students through friendship and personal characteristics is to convey your recognition and approval of students' positive qualities, thereby strengthening confidence in themselves as valued people with qualities admired by you, other adults, and other students. It is a form of influence that promotes qualities like helpfulness, fairness, kindness, leadership, and honesty in students.

Students respond to adults who have characteristics that they admire and like. Examples of these adult characteristics are warmth, humor, enthusiasm, relaxed friendliness, fairness, helpfulness, and approval. Students also are influenced by mannerisms of admired role models. Because many students imitate these appealing adult characteristics, it is essential that adults provide such models of behavior if they want to influence these students. Praise and affirmation are other major strategies for this form of influence. Varying voice tone can be used to effectively convey genuine interest, support, caring, surprise, or to pique motivation for an activity. Such strategies are highly effective with students who identify with adults as role models and seek adult relationship.

Adults are using relationship as a way of influencing students when you hear them say,

> "Do this for me."
>
> "I really like it when you do that."
>
> "That is really good work!"
>
> "I think that's the best you've ever done."
>
> "I'm proud of you."

The effectiveness of this form of influence is evident when you hear students say,

> "Teacher, how do you like this?"
>
> "I knew you wouldn't like what they did."
>
> "I made this for you."
>
> "I did what you wanted me to do."

You also can see the results in nonverbal behavior: students looking toward adults at a time when others are misbehaving, listening quietly and intently to what an adult is saying, directing the bulk of their social communication toward adults, spontaneously recalling what an adult has said to them on previous occasions, and volunteering bits and pieces of their most private thoughts and feelings.

Strategies involving friendship and the adult's personal characteristics are most effective when they are used in combination with other forms of influence. Friendship strategies and those that rely heavily on the adult's characteristics provide positive social role models and a personal, caring dimension to adult-student relationships. But when they are used by an adult as the primary source for influence, such strategies may promote personal dependency that restricts a student's development toward independence. Such student-adult relationships tend to turn into person-to-person bonding where the adult does not exercise necessary authority and the student fails to assume independent responsibility. Separation from such a relationship is very difficult for both student and adult.

Shared Skills.   Shared skills is a category of strategies that convey the adult's ability to help a student solve a problem that the student cannot handle alone. The goal in using strategies of this type is to teach new skills for social problem solving by providing models in the adult's own behavior and sharing skill and insight about the problem. The adult uses strategies that support a student through the waves of self-doubt, recrimination, panic, and anger brought on during crisis. The adult affirms the student's strengths, reviews social interactions, and interprets the behavior of others for the student. Then, together they come up with a plan for use of new ways to behave to avoid such distress in the future (see Chapter 10 for further description of this process during LSI).

A high level of trust develops between a student and an adult who shares expertise to help the student understand events and solve a crisis. Students also recognize an adult as an expert when a crisis is managed successfully and calmly by the adult. Students respond to the expertise of the adult who decodes their behavior and that of others. Their responses to an adult's expertise may be reluctant at first, but when they begin to observe, listen, and think then the decoding is accurate.

Sharing skills and expertise is misinterpreted by some adults who present themselves to students as "skilled superiors" or who display their own particular accomplishments for the students to admire (and emulate?). Remember, the purpose is to help the student solve a crisis that the student cannot solve alone!

An adult is using strategies that share skills and expertise with students when the adult makes statements like these:

"You don't need to worry, teachers are in charge here."

"I've had other students with this problem, and this is how we handled it."

"I've been watching how you were doing in that difficult situation; you figured it out correctly."

"Your behavior tells me that something else is really bothering you."

When adults are effective in helping students solve problems, the result is trust, admiration, and a positive relationship with the adult. An adult who shares skills with a student in crisis also creates a climate of security where the student has a new sense of freedom to learn because energy is not misdirected into defensive behaviors.

Consequences. Fear of consequences underlies a category of strategies that influences students' behavior. Often, students respond to a reality consequence as if it were a personality issue between student and teacher, when consequences should be seen as a result of the student's behavior. Some forms of discipline, "contract" programs, and reward/reinforcment systems modify behavior through sophisticated versions of this type of influence. The underlying fear is that deprivation or punishment will result if the student fails to meet behavioral expectations. Examples of adult strategies that influence through reality consequences include privileges denied, withholding adult approval, time out, authoritative or physical proximity (standing over a student), hostile verbal confrontation, restriction, suspension, and exclusion.

When adults continually use strategies that are threatening, as opposed to giving students choice with reality consequences, students are left in a powerless position. Powerless, students either give up, which arrests further development; or begin to acquire negative attitudes, defensive behavior, and resistance to adults. When this happens, anger often is subverted to passive aggression or explodes in violent or cruel acts.

With a few exceptions, these strategies seldom contribute to creating the sort of environment conducive to successful LSI. We only use such strategies with students who are so entrenched in their own unacceptable behaviors and alibi systems that they have great difficulty accepting the idea that change in behavior is needed (see Chapter 8). Students who have been in environments where adults have used coercion and fear against them extensively often rely on these types of behavior themselves. With such students, you may need to begin as a powerful authority, using fear of consequences to maintain law and order. This may be the only type of authority they understand, intitially. All students must know that an adult (authority) is in control and that they are physically and psychologically safe. When you find it necessary to use such strategies to influence students' behavior, keep it short and look for ways to change to strategies based on other forms of influence. You will find the climate changing rapidly, in a positive direction, when you switch your strategies.

Summary

The answer to the first question raised in this chapter, ''Who's is charge here?'' is simple: It must be the adult! But this answer does not imply authoritarian control. Rather, it requires an adult to convey the type of adult role that will produce the greatest possible social and emotional development for the student.

Some students need clearly visible adult action that cares for, controls, and directs the student. Other students need a less central, less visible adult so that their independence and personal responsibility will develop. Using the chart presented in Table 4.1, you can identify the general type of adult role that each of your students needs for optimal development.

Social influence is an attempt by an adult to influence the behavior of a student. The effectiveness of your social influence with students will be the degree to which they do what you expect them to do. Most strategies for managing behavior can be grouped into one of these categories: motivation, friendship, shared skills, or fear of consequences. Adults exercise these forms of influence over students almost continually; sometimes intentionally, other times unknowingly. The extent to which an adult is effective depends upon an accurate fit between the strategies the adult uses and the student's present attitudes and past experience with adults. An effective adult must be able to understand what type of influence is needed for a particular student in a specific crisis and use the strategies that convey it, as the need arises.

Study the strategies you typically use to influence students. If you find that you are relying on only one adult role and using only one or two types of strategies to influence your students, broaden your repertoire. To be effective in LSI, you should be able to use strategies from all four types of influences to convey the adult role needed by your individual students.

"WHO'S GOING TO HANDLE THIS PROBLEM?"
(The Existential Crisis)

The previous discussion focused on adult authority and the absolute need for adults to influence the psychological and interpersonal climate in positive ways during crisis. An adult must always have control of the environment during crisis. But this does not imply that the adult is responsible for the solution to every crisis. In this section, we consider the second major question raised in this chapter—Who resolves the crisis? Is it the adult's or the student's responsibility?

Few students are capable of managing crisis resolution independently when LSI is needed. Yet in LSI students must begin to learn how to assume responsibility for their behavior and its consequences. The long-term goal of LSI is to teach students to regulate their own behavior and meet their own emotional needs in appropriate ways without dependence upon adults. We view this gradual transfer of authority and responsibility for problem solving from adult to student as occurring in three phases, centering around a major developmental event in the lives of all children and youth. We call this developmental event the Existential Crisis.

The Existential Crisis is a phase in social-emotional development when a student's belief in the absolute, omnipotent authority of adults begins to falter. It is called existential because it deals with new awareness about the limitations of adults for taking care of life events in satisfactory ways. This uncertainty about adults leads a student to question authority and to look for other sources of security. It is called a crisis because it raises concerns about alternatives to adults taking care of problems, with the possibility that the student must be responsible. Most students doubt their own ability. Yet they must change, learn new behaviors, and take on new consequences. The challenge is nearly overwhelming to some students.

As we begin each LSI, we have found it helpful to determine if the student is in the preexistential phase, going through the Existential Crisis, or in the postexistential phase. If the student is preexistential, the LSI will need to be essentially directed and controlled by the adult. If the student is actually going through the Existential Crisis, the LSI must provide abundant adult support and direction but also must begin to shift responsibility for crisis resolution to the student. For students in the postexistential phase, the LSI focus is on assisting them to handle crisis with independence and a focus on personal responsibility. To help you recognize which of these general phases your students are in, we briefly review the dynamics involved and the management strategies that are appropriate for each phase.

The Preexistential Phase

In the preexistential phase, a student's emotional need is for an expert and all-powerful adult who directs behavior, makes rules, judges violators, punishes the guilty, handles all problems, provides protection, and provides what is needed for satisfactions. When students conform, adults like them and therefore will reward them.

To preexistential students, adult authority is all-powerful. Every first-grade teacher knows about preexistential students:

> "Teacher! Teacher! . . ."
>
> "He's not playing by the rules."
>
> "Make her give it to me."
>
> "He's bothering me."
>
> "I'm going to tell my mother."

Notice how each remark is directed to the adult, even though the issue involves peers. Such responses are normal for students under age 9. Adults are seen as benevolent authority, dictating and enforcing standards of behavior. In the view of preexistential students, adults are responsible. These students look to them for protection, solving problems, and maintenance of law and order.

Adults use many behavior management strategies that suit this stage of development and that are useful during the problem-solving steps of LSI with preexistential students. Here are several examples:

- Token systems
- Fixed classroom procedures
- Contingencies for adult approval
- Systems of concrete rewards (smiling faces, grades, treats, privileges, selection by teacher as "leader," checks)

- Rules and prescribed punishments
- Adults as behavioral role models

When you are using LSI with older students who are still preexistential in their development, these management strategies can provide only short-term solutions to crisis. These students expect to be controlled by outside force. However, if you continue to rely only on these approaches, the students will remain preexistential; that is, they will continue to look to you as the enforcer, the one responsible for finding solutions to their problems. At some point you will have to take these students through the Existential Crisis and into the postexistential period. The management issue here is one of gradually shifting the source of authority and responsibility from external control to controls from within. It will *not* happen if you continue to maintain external control at all times; students will simply continue to rely on you to set the standard and then will test and defy you as they experiment with independence.

During LSI, there are many opportunities for learning self-regulation. Much self-regulation comes through the opportunity to explore alternatives and make choices. Throughout LSI there are opportunities for individual input, exploration of ''what happens when . . . ,'' alternative rules to live by, and alternative behaviors to choose. Every successful choice puts responsible, independent development one step toward self-governance and responsibility. Every failed choice halts the progress and leaves authority, responsibility, and control with the adult.

The Existential Crisis

Normally, this period starts about the time children begin school and continues for several years, until the student has completed the shift from relying on adults as the sole source of authority and behavior control (preexistential) to a view of authority as coming from more than one source, including oneself, and assuming responsibility for control of one's own behavior.

During the Existential Crisis, students have doubts and uncertainty about adults. This leads to testing adult authority and credibility, defying directions, ignoring rules, denying responsibility, and shifting blame to others. During this phase, students vacillate unpredictably, one minute trying to conform to adults' expectations and then defying these same standards, losing control, regressing, or assaulting others.

Examples of strategies that are useful during LSI with students who are in the Existential Crisis phase include:

- Positive feedback and praise from adults
- Verbal reflection of words, actions, and feelings
- Interpretation of meanings behind behavior (decoding)
- Adult responsibility for rule maintenance and consequences

The volatility and lack of stability in student-adult relationships make this period particularly difficult for many adults. It requires us to use "preexistential" strategies (adult authority and control) at one time and then switch to "postexistential" strategies (adult guidance but not control) when a student begins to show control and responsibility for self-management. By recognizing and understanding the dynamics of the Existential Crisis, you can be prepared to provide changes in your own strategies, loosening and tightening control as needed.

Should we be concerned about younger students who go through the Existential Crisis too soon because of life experiences that have thrown them out on their own at an early age? What about older students who have failed to resolve the Existential Crisis satisfactorily? Will such deviations in the normal sequence of development affect their behavior and the way they are managed in the classroom? The answers are clearly "Yes!" These will be controlling, testing, untrusting, manipulative students who do not take adult direction yet avoid responsibility for their own behavior. They cannot accept authority and cannot allow adults to direct them. They seek reassurance and nurturance while lashing out at the adults from whom they seek approval and support.

The Postexistential Phase

As students turn increasingly to peers for behavioral models and affirmations of themselves, there is gradual detachment from psychological dependency on adults. Adults now become mirrors for students to see themselves as others see them and encourage students in independent, self-regulated behavior. As this course to independence is charted, adults are needed as backstops when challenges become overwhelming and behavioral skills are insufficient for successful navigation through crisis.

Effective teachers in upper elementary school and high school know about postexistential students and have teaching styles that allow independence to have its day, while providing sufficient direction for students to achieve success in independent personal and educational endeavors. These adults serve as role models for effective interpersonal relationships and independence in problem solving. Postexistential students know they have responsibility for self-regulation and generally exercise some degree of responsibility for crisis resolution. In LSIs with postexistential students, you will see them making an effort to resolve the crisis, making suggestions that show a sense of responsibility for the crisis and the resolution.

Summary

To summarize the answer to the second question in this chapter, "Who's going to handle this problem?": It must be the student, with adult support, to the extent that the student needs an adult in order to have a satisfactory resolution. To determine this need, we recommend using the three-phase guide summarized in this section. Begin every LSI with a quick review of how your student views adults. Some students will have the preexistential view that adults must solve problems and are responsible for crisis resolution because they are in charge. Other students will be experiencing the Existential Crisis, where they will swing between attempts to solve a crisis on their own and retreat to a preexistential position, defer to adult authority, and deny responsibility. Postexistential students will make attempts to resolve a crisis, even if their skills and judgment are seriously lacking. From this information, you will be able to balance the extent to which you assist and direct a student during each LSI or leave the selection of a solution and implementation plan to the student.

"HOW DO I KNOW IT'S GOING TO BE ALL RIGHT FOR ME?" (Motivation to Be Responsible)

The previous discussions focused on adult authority and the absolute need for adults to influence the psychological and interpersonal climate of LSI. We also described the changes students go through during the Existential Crisis as they begin to doubt the absolute nature of adult authority and learn that problems must be faced by themselves (with adult support). The question we raise in this section considers a student's motivation to be responsible for personal behavior and crisis resolution.

While we are concerned with making the transfer from adult authority to student responsibility, the student is concerned that personal needs will be met. Students behave in ways that produce results they value. They also try to avoid causing results they do not value. Values are the internal rules students (and adults) live by. They are the fixed point of reference for daily behavior—the glue that holds together the structure we know as personality and character. What students care about is what they value. Their values shape their choice of activities, influence their reactions to stress, and regulate their behavior.

From a student's viewpoint, what the student cares about the most during crisis is to avoid the unpleasant feelings of stress. This is the fundamental value. The Conflict Cycle we presented in Chapter 2 is fueled by this motivation for self-protection. Any attempt to break the cycle must guarantee that emotional hurt will not continue. Satisfactory resolution of crisis must be constructed from this fundamental value. Unless emotional protection is provided, students will not be able to change established patterns of defensive and unacceptable behavior. Students must believe that, in the end, they will be all right. When they believe this, they will participate in crisis resolution. If they doubt that their own interests are being considered, they will raise their defenses.

But assurance of emotional protection is only the first element and not the full story in successful crisis resolution. Central to every goal in LSI is the idea of learning independent responsibility for behavior. This gradual transition from guaranteed protection to independent, self-regulated behavior is not an easy task. To exchange dependence on others for independent responsibility requires a powerful motivation.

How do we provide this motivation for behavior change? Think of it as building solutions to crisis based on (a) a gradual change in a student's values and (b) a gradual shift away from emotional dependency on adults toward emotional independence. Following is a review of these ideas in a framework that should be helpful in identifying your own students' values and level of emotional dependency. From such information, you will be able to build motivations during LSI for them to change their own behavior.

Identifying Students' Values

Students' values develop in a natural sequence that can be summarized in five general stages: own needs, adult approval, fairness, responsibility for self, and responsibility for others. The basic value begins with the premise that one's own needs are paramount.

Any behavior that meets personal needs is satisfactory. Gradually, students move from that orientation to a belief that adults' standards are the ones that bring personal benefits; therefore, their behavior should conform, to please adults and avoid punishment. As students develop further, their views broaden to include justice and fairness as valuable guidelines for regulating their own behavior. When fairness first emerges as a value for regulating behavior, students are concerned that they receive fair treatment from others. Gradually, their view of fairness expands to include fairness for others. Finally, students embrace society's values of responsibility for self-regulation, justice, and care of others. Empathy and altruism also are added as values that regulate behavior.

Figure 4.1 summarizes this sequence of values and provides examples of solutions to crisis that would be considered satisfactory to students holding these particular values. (We refer to this sequence again in Part Two, Step 4 of LSI.) In this illustration, the band of dotted lines indicates the approximate place in the developmental sequence when the Existential Crisis occurs (discussed in the previous section). The two concepts tie together—students' changing values and their changing views of authority. Concern about personal needs begins to blend into the notion that satisfactory interactions with others bring satisfactory results for oneself. With this development, a student's sense of responsibility begins to broaden to include the possibility of giving up in order to get in return.

The arrows in Figure 4.1 indicate ways to challenge students to consider solutions at the next step in the sequence of values. As you bring LSI to the solution steps, you have two basic choices: (a) encourage students to choose solutions at their present stage of values or (b) challenge students to consider solutions at a higher stage. Sometimes it is necessary to guide students to choose solutions to crisis that are a part of their current value system. These solutions may include results that will bring specific benefits or better feelings, concrete incentives, adults providing rules and highlighting consequences, or adult affirmation and approval for "the right choice." These are preexistential solutions, necessary because the student may have tenuous confidence in making

Figure 4.1 The Sequence of Values Used to Determine "Satisfactory" Results

Types of Values	Examples of "Satisfactory" Solutions	Challenge Student to a Higher Level Value
Own Needs	"What I want." "What feels good." "Look after yourself first."	

Adult approves of student.
Rules to follow.
Leave it to adults.

Adult Approval

"I do what I'm told so adults will approve of me."
"Good people obey; bad people deserve to be punished."
"If you break rules, you get punished."

Reasons for rules.
Right vs. wrong.
Respect others' feelings.
Adults model fairness and value what's right and fair.

- -

Fairness

"Fairness is right for me."
"Kids have rights."
"Doing things with others brings rewards to self."
"Do to them what they do to you."

Existential Phase

- -

Expand values to "kindness,"
"leadership," "friendship."
Live up to others' expectations.
Consider others' feelings.

Responsibility for Self

"What do others think of me?"
"Everybody does it."
"I don't want to be left out."
"People who like you will help you."
You can have friends by being a friend."
"It's important for people to like you."

Expand personal goals.
Emphasize personal goals.
Personal responsibility.
Social responsibility.
Fairness and justice for everyone.

Responsibility for Others

"How does my behavior affect others?"
"The system should protect everyone."
"Be a responsible person for society."
"Live by a creed."
"Fulfill obligations."

Respect self as a moral person.
Actions based on principles.
Think for yourself about others' needs.
Follow others who act on moral principles.

choices or doubts about your reliability as an adult authority. Such students will resist making choices, leaving it up to adults instead. However, if you continue to fill the adult-as-authority role too long, you may restrict the student's development, and responsible independence will not be forthcoming. When you notice themes of fairness in a student's remarks, you can be confident that the student is moving beyond the Existential Crisis and may be ready for some self-responsibility. Then it is time to challenge the student to consider solutions at the next stage. You probably will find the student receptive to new ideas for solutions with themes of friendship, leadership, consideration, responsibility, and kindness.

One caution about "challenges": Sometimes adults tend to foster solutions to crisis that are beyond a student's current ability to handle successfully. If a planned solution has no value to the student or the expectation requires too great a leap forward for the student to accomplish, the desired behavior will not be forthcoming and the plan will fail.

Identifying Students' Emotional Dependence

A goal of most intervention programs, like the goal for LSI, is learning independent control of behavior. And when a student exits a program, the expectation is that the learning will generalize (independently) into other settings. It is enormously frustrating to most adults when students fail to achieve this goal. This failure may indicate chronic emotional dependence. Emotional dependence sustains the preexistential values held by some students about adults (described in a previous part of this chapter). Students who are emotionally dependent still fear that their own needs may not be met by significant adults. As long as a student clings to this view, the student will not be able to assume independent regulation of personal behavior. The student will be preoccupied with meeting emotional needs and protecting feelings.

Another form of emotional dependency results when a student lacks or has failed to form primary attachments to an adult or has been deprived of emotional care and support. Attachment is a concept that describes the essential human need to be connected and cared for. This fundamental need for emotional attachment may supersede all other behavioral motivators. It provides children and youth with confidence that they are valued by others. Normally an infant's caregiver is the first source of emotional significance. When students' basic need for emotional attachment has been satisfied, they do not have to struggle to attain it. They use it as a base of confidence to venture beyond the attachment to pursue new relationships and independence.

If attachments fail or never develop, emotional dependence results. A student has to put major effort into seeking and holding on to substitutes. Failed attachment is a major part of the first developmental anxiety, abandonment (discussed in Chapter 3). Unresolved, it seeks many substitute forms of gratification (e.g., work, hoarding, overeating, addictions, etc.) that erode interpersonal relationships and can severely modify the course of an individual's personality development. Such students often lack empathy and sensitivity to the feelings of others. They bend rules to gratify their own needs, operating from a "Look out for number 1" base.

The developmental counterforce to attachment is separation, a drive for independence. Students show a willingness to forgo primary attachments as they mature, seeking independence. The success of the separation process depends upon the previous success during the attachment phase of emotional development. Successful separation centers around the balance adults achieve between providing security and comfort (dependency needs) and simultaneously allowing freedom (independence needs). This is the daily challenge for children, youth, and adults.

If the separation process is experienced without success, psychological independence may not be achieved. Fear of failure, fear of inadequacy, and fear of not measuring up to the expectations of others spiral into a restricted personality. Such students fail to take initiative, resist change and new experiences, pander to others, and follow those who will control and direct them (the developmental anxiety, inadequacy, described in Chapter 3). These students also may be selected as scapegoats or isolates by their peers.

Whenever the attachment or the separation process fails, anger is a major by-product. A student feels anger toward family members who have failed to nurture and protect; anger at those who have ridiculed and criticized the student's attempts at independence. These failed relationships from home and community permeate the student's view of adults at school. They bring this emotional baggage with them. And their teachers become the substitute adults, the recipients of the anger these students hold toward adults outside of school.

In contrast, resolution of the conflict between attachment and separation produces emotional independence. It is clearly observable in most students in upper elementary school. These students believe that the important adults really care and will provide emotional protection when it is needed, while permitting them freedom to be successfully independent. Such students have no need to be preoccupied with protecting themselves emotionally. Keep in mind, however, that all students continue to experience stress, and they never will be totally independent of their need for emotional support and relationships with adults they admire. Emotional independence increases in direct correspondence to the extent to which dependency needs (attachment) are met and independent experiences are successful.

Here is a simple example of failure to maintain this delicate balance between the dependence and independence needs of a middle school student. This brief exchange was overheard at an office telephone:

> **Jack:** Mom, when are you coming home?
>
> **Mom:** You'll know when I walk through the door.

In this simple exchange is a depth of implicit emotional need. These two lines are not sufficient to tell us exactly what the need is, but it is there. Consider the many possible needs. Jack needs to connect to his important adult. This may be an exchange that fosters independent behavior, or it may erode it. There may be connections and reassurance between mother and son, or it may be an expression of Jack's insecurity or Mom's insensitivity. Could it simply be a call made for something to do when things are boring at home? Permission to leave? For transportation elsewhere? To be relieved of baby-sitting a younger child? To talk about something important that has happened? For reassurance that Mom is all right? To reassure Mom that he is all right? Does she want to know? For the hope of protection against real or imagined terror at home?

And what about Mom's response? We do not know if this is her "style" or if she is conveying some of her genuine feelings about Jack. On her part, could it be a simple put-off because she thinks Jack should not bother her at work? Maybe she thinks Jack should learn to be on his own and not cling to her. She may resent Jack for reminding her of her burdens at home. Could there be remnants of her own conflict between emotional attachment and independence in Mom's response?

Jack's reaction to Mom's abrupt remark will also be drawn from the same attachment-independence needs. He will interpret and respond to Mom from either need, whichever is paramount for Jack. Does he hear it as a put-down? Rejection? Dislike? Criticism? Could he resent the restriction of staying home? Perhaps he resents adult control? Or perhaps this is such a standard exchange that neither Jack nor Mom thinks anything about it. If this exchange is typical between Jack and his mother, it is a legacy that he will carry with him—adults are indifferent to his needs and so, for emotional insulation, he will be indifferent to them. This behavioral defense comes to school with Jack. Then we have a style of teacher-student interaction that mimics the indifference in the parent-child exchange:

> **Teacher:** Jack, you must hurry along to finish your assignment before the end of this period.
>
> **Jack:** I'll do it when I feel like it. And maybe I won't do it at all.

Jack's behavior comes from his history with adults—the models they have presented, the feelings they have aroused, and the developmental anxiety that has become a part of his personality as he struggles to resolve the conflict between his growing independence with his latent need to be cared for by others. The strength of his feelings about adults will play out in LSI between Jack and his teacher.

Summary

To summarize the answers to the third question in this chapter, "How do I know it's going to be all right for me?": Students must believe that their emotional needs will be met. Until they are secure in this belief, they will not move far toward appropriate, independently regulated behavior. Students will take "ownership" of a solution to crisis when they can see it to be of value to themselves. These values are the internal regulators of students' behavior. To have an effective outcome of LSI, a resolution to the crisis must be framed within the context of current values held by the student. This belief that something beneficial will occur, in exchange for using a new behavior, is a fundamental motivation for change.

During LSI it is essential to determine what a student really values. This information then can guide you as the LSI shifts focus to solutions. A solution to crisis that takes into account a student's values will be accepted as satisfactory. When a student believes that a solution is possible and desirable, the student will have "ownership." To produce lasting change from LSI, solutions must be owned by the student, not the adult.

The gradual transfer of responsibility from external control to appropriate, independent, self-regulated behavior also depends upon successful transition through the Existential Crisis, from emotional dependence to emotional independence. Emotional dependency is a major roadblock toward independent responsibility. Students must experience freedom from concern about protecting their feelings and meeting their own emotional needs, if they are to achieve emotional independence. When you use LSI, awareness of a student's level of emotional dependency is essential. It is a major force in the eventual progress made toward accomplishing LSI goals.

SUMMARY OF AUTHORITY, RESPONSIBILITY, AND MOTIVATION

The topics covered in this chapter are sometimes difficult to apply at first reading. Although we included a summary at the end of each topical section, it may be helpful to integrate the information by showing an application. The concepts we are dealing with are adults' roles, influence of adults, students' existential phases as they change from outer to inner directed behavior, personal values that regulate students' behavior, and the extent of students' emotional dependency. The example of Harry below illustrates how you can use this material. We have identified Harry's shifting values, using Figure 4.1; the three-phase guidelines for Harry's position in relation to the Existential Crisis; the various adult roles and influence, using Table 4.1; and the extent of Harry's emotional dependence, using observations of Harry's behavior and remarks.

It is not surprising to find that age does not necessarily determine views about authority or the values students use. We see numbers of older students responding to crisis with simple, self-protective values. We also see students who attempt to use more complex values. When they are not successful in their independent attempts to solve a crisis with these values, they tend to slip back to more primitive values. Unfortunately, it often is an adult, using inappropriate values and ineffective forms of influence, that provides the inappropriate model. This is a brief example of such a student and the adults who are involved in his crisis. Here is what happened.

HARRY

A fight breaks out on the middle school bus. Harry is a known instigator and has a history of fights and suspensions. His view is that he is being picked on by the others and is simply defending himself. The scenario unfolds in a typical way. The angry bus driver reports Harry to the assistant principal. Harry cools his heels in the outer office for an hour or so until the assistant principal returns for a talk. The result is a call to Harry's mother and another suspension. Harry (H) explains all of this to the coach (C) as he prepares to leave school.

> **H:** I got busted again.
>
> **C:** The bus?
>
> **H:** It wasn't my fault! Leon pulled a knife and I told him to put it away.

**VALUE: Modeling adult standards of right versus wrong.
EXISTENTIAL PHASE: Postexistential. Student tries to handle the problem independently.**

> **C:** Then what happened?
>
> **H:** (Defensively.) It's my right to protect myself!

VALUE: Rights for self. EXISTENTIAL PHASE: Postexistential.

> **C:** You thought Leon might really be serious?
>
> **H:** You can't ever tell about Leon!
>
> **C:** So, what did you do?
>
> **H:** I told the driver that Leon had a knife.

**VALUE: Adults are powerful and should protect students.
EXISTENTIAL PHASE: Crisis. Independence was not working, so adult authority was needed.**

> **C:** What did she do?

H: She told us to get off the bus or settle down. Then she turned us both in.

C: Now you're suspended! What will happen when you get home?

H: I guess my mother will get mad about them throwing me out of school when it wasn't my fault. So, she'll come down here and straighten out that principal. She won't let them pick on me!

> VALUE: Adults are powerful. EXISTENTIAL PHASE: Preexistential.

C: Will she punish you?

H: Naw, she'll wait for my dad.

C: What will he do?

H: He's always telling me to stick up for myself. He says there's only one way to settle some scores. He'll tell me, "Next time, Harry, just beat the shit out of Leon, so he'll know not to pick on you again!"

> VALUE: Adult authorizes punishment for bad kids. EXISTENTIAL PHASE: Preexistential.

Harry tries to solve this problem on his own, at a fairly advanced value level. Without success from his first attempts to handle the problem himself (postexistential), Harry gradually regresses to a preexistential attitude that it is someone else's problem (his mother's) and he will follow the authority of the powerful adult in his life (his dad).

Track back through this scenario again to identify the vacillations in adult roles and types of influence conveyed to Harry:

Bus driver [Sent to office for punishment.]
Role: Upholder of law and order
Influence: Fear of consequences

Assistant principal [Sent home for punishment.]
Role: Upholder of law and order
Influence: Fear of consequences

Coach	[Not shared skills; perhaps friendship?]
	Role: Counselor, friend
	Influence: Not clear
Mother	[Wait for father.]
	Role: Advocate
	Influence: Shared skills [taking care of basic need for protection] and fear of consequences
Father	*Role:* Teacher of behavior
	Influence: Fear of consequences

Harry's attempts at emotional and behavioral independence crumble under pressure. We can be fairly confident that Harry is still emotionally dependent on the approval, power, and authority of adults.

Compare this analysis with Table 4.1, which outlines the general sequence of adult roles needed by students to develop social and emotional maturity. It is clear that each adult is responding to Harry from a personal "style," with no thought of Harry's potential for social problem solving and what he needs in order to learn to regulate his own behavior and resolve the crisis independently. The coach was a key. He could have used this crisis to reinforce Harry's emerging attempts to solve problems in appropriate ways. It was a time for LSI!

If you had been the coach and carried the LSI forward, what role would you have taken? What type of strategies would have been helpful in influencing Harry to change his behavior? What existential phase should you emphasize? What value level should be used as you and Harry approach the resolution phase of LSI? These are the same questions to be answered as you begin every LSI.

EXAMPLE OF ADULT ROLES AND STUDENT VALUES

By now you should be quite comfortable with the LSI between student (S) and homeroom teacher (T) in "The Cap and Gown Crisis" at the end of Chapter 1. Let's go back to that example to see how the concepts from this chapter are applied in a full LSI:

STUDENT'S VALUES: Cares about being recognized for personal initiative, for being responsible, and for people to see student as mature. Student is ready for challenge to the top value stage for adolescents: being a responsible citizen and contributing to the group well-being.

VIEW OF ADULTS: Postexistential. Student tries to handle the responsibility of securing his brother's cap and gown. The LSI reinforces this level of responsibility while expanding the student's understanding of competing values (his own and that of authorities.)

ADULT	ROLE OF ADULT	INFLUENCE STRATEGIES
Security guard	Upholder of law and order	Fear (threat of punishment)
Substitute	Upholder of law and order	Fear (threat of suspension)
Principal	Upholder of law and order	Fear and shared skills
Mother	Advocate	Shared skills
Mr. Smith	Teacher	Motivation (successful class work)
LSI teacher	Counselor	Shared skills (knows how to help student)

In contrast to Harry, this student from the LSI in Chapter 1, struggling with his brother's cap and gown, gives many indications of emotional independence. This student does not seem to have the emotional pull to slide back as Harry did from his postexistential position. This student does not show that he will rely on adults to solve his problem, even though he seems to value what adults think of him. What is clearly evident is that this student needs affirmation for his positive attempts to be independent and responsible, and he needs support as he learns that self-directed behavior must be restrained when results may work against him. He is clearly ready for this insight, and the LSI in Chapter 1 is on target.

WHAT WE HOPE YOU HAVE LEARNED
FROM THIS CHAPTER

- The five general types of adult roles, and the social and emotional support these roles provide to students.

- A definition of an adult's social influence from a student's viewpoint.

- Four general types of influence used by adults in their daily work with students.

- The adult role and type of influence you typically use, as you work with students in crisis.

- The three-phase guideline for identifying a student's attitude about who should solve problems.

- Examples of intervention strategies to be used with students in the preexistential, postexistential, and Existential Crisis phases in their relationships with adults.

- The sequence of five general types of values that can influence a student's ownership of a solution to a crisis.

- How emotional dependence can keep a student from learning self-regulated behavior.

PART TWO

In Part Two, we describe each step of LSI with an introduction to the purpose and content of the step, a description of what happens in each step, a synopsis of what the adult should do and say, and suggestions about how to end a step and bridge into the next one. By dissecting the six steps of LSI, we hope we have given you enough detail to learn a basic structure for using LSI in any crisis, with any student, of any age.

In actual practice, you will find the LSI steps blending together. Often you will recycle back to repeat a step as new information or insight is brought out. As you begin your first LSIs, make copies of the charts for each step found in Appendix C, to keep with you "on-site." Gradually, as the steps become familiar, you will see many more ways to use LSI than we are able to describe in a single book.

The last chapter in this section contains our perspective on the entire "helping process." We believe it is important to view every LSI in this framework—as a helping process. We include a checklist of skills adults need for using LSI effectively. (This checklist also is included in Appendix B.) To illustrate how small errors in strategy can cause adults to lose the helping perspective, we also include a review of potential pitfalls.

We hope that you will find LSI such an effective strategy for helping students through a crisis that it will become one of your most frequently used interventions. Throughout your LSI experiences, keep focused on the end results you can obtain with this process: positive change in a student's views of self, ownership of solutions to resolve crisis, responsibility for changing behavior, and greater understanding of others' reactions. Gradually, over a series of LSIs, there should be a reduction in a student's emotional memory bank of negative thoughts and feelings. The student's changing insights, behavior, and underlying feelings begin to build self-esteem because of greater insight, better feelings, more impulse control, more positive feelings about self and others, and more positive self-regulated behavior.

"Talking about my past is like picking up a dead dog out of the road, that's been hit by a car."

(from a student)

Chapter 5

A GUIDE TO LSI BASICS

When you work closely with children and youth in crisis, their personality characteristics and patterns of behavior for maintaining an emotional comfort level become apparent. Some students become aggressive and willful, blaming others for their misfortunes. Others put gratification of needs first and feel little guilt. Still others become confused and disoriented under stress; while others withdraw from involvement, develop debilitating anxieties, experience profound guilt, or spend their energies defending against real or imagined inadequacies in themselves or others. Most of these students are unaware of their ways of dealing with stress and crisis. Unless they learn to recognize these behaviors and feelings in themselves, and change their responses to stress, the Conflict Cycle (see Chapter 2) will go on unabated. For many students, it is safer to believe that the world picks on them than to feel that they are "no good," "a misfit," or "stupid."

Students often are in crisis because they failed to recognize the meaning of events that led to the crisis. Their perceptions and insights about an event are distorted by their defenses and the intensity of their emotional reactions to stress. They fail to anticipate the reactions of others and seldom see how their own behavior contributed to the crisis.

Crisis is the result of students' unsuccessful attempts to cope with stress at a specific moment. Usually, their reactions to stress are emotional and defensive rather than rational. The incident touches their vulnerable private storehouse of thoughts and feelings; and they react with more feelings, words that upset others, and counterproductive behaviors. The result is the Conflict Cycle we describe in Chapter 2: Reaction to stress evokes negative reactions from others, which, in turn, cause more uncomfortable feelings in the student, more unacceptable behavior, and more negative reactions from others. From this cycle, a crisis is born.

LSI is a verbal strategy for providing active intervention in students' lives in times of crisis. It assists students to understand and cope with the specific crisis that they could not handle on their own. It is conducted by someone who is part of their natural environment and takes place as soon as possible after the incident occurs. The time shortly after crisis is a productive time for students to learn new social skills and gain new insights into their own feelings and behavior and that of others. Because current behaviors have failed them and emotional distress has intensified, students are more receptive to change immediately following crisis than at any other time. By using LSI, the adult conveys to a student that (a) problems can be solved, (b) the student has an advocate in the adult, (c) the adult sees the good qualities in the student, (d) the student has the skill to solve the crisis, and (e) even though the student must exercise constraints or curtailments, he or she is still important as a valued individual.

USING LSI WITH A GROUP

The focus of this chapter is on the basic LSI steps used with an individual student, but the application for use with a small group of students is similar. We use this same LSI process frequently with two students, or a group, when they have been party to the crisis. Group LSI sometimes follows an individual LSI when one student has played a central role in a crisis but the group has been involved. Bringing the issue back to the group emphasizes group responsibility and individual responsibility to maintain a satisfactory group. Sometimes we begin with group LSI, when a crisis is clouded by the involvement of everyone. As we go through the LSI steps with the group, several central issues begin to emerge. Usually, this is followed by one or more individual LSIs, as particular perpetrators, instigators, and victims are identified.

When using LSI with a group, the adult must take the central role and maintain control. Relationships in a group are usually tentative, volatile, and can be destructive to individuals if care is not exercised to provide psychological protection for everyone participating. Face-saving and posturing for peers is a major element that you will find in most group LSI. Individuals under stress must be supported when emotionally charged discussions begin. Individual group members also must learn how to participate in group discussions constructively and be respectful of each other's "space."

Morgan (1981) summarizes the group LSI process vividly in his illustrations of its use in therapeutic camp settings. He describes five applications for group LSI: organizing group functions, solving group problems, reinforcing group values and expectations, reinforcing a healthy group perspective, and creating awareness of group progress. Morgan also emphasizes the skills needed by an adult using LSI with a group: "The teacher must be aware of the prevailing group code, roles, and the power structure as basic conditions of leading a group LSI" (p. 37).

Similarly, Wood (1979, 1981) describes important developmental milestones for social and emotional growth that can be taught with group LSI. Here is a summary:

- Participating in group discussion appropriately.

- Interacting with peers in socially acceptable ways.

- Expressing pride in group achievement.

- Asserting oneself in a group.

- Responding appropriately to the group's leadership choices.

- Participating in group problem solving.

- Expressing feeling appropriately in the group.

- Responding to group suggestions.

- Respecting others' opinions.

- Identifying others' differing values.

- Drawing social inferences from group situations.

Just as individual students must learn how to discuss feelings and actions, students in groups must learn these group communication and social skills if they are to deal with crisis more successfully in the future. Following the basic steps we describe in this chapter, you can teach these group skills as you also teach your individual students more successful ways to respond to crisis.

ESSENTIAL STRATEGIES

Redl (1959a), Fagen (1981), and Heuchert and Long (1981) describe several essential processes to consider while using LSI. Throughout this chapter we expand on these ideas, as summarized here:

- Selecting the pertinent issue.

- Focusing on the aspects of the issue that clarify and simplify it for the student.

- Conveying an adult role compatible with the student's view of authority and the therapeutic needs of the student.

- Responding realistically to the manageability of the mood of both student and adult in a specific crisis—when LSI may have to be cut short because emotions or fatigue interfere with the potential for successful LSI.

- Timing LSI to follow as soon after crisis as a student can participate without being so upset that talk would be useless, and allowing sufficient time to complete the process without interruption from other mandatory activities, such as the bus leaving, or recess.

- Selecting a place for conducting LSI that provides what is needed where the event occurred, lending authenticity to the discussion of the event; or away from daily traffic, to provide privacy for freedom to talk without distractions or betrayal of confidentiality.

Some crises are so specifically related to events that you can track the sequence, during the first two steps of LSI, in a straightforward time line from antecedents to the crisis. From that point on, the problem-solving steps of LSI address the incident that provoked the crisis. Resolution of the crisis becomes the primary outcome.

However, many crises have little or no significance in themselves. They explode as ways to express thoughts and feelings in disguised forms. A student who walks into class with "emotional baggage" or as "a crisis about to happen" is a student dominated by anxiety and full of defensive behaviors. When a crisis results, there is an agenda behind the incident that may not be easily spotted. The student perceives events through a filter of private thoughts and intense emotions, using the incident as a vehicle to discharge greater concerns. In such situations, the incident may not be the real issue, so the LSI focus shifts from the incident to the underlying issue.

Determining the central issue for a particular LSI may be the hardest thing to learn. Your role as a significant adult in the student's life is put to the test as you become the mediator among the student, the feelings behind the student's behavior, and the central issue embedded in the crisis. Your purpose in using LSI is to expand the student's insight about the central issue and how feelings and behavior of self and others played into producing the crisis. With changes in insight, students can change behavior for more satisfying results.

Lasting change seldom occurs with a single LSI. Think of LSI as a series of interactions between you and the student. Because trust, acceptance, and developing relationship are all necessary ingredients for effective LSI, it is important that consistency be maintained in dealing with each student. A student cannot be shifted from adult to adult for LSIs and then be expected to achieve the same amount of progress that would be achieved through a consistent relationship with one adult over a series of LSIs. Chapter 1 contains a discussion about short-term and long-term outcomes of LSI. The end results can be summarized as positive change in students' views of self, ownership of solutions to resolve crisis, responsibility for changing behavior, and greater understanding of others' reactions to themselves. Gradually over a series of LSIs a reduction should occur in a student's emotional memory bank of negative thoughts and feelings. The student's words, behavior, and underlying feelings begin to change because of greater insight, better feelings, more impulse control, more positive feelings about self and others, and more positive self-regulated behavior.

THE SIX LSI STEPS

As you may recall from Chapter 1, the six steps of LSI can be divided into "three and three"—the first three steps focus on the crisis and the central issue, while the last three steps address the solution. In the remainder of this chapter, we describe the basics of each step in detail. For each step, we present a synopsis of major points to summarize the purpose of the step, students' characteristic responses, the adult's major task, and the way to tell when a particular step should end and bridge into the next step.

The charts illustrate major points and dynamics of each step. In these charts, solid lines convey the use of words and rational, cognitive processes. Dotted lines indicate emotions, drives, feelings, and underlying anxieties. The solid, horizontal line through the center of each chart is used to designate the separation between "public reality," evident to all who are involved, and a student's "private reality," seldom evident except through decoding. These charts are included at the end of the book in a form that can be photocopied to assist in studying.

Throughout the steps, we recommend abundant use of affirmations, shown in the charts as "A." With some students, you may need to begin LSI with an affirmation, conveying a positive view of the student and reflecting on some positive quality to defuse intense defensiveness, emotional flooding, or resistance. When you begin to decode, connecting behavior to feelings, you also will need to provide affirmations. Affirmations offer support for students, communicating understanding of what they feel without demeaning or disapproving. Perhaps the most important results from abundant use of affirmations are changes in a student's attitude toward the adult who affirms the student. Messages are communicated that the adult likes the student, believes that there is a better side of the student, and is an ally for bringing out better qualities. Almost all students are hungry for this mirror of a positive image. It is their only hope to be more than they have been before!

STEP 1: FOCUS ON THE INCIDENT

SYNOPSIS OF STEP 1

Since many students are not eager to talk immediately after a crisis, the first step in LSI is used to convey support and understanding of the student's stress and to drain off emotions to the point where the student can begin to talk about the incident in a rational way. Intensely angry feelings, "shock" words, and inappropriate actions are almost always present as you intervene in crisis and begin LSI. When a student believes that someone will help and when the student can begin to use words instead of being dominated by emotion, LSI can start and therapeutic gains can be made.

Hopefully, Step 1 is brief. Sometimes it takes only a sentence or two. With other LSIs, it may take as much as half an hour to diminish emotional flooding so that a student can begin to talk. If emotional intensity does not diminish to a point where the student can use words, do not go further into LSI. Instead, try Emotional First Aid, which we describe in Chapter 6 as part of the helping process.

Figure 5.1 illustrates the dynamics and process in Step 1. A student is ready for the next step when (a) the student begins to use words to talk to you, and (b) the reason for the talk has been brought out in the simplest words possible.

What Happens in Step 1?

When a crisis occurs, emotions are near the surface or already spilling over. When a student is so overwhelmed by emotional flooding that speech is not possible, *wait!* Nothing can be gained by beginning the LSI when a student is not able to use words.

Support the Student. While waiting, convey calm support. Communicate that you hear and see that the student is upset but the situation is not beyond solution, and that you are not emotional about it. Sometimes these messages are communicated nonverbally through your body language and facial expression; sometimes by holding (dispassionate restraint); and sometimes by a few simple words, sounds, or statements such as these:

> "You are really upset about . . ."

> "It's going to be okay."

> "We're going to work this out and you'll feel better."

> "Teachers can help . . . (take care of you)."

> "When you calm down, we can work this out."

> "I can tell you're almost ready to talk. You've stopped all that noise."

Figure 5.1 *Step 1:* **Focus on the Incident**

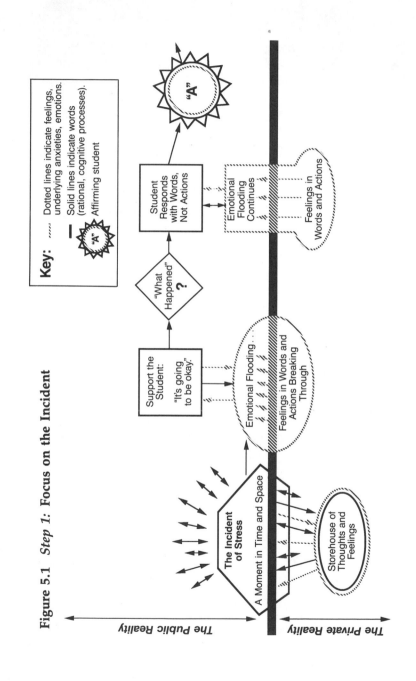

Key:
..... Dotted lines indicate feelings, underlying anxieties, emotions.
— Solid lines indicate words (rational, cognitive processes).
"A" Affirming student

The Public Reality

The Incident of Stress
A Moment in Time and Space

Storehouse of Thoughts and Feelings

Emotional Flooding...
Feelings in Words and Actions Breaking Through

Support the Student: "It's going to be okay."

"What Happened"?

Student Responds with Words, Not Actions

Emotional Flooding Continues

Feelings in Words and Actions

"A"

The Private Reality

When the student begins to give some verbal response, it usually means that a shift is taking place from an emotional mode to a rational mode for dealing with the problem. Now you have a way to begin the LSI process. By converting emotions, difficult feelings, and destructive behaviors into words, you and the student have the tools for problem solving.

As Talk Begins. When a student starts to talk about the incident, you probably will not know exactly what your therapeutic goal will be, what type LSI will be needed, what you will say, or what the outcome will be. You will build the LSI around the specific incident, shaping the direction from the responses given by the student. Here are examples of questions that often help get LSI started:

> "It seems you are ready to tell me what happened."
>
> "Why are we here?"
>
> "Let's see if we can make sense of what happened to you."
>
> "Tell me where the trouble began."
>
> "Who was there?"
>
> "When you are upset it is difficult to remember what happened, but let's give it a try."
>
> "Let's talk. It's important for me to understand why you are so mad (scared, frightened, angry)."

If a student is eager to talk, you have moved automatically into the next LSI step. But if words are slow in coming, it may require considerable effort to begin (see Your Role in Step 1, below). The important point is to *begin* a dialogue about the incident. It is not important to confront lies or straighten out distortions in perceptions during this first step. Nor is it necessary for you to get your own position on the matter out in the open.

If a student continues to be dominated by emotion or has not begun to center on the incident, don't go further into the LSI. Continue to focus on the incident, on your ability to help, and on the student's potential to talk about it. (Chapter 6 contains a discussion about emotional flooding and how to respond to severely resistant students who will not talk.)

Your Role in Step 1

Begin Step 1 by being a representative of "fairness for all"—an accepting, impartial adult who shows concern for the student's situation (the crisis, feelings, or dilemma). You are setting the tone for the relationship that will follow throughout the six LSI steps. Questions and statements like the following help to convey the idea that adults can listen, understand, and be objective and fair:

> "This is a difficult situation, but we can work it out together."
>
> "Sometimes things are terrible, but there are always ways to make it better."
>
> "When you lose control, everything seems to go wrong."
>
> "This is a serious situation. It calls for some serious thinking about what to do."

The student must have confidence and trust in you. Provide support and a sense of optimism that the problem can be handled. Convey your belief in the importance of being fair. Communicate your confidence in the student. Let the student know that you understand and respect the student's feelings even though you cannot accept the behavior. Focus on the student's positive attributes that can be used in solving the crisis. With practice, you will develop increasing verbal skill for even the most difficult problem. In time, you will find few, if any, students failing to respond to your strategies for getting the LSI rolling in Step 1.

As You End Step 1

The first step is over when the student responds to you with words (or in some instances, positive gestures or grunts) that acknowledge willingness to go along with you on the topic. Here are examples of ways to end Step 1, affirming your confidence in the student, your support, and confidence in a good outcome:

> "When you talk about a problem like this, it can be straightened out." [Affirmation of the student's ability]

"This may be a bad situation, but we're going to work it out." [Affirmation of your support for the student]

"Even the worst problem can be solved." [Affirmation of hope]

STEP 2: STUDENTS IN CRISIS NEED TO TALK

SYNOPSIS OF STEP 2

This is an intensely interactive step: The student talks and the adult questions, clarifies, and decodes. The exchange between adult and student during this step has two objectives. For the student, there must be a decrease in emotion and an increase in rational words and ideas, organized around the sequence of events—a time line. If this does not occur, the student cannot benefit from the problem-solving phases later in the LSI. For the adult, the objectives are to expand and clarify details about the student's perception of the incident, begin to decode behavior, and begin to construct an answer to the question, "Is the incident the issue, or is some underlying anxiety the real issue?" The answer will shape the rest of the LSI.

Figure 5.2 illustrates the dynamics of Step 2, showing how the decoding process is used along with questions to the student to clarify the student's perceptions about the "When," "Where," and "Who" of the incident and the events leading up to it. Step 2 ends when the talk has produced a review of time, place, and people involved in the incident. The adult has a sufficient understanding of the student's reactions and point of view to begin to focus the following step on a central issue—the incident or an underlying anxiety.

What Happens in Step 2?

Step 2 is a student's chance to "tell it like it is" or, at least, to tell how the student views the incident. For the student, Step 2 is an exercise in using rational words to deal with a problem. This is not easy for most troubled students. Typical reactions to an incident are anger, denial, evasion, distortion, or verbal counterattack. Because an incident taps a student's private storehouse of intense feelings, it has infringed on personal space. So,

Figure 5.2 *Step 2:* **Students in Crisis Need to Talk**

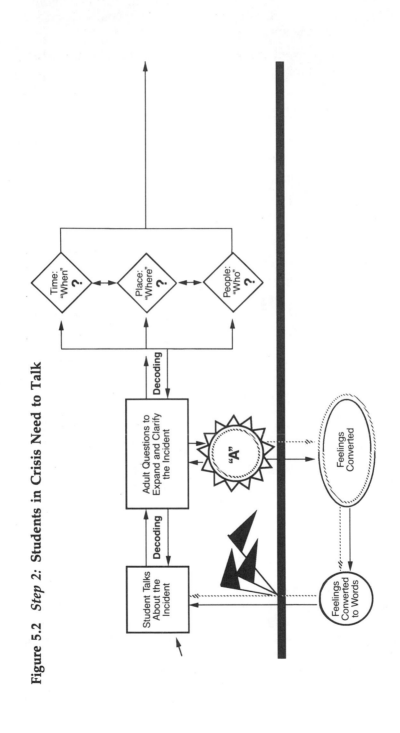

talking about the incident also should provide opportunity to bring out deeper or different issues and concerns that may be preoccupying a student's thoughts and energy.

Student Talks About the Incident. When the incident is first identified during Step 1, a student's reaction usually is loaded with defensiveness and intense denial of concern or responsibility. An emotional intensity exists that obscures objectivity. The student tries to justify what has happened by blaming you or others, insisting that the incident occurred in a specific way, or initiating talk that is off the topic.

As the student talks about the incident in Step 2, he or she should cool down considerably. You will notice a change in the tone of the exchange between you—from a predominance of emotion to a more rational tone. When this happens, don't be misled into thinking that LSI is nearing an end because the student is using rational behavior. Cooling down and talking rationally are *not* the only objectives in this step. Until you make a preliminary choice about whether or not the incident is really the issue, the LSI should not move to the next step.

As students become more comfortable with you and more spontaneous in their talk, they occasionally drop small comments or make aside remarks that seem to have no bearing on the incident or the discussion. We call these red flags. Unless you are actively listening and looking for them, you may let them pass as unrelated comments or as attempts by the student to get off the subject. It is difficult to discern a student's intent when he or she goes off the subject. These red flags often represent the first attempt by a student to let you know that there is another, real issue that goes beyond the incident. A red flag is a representative of the private reality we talked about in the introduction to this section. When you spot one, it may or may not be the right time to respond to it. If you choose to respond, a simple verbal reflection of the student's words is usually enough to highlight the fact that you hear it and recognize it as important. This response usually is enough to get the topic under way. If the student does not respond verbally but body language says there is an interest in further talk about the topic, try a follow-up question. If students are not ready to talk about a private issue, they usually initiate a shift in the conversation or show such agitation that it is apparent the topic is too uncomfortable to pursue for the moment. In which case, let it go but remember it for later use. (You will find an expanded discussion and examples of red flags in Chapter 6.)

Adult Questions to Expand and Clarify the Incident. For the adult, Step 2 is used to expand the student's awareness of the complexities of the incident. This is the step for teaching the student to look at an incident as a sequence in time. Since LSI usually takes place because of a situation that the student could not handle, the first focus is on the "here and now."

"When did this happen?"

"What happened before that?"

"What happened next?"

"Did this happen before or after you got up to get your supplies?"

After confirming the time line, explore the setting:

"Where did this happen?"

"Where were you when he said that to you?"

"Did this happen at the . . . ?"

"Has it happened at that same place before?"

"What was going on?"

These types of questions easily flow into questions about the people involved, and who said what to whom, when.

"Who made the first move?"

"Who was there?"

"Who else was there?"

"What did they say (do)?"

"Then what did you say?"

"What did they do after you said that?"

"Could you repeat what you just said?"

"I'm not sure I understand what happened. Please tell me again!"

"Let me summarize what I heard, and you correct anything I didn't get right."

It sometimes helps to use words that evoke specific visual images about the incident. Such graphic imagery aids many students in making the shift from raw emotion to using words in thinking and talking about the incident. Here are examples of ways to spice your questions with imagery:

"Was he wearing a red shirt?"

"Was he that great big guy that looks like a wrestler?"

"Did she scream at you?"

"Where did they run after that?"

"Were you sitting in the front of the bus?"

"Can you hear that noise when it's dark?"

As Step 2 gets under way, use the discussion to separate the various issues embedded in the incident. From each remark the student makes and each accompanying behavior, you must choose to respond or not. Respond to those remarks that seem to hold promise for greater understanding and insight. Also respond to remarks that seem to hold significance for the student, even if not for you. By this selective process, you expand and clarify the incident. You also increase the student's investment in the process. It is a balancing act that requires considerable mental agility.

Your Role in Step 2

Your role during Step 2 is to convey a sincere interest in hearing the student's perception about what happened and obtaining more facts about the incident. The more information you have from the student, the more opportunities you have for decoding and understanding. To do this, we find these strategies useful: questioning, listening, observing, reflecting, and interpreting (decoding).

Questioning. Questions are the most direct way to expand the information you have about the incident and the extent to which the student has invested emotionally in it. (How much of the private reality is embedded in the incident?)

Generally, the first questions ask for facts about the incident and explore the sequence of events—the time line. Illustrations of these basic questions are in the previous section. Use more complex questions when you begin to explore the extent to which the student can view the incident from the perspectives of others. You want to know if the student can associate personal behavior with the resulting reactions from others. Questions also can reveal the extent to which a student is relying on adults to provide external control, and what kind of issues the student sees as important in the incident. Here are examples of questions that expand the information base:

"How would he describe what just happened?" [Perspective of others]

"Could there have been something that set him off?" [Cause and effect]

"Is this a problem for kids to handle by themselves?" [Source of authority]

"Is fairness (kindness, leadership, friendship) an issue here?" [Values]

Active Listening (and Active Remembering). Listen carefully to what a student says. Key words and ideas are the cues you use to form your own responses. The student's view is the starting point, giving you a base from which to structure the remaining LSI steps. It also is important to remember what the student says. In later phases of an interview, or in subsequent interviews, you should be able to recall and reuse important or vivid points made by the student during these earlier steps.

Observing. Active observing is an important part of active listening. The body language a student uses to accompany a statement often tells more than words. A student's behavior also tells a lot about the impact you are making.

You can be confident that the talk has validity if a student participates, responds with vigorous denial, or shows a pronounced increase in body movement, even if these reactions are negative. Occasionally, a student will be very still, making no remarks and showing no behavioral response. This particular reaction almost always indicates interest in hearing an elaboration of the idea you just brought up.

If the talk is not on track, a student will show boredom or disinterest, and may bring up another topic. This same strategy also may be used by some students when content is too close to the private reality. You can usually tell the difference by the amount of agitation or emotion accompanying the response.

Reflecting. Words that mirror an idea or an observed behavior back to a student convey that the student's idea or behavior is worthy of note (or in the case of a negative point, that it has not escaped your notice). Reflections require no responses from students because they are statements of fact. A reflection does not intrude into a student's space and is not judgmental. It is among the best strategies for minimizing defensiveness and keeping a sensitive topic going. It also is useful with students who have problems with authority because it does not require a defensive response. Typically, a student will respond to a reflection by elaborating the details and expanding the content spontaneously. If not, go back to questions again; more facts and observation are needed.

If you have not used reflections before, you may feel a bit awkward simply turning an observation or a remark back to a student without adding your own ideas. Here are examples of simple reflections, where the student has made the first statement and the teacher has simply rephrased it:

> "This wasn't fair."
>
> "You were sitting there minding your own business."
>
> "You didn't mean to make her mad."
>
> "You think he is the one I should talk with."
>
> "You were feeling good about what you had done until he made that remark."

Did you notice that these examples are not stated as questions, although they could be? Reflecting involves no question, so a response is not necessary. These examples do not reflect negative feelings or negative behavior. Also, they are nonjudgmental, without innuendo that the student somehow really might need to reconsider. This technique is very effective during the first steps of LSI when you want to encourage students to freely express their views and feelings about the incident.

Interpreting (Decoding).   When you make a statement that connects meaning or feeling to a student's surface behavior or words, you are interpreting, which is a form of decoding (see Chapter 3). Interpretation expands a student's awareness of connections between behavior and private thoughts and feelings. Construct your first interpretations from reflections about the student's behavior and words. The reflection provides credibility to the interpretation, because it gives observed evidence. This is a neutral way to link what is publicly observed to what may be private. Here are examples of the way reflections and interpretations are used together in decoding behavior:

> "When a student puts his hands over his ears [reflection of behavior], he is telling me that this topic is uncomfortable for him to hear [interpretation of feeling]."

> "I've noticed, each time you get around him, you end up in trouble [reflection]. Maybe he enjoys seeing you out of control [interpretation]."

In the first example, the third person pronoun is used to provide psychological "space" for the student. Because the interpretation does not directly address the student, there is distance enough for the student to hear the message without having to be defensive. The student is not forced into a position of having to respond. The reflection and interpretation are simply matter-of-fact.

When a student's response suggests that decoding has been accepted, follow it with a question. This helps the student begin to deal with the new material. Usually a yes- or no-type question is the easiest to answer, particularly if the interpretation is heavily ladened with private concerns.

With sophisticated students and those who have had previous LSIs, you can begin to shift responsibility for decoding and making interpretations to the student, using statements and questions like these:

> "That must have hurt your feelings."

> "What went through your mind when he said that?"

> "Why did that set her off?"

"Do you think your comment had anything to do with the way you were treated by them last week?"

"Some people really know how to make other people feel bad."

While you or a student are decoding, you are seeking answers to the question, "What is the real issue here?" Decoding is the simultaneous mental process used to translate interpretations of behavior into a broader network of meaning, connecting the obvious to the not-so-obvious. Decoding should shape the direction of your LSI, moving it toward a focus on the surface incident or toward a more private, underlying issue. (If you are still hesitant about decoding, we suggest going back to Chapter 3 for review. Decoding is a skill that must be used for effective LSI.)

As You End Step 2

Step 2 begins to merge into the next step when the student has reviewed the incident and you have used interpretations, decoding, and affirmation of the student to forge links of confidence between you and the student. Inaccurate interpretations, failure to decode, or a shortage of strong affirmations will dilute confidence the student may have in you: "This adult is like all the others and doesn't really understand (or care)!" In contrast, with accurate interpretations and sensitivity to the student's private space, the student begins to believe, "This adult really does understand!" When this happens, the student will be ready to go with you into the next LSI step. Here are examples of ways to end Step 2 and bridge into the next step:

"This is what I hear you saying . . ."

"You've described the situation clearly. Let's review what you have said."

"It's interesting how something that you meant to be nice backfired and led to all this trouble."

"It seems like the problem that got you in here to talk isn't the real problem after all."

STEP 3: FIND THE CENTRAL ISSUE AND SELECT A THERAPEUTIC GOAL

SYNOPSIS OF STEP 3

Step 3 builds a conceptual structure about what is really significant in the incident. At the moment of the incident the student's life space includes both public and private realities. It is the adult's job in Step 3 to sort it all out and make decisions about which reality predominates and what is needed. Step 3 is the decision-making point for selecting a therapeutic goal.

The key processes in Step 3 are to explore the student's view of the incident and underlying anxieties until you have sufficient understanding to (a) concisely state the central issue; (b) assess the student's perceptions, insights, and motivation to change behavior; and (c) decide what the therapeutic goal should be for this particular LSI.

Figure 5.3 illustrates these processes. When they are completed, a logical outcome will begin to take shape and the information is used to structure the remainder of the LSI. Step 3 ends when the issue has been stated concisely by the adult or student and the adult has chosen a therapeutic goal. (The first five chapters in Part Three provide in-depth discussions of how to conduct LSI specifically for each therapeutic goal.)

What Happens in Step 3?

Preliminary information needed to identify the central issue is obtained in the previous steps by decoding the student's words and behavior. As Step 3 begins, you have an initial idea about whether the incident or underlying anxiety is the central issue. During Step 3, this idea is tested and expanded by obtaining information in greater depth, looping back for more verbal exchanges and more decoding.

Find the Central Issue and Form a Concise Statement. The first decision in Step 3 is whether to focus on the incident or to put your effort into unfolding some of the underlying anxieties associated with the incident. A focus on the incident assumes that relevant antecedents (the chain of events that triggered the incident) are observable and are major ingredients of the incident.

Figure 5.3 *Step 3:* **Find the Central Issue and Select a Therapeutic Goal**

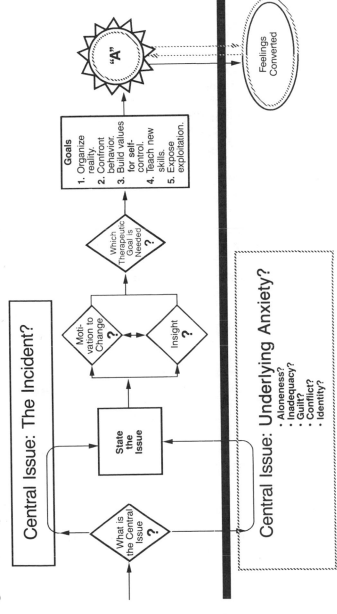

A focus on underlying, developmental anxieties assumes that the incident is not actually the issue but serves only as a means to express a hidden concern. Such issues often are obscured by behavior or words used by students for protection against invasion of their psychological space or from being overwhelmed by private thoughts or feelings that cannot be expressed or recognized openly.

To focus on underlying anxiety is difficult and complicated. Yet, unless this is done, a student's private reality remains untouched and unresolved. Emotional volatility builds up as a student attempts to function while avoiding a private burden. If the underlying anxiety can be brought out and dealt with in a satisfactory way, emotional tension is reduced. Then, rational problem solving becomes the bridge for management of these private feelings and thoughts.

With increasing information comes greater clarity about the incident and underlying anxiety. This information must be distilled into a single statement that represents the central issue. The central issue should be clearly stated in one or two simple sentences before continuing further into the solution steps.

In some interviews, the adult summarizes the central issue; in others, the student will be able to blurt out some version of the central issue. Here are examples of statements about central issues. The first examples are of the incident as the central issue:

"It seems that messing with others leads to problems; and here you are, having the problem."

"So, you are the one who is always picked on. He called you a name and then you hit him. . . . And here you are."

"You tried to follow the rules, but they wouldn't let you. It's really hard to do right when everyone else is breaking rules."

"You think he deserves cussing because he cusses you all the time."

"She thinks she knows everything and that gets to you. Is this another example of getting set up by someone else?"

"You really didn't mean to hurt him. It's hard to stop yourself."

A part of every incident touches a student's underlying anxiety. Sometimes the incident is played down in order to explore underlying issues. The following statements were made by teachers responding to fights in their classrooms or on the playground. In each instance, the teachers built their responses from what the students brought up as they discussed the fights. In these examples, the teachers chose to form the central issues around underlying anxieties.

"It's going to be hard to wait for 10 days to see if the judge is going to let your (foster) parents adopt you." [Issue is abandonment.]

"Tearing up the room may not be the best way to get the other guys to be friends with you." [Issue is inadequacy with peers.]

"You said you were an 'idiot,' but what I hear between the lines is a message that maybe with some help from teachers things could be better." [Issue is guilt, self-judged.]

"When your dad told you to turn off the TV last night and you went on a rampage, just like the one in class today, it was like the little kid in you. Isn't this what we've been working on, how to handle feelings without having a 'royal fit'?" [Issue is independence-dependence conflict.]

"You're very concerned about Ron being a fag and how you're going to act around him." [Issue is adolescent identity.]

In these examples, we used each of the basic developmental anxieties discussed previously in Chapter 3. If you still feel uncertain about them, a review of that chapter should be helpful. It is important at this step in LSI that you are able to identify these developmental anxieties as they emerge during the discussion and decoding processes that have occurred.

Assess the Student's Perception, Insight, and Motivation to Change. Following a concise statement of the issue, think through the preceding dialogue to summarize, in your own mind, how the student views the incident and the extent of the student's sense of responsibility. From the previous steps, you should be able to see the events as the student has seen and felt them. Some students have extremely restricted social perceptions, while others are clearly able to view an incident from another's point of view. It is essential to understand how your particular student views the event. You also should have an idea about the student's view of adult authority and who should solve the problem—the student or the adult (see Chapter 4, concerning the Existential Crisis). This information will help you evaluate the student's motivation to change and assume responsibility for the crisis. It also will give you an indication of the student's readiness to give up old alibis and rationalizations for new or different behavior in the future. With this information, you and the student are ready to move on to the last process in this step—selecting a therapeutic goal, based on the student's current perceptions.

If you still feel unsure about how the student views the event and his or her personal role in it, then it is necessary to return to Step 2 and further expand and clarify the student's perceptions and motivations to change. Decoding a student's words and actions is the single most helpful process you can use to accomplish this.

If it is clear to you that the student continues to deny responsibility and has shown resistance to change, one of the first two therapeutic goals will be appropriate (outlined below). If the student has shown remorse, willingness to change, or good intentions (even though failing to do what was known to be acceptable behavior), the student is ready for one of the other forms of therapeutic goals. In the following material we summarize these goals and the typical student perceptions that lead to each goal.

Select a Therapeutic Goal. Once the issue has been clearly stated, and you have assessed the student's perceptions and motivation to change, the question becomes, ''What do I want to happen, now that the student and I have agreed upon the problem?''

Almost every central issue can be managed within one of these five general types of therapeutic goals. The labels we use emphasize the adult's roles and are Fritz Redl's original therapeutic goals restated. The original names given by Redl (1959b) are in parentheses. Selection of a particular goal is based on the student's current perception of the incident and issue and your assessment of the student's motivation to change. Each goal also has a new, broader insight for the student as its objective. Here is a summary of each type of therapeutic goal:

ORGANIZE REALITY ("Reality Rub-In")

Student's Perception: "I have been treated unfairly, and gypped, and now they are blaming me for the problem."

Uses: With students unable to interpret or perceive events accurately; those unaware of their own behavior or the reactions of others; or with mentally disorganized students, those with thought disorders or with limited ability for abstract thought.

Goal: To interpret events in an accurate way, correcting distortions, social myopia, or misperceptions about the incident.

Focus: Organizes students' mental perceptions and sequence of time and events; developmentally the most rudimentary of the therapeutic goals.

Student's New Insight: "I only saw and remembered part of the problem. Now I see the part my own behavior played in the crisis."

CONFRONT UNACCEPTABLE BEHAVIOR ("Symptom Estrangement")

Student's Perceptions: "I do what I have to do even if it hurts others." "I have a right to be free." "I have a reputation to maintain." "I have no need to change."

Uses: With students who are too comfortable with their deviant behavior, receiving too much gratification; those who rely excessively on aggression, passive aggression, manipulation, or exploitation of others.

Goal: To make a particular behavior uncomfortable, by confronting the rationalizations and decoding the self-serving narcissism and distorted pleasure the student receives from the unacceptable behavior.

Focus: Uses external adult authority, power, and judgment to help student obtain new insight into deviancy and the inevitability of dire consequences.

Student's New Insight: "Maybe I'm not as smart as I tell myself." "Maybe I've been cruel." "Maybe I've been tricking myself."

BUILD VALUES TO STRENGTHEN SELF-CONTROL ("Massaging Numb Values")

Student's Perception: "Even when I'm upset, a part of me is saying, 'Control! Stop yourself' . . . but I don't."

Uses: With students who, after acting out, are burdened by remorse, shame, inadequacy, or guilt about their own failures or unworthiness; those with a destructive self-image; and those who have a negative social role.

Goal: To emphasize a student's positive qualities; strengthen self-control and self-confidence as an able and valued person with qualities like fairness, kindness, friendship, or leadership potential.

Focus: Expands student's self-control and confidence by abundant affirmations and reflections about existing socially desirable attributes and potential for future acclaim by peers; developmentally this goal requires a shift in source of responsibility—from adult to student.

Student's New Insight: "Even under tempting situations or group pressure, I have the capacity to control myself."

TEACH NEW SOCIAL SKILLS ("New Tools Salesmanship")

Student's Perception: "I want to do the right thing, but it always comes out wrong."

Uses: With students seeking approval of adults or peers but lacking appropriate social behaviors to accomplish this.

Goal: To teach new social behaviors that student can use for immediate positive gain.

Focus: Instructs in how to do specific behaviors that will have immediate payback in desired responses from others; developmentally reflects emerging independence and responsibility.

Student's New Insight: "I have the right intention, but I need help to learn the skills that will help me make friends, achieve, and get along with adults."

EXPOSE EXPLOITATION ("Manipulation of the Boundaries of the Self")

Student's Perception: "It's important to have a friend even if the friend gets me into trouble."

Uses: With students who are neglected, abused, scapegoated, isolated, or who seek out destructive friendships by acting out for them.

Goal: To help a student see that another student (or adult) is manipulating events in a way that is working against the student's best interest.

Focus: Provides insight into reasons for the behavior of others; views social interactions from the perspective of motivations and behaviors of others; developmentally, this goal requires considerable maturity on the student's part, as the student learns to understand how others think, feel, and behave.

Student's New Insight: "A friend is someone who helps you solve problems and feel good rather than someone who gets you into trouble."

Examples of LSIs using each of these therapeutic goals can be found in Chapters 7 through 11. These chapters illustrate how each of the therapeutic goals uniquely shapes the strategies and content of the remaining steps in every LSI.

You may be asking, "Why put off selecting a therapeutic goal until Step 3?" Sometimes, with students we know well, we have an idea from the very beginning about what the student's characteristic perception is, what central issue is involved, and what therapeutic goal is needed. Even so, it is necessary to establish the facts and observations about the crisis and obtain the student's current perception. Without this, the student will not become sufficiently involved to "own" a solution as it is developed.

Because LSI is a dynamic tool and is responsive to inner concerns as well as to the incident, sometimes you may select more than one therapeutic goal. For example, there may be a need early in LSI to organize reality, then it becomes clear that the student also needs a great amount of bolstering of self-esteem to make any further progress. Or the student may be ready for some direct teaching of social skills. Having several therapeutic goals should not be a problem, if you keep the central issue and the goals clearly in mind during the problem-solving steps of LSI.

Your Role in Step 3

Step 3 is perhaps the most difficult for the adult. It requires a good bit of head work, a lot of decoding, thoughtful observing

and listening, and good verbal skills for questioning and interpreting behavior. We described these processes in the previous
steps, but it is necessary to return again and again to the cycle
of Step 2 as you form a statement of the central issue in Step 3.
You will find yourself in an information loop—questioning, interpreting, decoding, and hypothesis building. While sifting through
the maze of responses you hear and observe, you also must form
your own questions and responses in ways that begin to shape
the direction of the LSI. Your task is to decide upon a particular
therapeutic goal.

In Chapter 4, we discussed styles of adult influence and their
importance in conveying the type of adult role needed by a particular student. Here in Step 3, a clear view is essential about
which type of strategies you should use to influence the student.
You may recall that the five types of therapeutic goals are labeled
according to the type of action required by the adult. These labels
also suggest the style of adult influence needed to accomplish a
particular goal. They are helpful guidelines for the roles you might
take: organizing a student, confronting behavior, building values
for self-control, teaching social behaviors, or providing insight
into the behavior of others. Here are examples of statements made
by teachers illustrating each of these adult roles:

The organizer:	"You were standing there on the playground with the ball. Were you holding it with both hands?"
The benign confronter:	"You're trying to tease Thomas, hoping he would react to you. But look at what really happened. He is still having fun, and you are in here. Tell me, who has the problem, Thomas or you?"
The builder of values for self-control:	"I can see leadership qualities in you. The others pay attention when you have an idea for the group."
The teacher of social behaviors:	"Maybe if you had a friendly sound to your voice, he might let you use his equipment. Try it out with me."
The provider of insight:	"He must have been really mad to have hauled off and hit you after what you said. Why do you think he got so mad about that?"

The goal and the adult role are closely bound to each other. And it is the student's responses that confirm or negate the validity of the choices you make. As you convey the type of adult needed to accomplish the therapeutic goal, the flow of the LSI distinctly turns toward possible solutions.

As You End Step 3

The third step is over when a statement has been made about the central issue and you have chosen a therapeutic goal. With the end of Step 3, the problem phase ends and the solution steps begin. Here are examples of ways to end this step and bridge into the next step:

> "This situation is beginning to make sense. When a person understands a problem, it can be solved."

> "This has not been a good day for you, but we know what's wrong now, so we can do something about it."

> "We know what the problem is, but what are you (we) going to do about it?"

STEP 4: CHOOSE A SOLUTION BASED ON VALUES

SYNOPSIS OF STEP 4

This step begins the "how-to-fix-it" part of LSI. Three questions are to be answered during this step:

- Will the solution address the incident or the underlying anxiety?
- Will the adult or the student provide the solution?
- What will the student see as a satisfactory solution that can be "owned"?

Often there are two solutions to be dealt with: One is the reality issue surrounding the incident; the other is an underlying issue that may have emerged during the previous steps.

When possible, the student should come up with alternative solutions and then select the course of action that seems best (4a in the chart in Figure 5.4). If the student is not able to do this, the adult helps the student by providing guidelines, values, or rules (4b in the chart).

To have genuine ownership of solutions by students, solutions should be ones that students value as beneficial. We use a framework of developing values to identify which general stage of values will be suitable for a particular student. These are listed in Figure 5.4 and described in the following section. Step 4 ends when a solution is chosen and the student puts the solution into words.

What Happens in Step 4?

During this step, the nature of LSI changes distinctly. The tone is upbeat; the focus is on solutions—making things better. Before this point, LSI is diverging, exploring, and expanding for a full understanding of the incident and the central issue behind it. Sometimes it has been painful or confrontational; but now, with a concise statement about the issue out in the open, solutions become the focus.

Should the Solution Focus on the Incident or the Anxiety? The solution should address the stated issue. If the previous step has been well done, the central issue has been clearly stated and Step 4 can now focus on a solution to that issue. In some LSIs, you will choose to focus solutions on the incident, even though it may have been evident that developmental anxieties are present. In such cases, leave the problem of underlying anxieties for another LSI and focus on a solution to the incident.

In other LSIs, the issue statement may have left the incident to focus instead on a pervasive developmental anxiety. In these LSIs, solutions deal with resolution of the anxiety and not the incident. However, the incident itself is not entirely neglected. It is reintroduced later, in Steps 5 and 6, as you and the student plan ways to implement the solution or as you prepare the student to reenter the group. However, the original incident still requires some resolution. In these situations, it is treated as part of the milieu and is resolved in the general plan for reentry into the group.

Figure 5.4 *Step 4:* **Choose a Solution Based on Values**

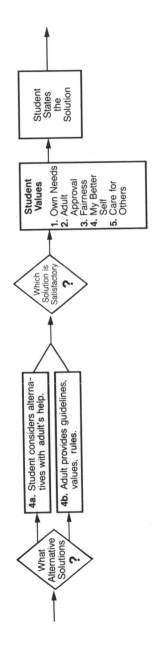

Consider Alternative Solutions. Encourage the student to consider as many alternative solutions to the issue as possible. Some alternatives will be outrageous and inappropriate; others will be self-serving; and others, totally counterproductive. But some alternatives will be on the right track. Discuss the pros and cons of each possible choice. By considering every alternative to a problem, no matter how farfetched, the student's view of ways to solve a problem will be broadened, and the student will be learning a fundamental problem-solving approach.

During this process, guide the student into making an objective evaluation of each alternative, considering what would be gained by a particular solution and what might be lost or harmful. Here are examples of statements made by adults helping students to evaluate possible solutions:

"Would that be fair?"

"If you handled it that way, what would she do?"

"You've seen guys handle it that way before; you know what happened afterwards?"

"If you did that, what would he do then?"

"Would the others respect you as a person they could look up to?"

"That would be really hard to do. Would it be worth the trouble?"

"That sounds like an idea that might work."

"Do you think you could pull it off when . . . ?"

What Is a Satisfactory Solution? Satisfactory solutions are those that students believe will work to their benefit. The most acceptable solution from an adult viewpoint may not be seen as satisfactory by a student. If a student does not really buy into a solution, the time has been wasted.

There is a general sequence in the development of social problem solving for most students that is based on what they really value. In Chapter 4, Figure 4.1 outlines this sequence. It begins with an orientation that one's own needs are paramount. Any solution that meets these needs is satisfactory. At this beginning level, service to self is the satisfactory solution. Any solution that will work to the student's immediate benefit will be "owned."

Here are several guidelines for satisfactory solutions at this beginning level of social values that provide for the student's *own needs*:

- Pleasure-producing results
- Concrete incentives
- Tangible rewards
- Privileges

Gradually, students move from that orientation to a belief that adults' standards are the ones that will bring benefits. At this stage, students believe that service to adult authority will protect self and bring benefits. Here are guides to solutions that seek *adult approval*:

- Accept rules and punishment as inevitable.
- Acquiesce to adults as the ultimate power.
- Do what gains adult approval.
- Accept punishment that restores reputation as a "good" person in the eyes of adults.

Development then broadens into the view that justice and fairness are valuable guidelines for problem solving. Social values at this stage are oriented to fairness and justice for like-minded individuals. Here are examples of satisfactory solutions involving fairness and similar social values:

- Look out for friends.
- Do things for others so they will do things for you.
- Demand fair treatment for yourself.
- Do what you think will get you a fair deal.

Gradually, students embrace values that reflect society's standards for responsibility. Satisfactory crisis resolution for students at this stage of social values is derived from concepts about responsibility, reciprocity, and cooperation in a larger, valued community. Here are examples of solutions that reflect this *better self*:

- Do things to have friends.

- Be a good person, so people will like (help, admire) you.

- What others think of you is important.

Finally, some adolescents may reach the stage of social values where responsibility for asserting individuality and independence is blended with responsibility for responding to the needs of others. It requires students to understand the perspective and feelings of others. It also requires empathy and altruism, sometimes putting the needs of others first. Few, if any, students we work with are at this stage of values. When they are, they often can work through a crisis to a satisfactory solution with minimal adult assistance. Here are examples of solutions at this stage of *care for others*:

- Show what you can do to contribute.

- Be a responsible person.

- Keep your eye on where you are going.

- Be someone who helps others.

- Make your life worth something.

These five stages embody most of the social value orientations you will find among solutions that students develop. If you can identify the general stage that a student operates in, you can be fairly confident that the student will "own" a solution that fits into the current stage.

Sometimes it is useful to go beyond a solution that is within a student's current stage and challenge the student to consider solutions typical of the next stage. This push into a higher stage is useful when a student begins to choose solutions in a manner that seems superficial, trivial, or produces little real motivation or commitment to change.

Your Role in Step 4

Adult responsibility during Step 4 is to guide. How much guidance you provide should be adjusted to the student's view of adults and who has responsibility for solving problems. (You may find it helpful to review the discussion about the existential phases and styles of adult influence in Chapter 4.) Your influence can widen a student's consideration of alternative ways to solve a problem, or you can restrict it without intending to do so. Here are examples of ways to encourage a discussion of alternatives:

> "There are lots of ways to go about solving this problem."
>
> "That's a solution that might work. Can you think of another way?"
>
> "Let's count how many different ways you can think of to handle this situation."

The most important part of being an effective guide in this step is to see that the student describes as many alternative solutions as possible. The benefits and problems associated with each alternative should be considered. To do this, you will need to ask questions and reflect real consequences. The best help you can provide during this step is to suggest guidelines, rules, and values that are relevant to the incident and alternatives the student is struggling with. Through your responses you can influence the student to carefully consider solutions that the student might not otherwise consider but that are appropriate to the student's current stage of value development (refer to Chapter 4 for an outline of student values and a discussion about forms of adult influence). Consider motivating the student to solutions at the next higher level of values. Some students will be ready to make the step to the next value level, others will not.

Exercise caution that you do not jump in too soon and influence the student to a particular solution. If you provide the solution, or the student sees you as having already made up your mind, the chosen solution will be "owned" by you and *not* by the student.

As You End Step 4

This is the point in LSI when the adult and student begin to merge in agreement about a solution. Among the alternative solutions discussed, one will begin to take shape as being more satisfactory than others. The process is a series of trade-offs: The student is seeking relief from the stress and comfortable resolution of the incident. You are seeking the most constructive solution for the student's perceptions, stage of value development, and the therapeutic goal you have selected in the previous step. Keep in mind the desired change in student insight that each of the therapeutic goals should produce. New insight is an essential part of a satisfactory solution. The solution also should provide for enhancement of the student's self-concept and confidence to face the people and situations that were a part of the incident. On the other side, the solution must really be possible. It must fit into the real situation the student faces, be within the student's capacity to use successfully, and be ''owned'' by the student.

Step 4 ends when the student puts the chosen solution into words. Sometimes you may have to summarize it first and then ask the student to tell it again to you. With other students, you can end the step with requests like these:

> ''We've talked about a lot of ways to handle this. Which seems the best to you?''

> ''It seems pretty clear that you've considered all the choices. Which will work best for you?''

> ''You have thought this through carefully. Let's review the choices.''

STEP 5: PLAN FOR SUCCESS

SYNOPSIS OF STEP 5

This is the step for realistic planning and rehearsal of what will happen when the solution is actually used to resolve the central issue and to deal with the incident. The focus is on specific behaviors—what to do, when, . . . Both negative and positive aspects are anticipated, and a working plan is formed for putting the solution into action.

This step is used to expand the student's understanding of what the chosen solution will require. The question is, ''Will this solution work for the student?'' Behind this simple question is the major concern of intervention programs, that there be a carryover effect. Step 5 is the place to see that this occurs.

As indicated in Figure 5.5, affirming a student's ability to carry out the plan is essential for success at this step. It is almost always difficult for students to use new behaviors and a new approach to a problem. Without practice and abundant encouragement, they tend to fall back on old ways of behaving.

Figure 5.5 *Step 5:* **Plan for Success**

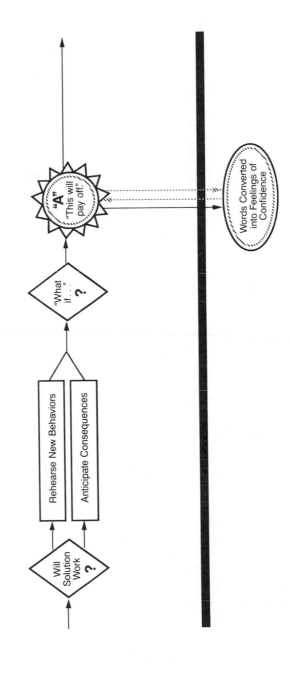

Step 5 ends when new behaviors have been rehearsed, future problem situations have been anticipated, and the student has confidence that it can be done successfully and will produce benefits.

What Happens in Step 5?

Step 5 is used to help students plan for resolving the current problem and avoiding repeats in the future. It teaches them to anticipate problems and expands their behavioral responses. This step could be called "rehearsal," for that is exactly what is done.

Rehearse New Behaviors. The adult usually begins the planning and rehearsal by asking questions like these:

> "You have a good idea here. What will you say to him when we leave here?"
>
> "This is a really good plan. Let's pretend I am that person; what will you say to me?"
>
> "It's not always easy to do what we've been talking about."
>
> "This is going to take a lot of courage on your part. How will you begin?"

During this step you are preparing the student to put the solution into action; to respond to events and people with new behaviors and new understanding. The plan for success may have several parts: One has to do with consequences that have come about as a result of the incident. Rules may have been broken, property destroyed, or offenses committed against another student or adult that must be handled. Preparing the student to accept consequences with responsibility and understanding is an essential part of this step.

Another part of the plan for success may involve more subtle changes in attitudes and behavior, resolving the central issue and developing long-range strategies for handling similar problems in the future. Be guided by the individual student, the nature of the original crisis, the nature of the central issue, real consequences that have resulted from the incident, and the therapeutic goal you have chosen.

With students who are limited in their mental ability or who are developmentally young, simple role play is an effective way to rehearse the plan. If the rehearsal is carried out carefully, it may bring out the fact that the chosen solution is simply not the right one for the student. You may notice that the student missed the point. Or the student simply cannot perform at that level. Or it may become clear that the solution is a version of an old behavior and fails to challenge the student to a more effective solution.

If you have doubts about the potential efficacy of the chosen solution, you may want to recycle back through Step 4 again. Maybe you and the student can come up with a more realistic solution and a set of behaviors that can really be done.

Anticipate Consequences.   Students often fail to anticipate that consequences may result from their behavior in an incident. They often resist thinking about what they have to do, how they will feel, how they will react, and how others will react to them. Such unanticipated problems need to be brought out during Step 5. Here are examples of statements that help a student prepare for the consequences resulting from an incident:

> "It's clear you have a good plan. When we go back into the room, he will be expecting you to apologize. What will you say?"

> "This talk shows your best qualities. You're going to be able to . . . (repair the damage, handle the suspension, explain it to your parent, face the judge, talk with the principal, apologize)."

> "I'm sorry you have to . . . (face the consequences), but once it's over, I know you're going to handle something like this in a better way next time it happens."

Such statements affirm your belief in the student's capacity to face the consequences in a responsible way. They also provide support for the most unpleasant part of an incident, enduring the consequences.

Students also fail to anticipate stress. The next stress may catch them unprepared and resorting to old coping strategies. Here are several examples of statements that help students anticipate problems:

"What will you do if . . . ?"

"What if he doesn't listen?"

"What if she tells you it doesn't matter that you had a reason for breaking the rule?"

"Do you think they'll believe you next time?"

"It will be hard to say no to your friends when they try that again. What will you do?"

"The next time they tell you to do that, you are going to feel just as angry. What will happen?"

"They may ignore you. What then?"

The difficult part of this step is to balance the amount of reality you expose the student to with the student's capacity to imagine future stress. If a student is just beginning to try new behaviors, it may not be the time to expand the horizon to foresee all possible problems. On the other hand, if a student has already participated in a series of LSIs with you and there does not seem to be carryover into real-life situations, you may need to spend more time on rehearsal.

Affirm Potential Benefits. The importance of repeatedly affirming and reaffirming a student during this step cannot be overemphasized. Your support and confidence in the student are the mirror in which the student sees a changed self, using new behaviors to deal with old problems in new ways. Troubled students cannot make this change in their view of themselves by themselves. Your affirmations and belief in the student make real change possible. And without such a change in a student's private thoughts and feelings about self, there will be no real, lasting benefit from LSI.

Your Role in Step 5

The adult role in Step 5 is to simultaneously provide a window of realism while conveying confidence in the student's ability to actually use the solution and conviction about the benefits that will come to the student for trying out the new solution.

Realism is conveyed through "What if . . . ?" questions. Beginning with a future scenario that repeats the present incident, lead the student through a replay while the student describes the new behaviors to be used. Then you can help the student imagine how others will respond. This effort should provide expanded insight into social exchanges, both positive and negative.

Confidence is the fuel for change. It is conveyed through statements such as, "I can see your maturity. . . ." Communicate the value of being responsible. Be convincing in communicating your belief in the good qualities of the student and in the capacity of the student to carry out the plan. Every troubled student is enmeshed in self-doubt. It is not likely that a student can come up with the self-confidence to make significant change in behavior unless important adults clearly convey confidence.

Benefits must be included in the plan for success. Benefits are not easily anticipated by most troubled students. They seem to live with a sense of immediacy that precludes planning. They also seem to operate under a cloud of pessimism, always ready to see the worst possible side of every person and event. To a large extent, experience has been the teacher, and their experiences have been downers.

Watch for these same qualities in yourself. If you do not convey confidence and approval, you probably will not be successful in helping students to see that desirable things can happen even as they face the consequences of their old behaviors, alter their strategies for dealing with problems, and change the way they view themselves. Teaching students to look at the future with optimism requires optimism and enthusiasm on the part of the adult. These are the types of statements that convey future benefits:

> "If you can do this, people will give you a break. They'll treat you with respect (fairness, love, kindness, understanding)."

> "When you start making these changes, people are going to start saying, 'Wow, I want to be around her.'"

> "When you handle things this way, people will say you are a person they can count on."

> "Your friends are going to look up to you."

> "People will see you as . . . (a leader, a really nice person, a friend, someone they enjoy being with)."

"This will give you the chance to show the really nice things about you."

As You End Step 5

When a student tells how the current issue will be handled and responds constructively to the idea that the issue will come up again in the future, Step 5 is finished. You know that the LSI has been successful if the student provides constructive and realistic responses to the present situation and future incidents. You will have a sense that the student is realistic about the current consequences and ready to try a new solution, recognizing that it will not be easy to implement but worth the effort in future benefits. Here are ways to end Step 5:

"This has been tough talk, but it's going to pay off for you!"

"You have a view about this that will really work for you."

"Things will be better when you handle it this way."

"It won't be easy, but I believe you can do it!"

"You can do this, even if it's hard to do."

"You'll see how good things will be."

STEP 6: GET READY TO RESUME THE ACTIVITY

SYNOPSIS OF STEP 6

The last step in LSI is the plan for the student's transition back into the group. This step helps the student prepare for reentering and participating in an ongoing activity. It also is important to use this step to close down private topics or feelings brought out during LSI.

The topic shifts in this last step. The incident and the central issue have been put to rest, at least temporarily. A solution has been worked out. Now the discussion centers on where the student will go now as LSI ends. Figure 5.6 illustrates the main topics to be covered. Discuss what has

Figure 5.6 *Step 6: Get Ready to Resume the Activity*

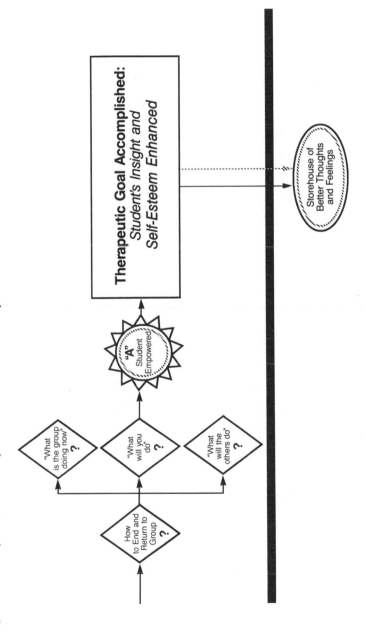

happened while the student has been away, what the student will do, and how others will react as the student rejoins the group. Affirmations of the student's new insights and competence for carrying out the new plan are essential if you want to assure that the student is empowered to be successful.

Step 6 ends when the student is under rational control, has described the behaviors needed for successful reentry, and conveys a responsible attitude for participating in the group.

What Happens in Step 6?

Step 6 generally begins with questions such as these:

> "What is your group doing now?"
>
> "When you left the group they were writing news reports. Do you think they are still at it?"
>
> "We've been in here for quite a while. Do you think your group is still working on that same project?"
>
> "When you go back to your room, what will they be doing?"

Such questions shift the focus from the student to another place and other people. The questions are designed to help a student begin to think about the peer group, the activity that has gone on while the student was away, and the expectations for participation and appropriate behavior.

Like the first LSI step, the last step often is brief because the student usually is eager to put the problem aside and get back to the group. This response is typical of students who believe that there has been a satisfactory solution. These students also have developed some additional confidence in themselves during the LSI.

The moment of reentry into a group is a powerful one. A student usually left the group because of crisis and conflict. Now, as the student is poised on the edge of reentry, there is a choice between defensive reactions or participation, with the best possible face put on for the group. An emphasis on participation and looking good in the eyes of peers can do the most for enhancing the reentry process.

The Expectation to Participate. To help a student accept responsibility for participating, review anticipated activities and expectations for participation in detail. We have seen students make substantial progress in highly successful LSIs and then lose it all because they had not been sufficiently prepared for the demands made on them when they reentered their group. The fewer surprises the better. Here are examples of ways a student is prepared for the expectation to participate:

> "They are keeping your work for you. You will have time to finish it before lunch."
>
> "The project will be over by now, and they will be working on something new. The teacher will tell you how to do it."
>
> "I know you are sorry to have missed the field trip because of our talk, but we have done something important in here today. They'll want to tell you about the trip when they get back. That will give you the chance to use what we have been practicing in here."

By reviewing the behavior needed for successful reentry and participation before a student returns to the group, the student has increased assurance that the reentry can be handled smoothly and without further embarrassment. When you include affirmations of the student, you increase self-confidence. Without feelings of confidence, students seldom make it back to the door before anxiety floods again and the student is out of control.

Regression and Resistance. Occasionally you may have a student who participates in LSI with cooperation and collaboration until the talk turns to returning to the group. If a student loses control, appears to suddenly reject everything that has gone on previously, seems reticent to return to the group, or tries to control the reentry process, something is producing anxiety about the reentry. The student:

1. may have more to say than has come out in the LSI,

2. lacks the confidence to make the shift back to being a participating group member, or

3. has difficulty separating from the interviewer.

Consider each possibility. You may find that you need to go back to an earlier step to bring out something that has been overlooked. Or you may have to review again the plan you and the student rehearsed in the previous step, perhaps modifying it to be more realistic about what the student can do successfully. Or you may have to examine the quality of the relationship that has developed between you and the student. The student may be feeling the anxiety that comes with separation from you. Your role in that relationship may be the significant force in fueling resistance to reentry or spiraling negative behavior. We discuss ways to handle this in the following chapter on the helping process and potential pitfalls.

Reactions of the Peer Group. The success of a student's reentry is directly related to how effectively someone deals with the peer group while the student is away. It requires close team work to reconnect the student and peers. Too much emphasis on the student's problem in front of peers tends to highlight the deviance, while ignoring the disruption to the group or the discomfort of group members communicates either the adult's desire to ignore it or inability to handle it.

In the complex dynamics of a peer group, the reentry of a member who has been away will almost always cause disturbance in other members of the group. This reaction is intensified dramatically when the absent member left the group under less than pleasant circumstances and may have transgressed against members of the group. Even if not directly involved, members watch a crisis with fascination, closely identifying with trouble and vicariously projecting themselves into the situation.

When the student reenters, it often excites the group again, and these reactions need to be anticipated. Sometimes the reactions will be directed against the reentering student, sometimes against the adult, and often against each other. Anticipating these reactions often heads off trouble.

The adult who remains with the group must deal with the reactions of group members during the time the student is away. Spiraling anxiety can be diluted or avoided by sensitive observation and redirection of each member to something personally successful. During this time, group members want to be reassured that they are "safe," the adult in charge is looking after their needs, and the absent student is being treated fairly. Here are examples of statements made by teachers and teacher aides to reassure the peer group:

"You're doing so well with this project that you should be able to show him the finished product by the time he gets back." [The message—Everything is under control, and the student will return to participate.]

"When we see a friend having to leave the room, we wonder what is happening to him. He is okay, and when he works out the problem with the teacher, he will be back." [The message—The student is not being brutalized; he is problem-solving.]

"When a problem gets out of hand, it is a good thing to go out with a teacher to talk." [The message—A talk with an adult is the way to solve a problem.]

Your Role in Step 6

The relationship between adult and student is significant in every step of LSI. However, this final step validates the relationship developed in the preceding steps. If you have not been on target, the student will present some of the problems we described above in the section on regression and resistance. If you have been accurate in identifying and conveying the type of adult the student needs, using the type of strategies that influence the student, the student will respond constructively to this last step. The relationship you have established through your adult role should provide the student with confidence that reentry is something that can be handled in a satisfactory way.

The view a student holds about adults changes as a result of increasing emotional maturity. (We discuss adult roles and styles of influence in Chapter 4 and again in Step 2 of this chapter.) Here are illustrations of how these changing roles are conveyed by adults during Step 6:

"If this trouble starts again when you go back, remember that you decided to tell the teacher before you get in trouble." [Adults provide protection, power, and approval.]

"I've noticed how you can control yourself even when others are losing it. When you go back, keep your attention on the assignment and you will be the winner!'' [Adults' attitudes and views are best and so should be imitated and internalized.]

"This is your chance to show the other students your style. When you go back in there cool and calm, they'll know you have the situation under control.'' [Adult mirrors the student's best self— one who can be independently responsible.]

As You End Step 6

The end of Step 6 usually signals the end of the LSI. The student understands that the next step is to rejoin the group and participate in a responsible way.

If the LSI has been effective, the student's insight has changed, anxiety has been reduced, and self-esteem enhanced. The therapeutic goal has been accomplished, and the student is left with changes in emotional memories—a modified private storehouse containing better thoughts and feelings. These are the ingredients of confidence. They carry over as the student leaves the privacy of the LSI setting and begins to put these changes into action in the peer group and classroom.

Here are statements that end LSI conveying confidence that the student can carry out the plan for responsible behavior and participation. We put them in a sequence of students' increasing emotional maturity.

"The next time he starts that stuff, what are you going to do so you won't be in trouble?''

"You've made a lot of progress today. When you go back to the group now, keep thinking about our plan and you will have a good day!''

"You know what to expect from the others when you go back to class. It's going to be tough, but I have confidence that you can handle it now without getting into trouble again. If you spot trouble beginning, let's come out here and talk again before it happens.''

"We have some things to continue talking about on another day. When we talk together like we did today, we can work out ways to handle this problem.''

At the end of LSI, you may find yourself wondering, ''How long before we have to go at it again?'' Don't be discouraged. Each LSI can be a building block for genuine change, but lasting change comes slowly and in small increments. View each LSI as one small step in a series. Each discussion that expands a student's self-confidence, insight, and ability to change behavior is a step toward eventual independent problem solving and emotional maturity.

WHAT WE HAVE LEARNED USING LSI

- Don't begin LSI if a student is out of control or emotionally flooding; use behavioral controls or Emotional First Aid instead. (We discuss this topic in the next chapter.)

- Don't begin LSI if you are overwhelmed by the crisis, the student, or your own feelings; wait until your own reactions are rational, objective, and completely under control.

- Don't begin LSI if you cannot take the time to go the distance, as needed, through the steps; 30 minutes is about average for a typical LSI (highly verbal or resistent students may need an hour).

- Create a sense of privacy in the place where you do your LSIs; what is said is not shared with teachers, parents, other students, or anyone else, unless you and the student agree to it as a part of a resolution plan.

- Maintain a mind-set that crisis is a time for teaching and learning, not judging and punishing; a time of instruction and not destruction.

- You (not the student) must control the pacing in LSI, spending as much time as is needed with each step; and *you* decide when it is over.

- Sometimes LSI will not occur in orderly steps; you may find yourself cycling back through Steps 2 and 3 several times, to clarify or expand an issue.

- Therapeutic goals can be combined; you may start with one goal in mind and then shift in response to the student's reactions.

- If you have some doubt about a central issue, deliberately go back through Steps 2 and 3 and try more decoding.

- Don't feel that you must complete all six steps each time you begin LSI; if necessary, go directly from Step 3 to Step 6 and close down the LSI with the message that you and the student will talk again.

- Keep in mind that behavior, feelings, and anxieties cannot be changed by a single LSI; it is by repeated use that lasting changes result.

- Remember the details of each LSI, so that you can build continuity into the next one with the same student.

WHAT WE HOPE YOU HAVE LEARNED IN THIS REVIEW OF BASICS

- The purpose in using LSI rather than an alternative behavior management strategy.

- What occurs to end Step 1.

- In which step a student is encouraged to talk even if what is being said is not particularly truthful or logical.

- In which steps you decode and probe for deeper meaning.

- The step in which you finally select a therapeutic goal.

- The point in LSI where you change from talk about the problem to talk about solutions.

- The particular statement that you must make before LSI can shift from discussion of the problem to discussion of solutions.

- The steps in which you use the most probes and interpretations as a part of decoding.

- When to encourage a student to review many alternative solutions, even those that you might see as inappropriate.

- What a student should say before a complete LSI ends.

- Why a cooperative student might begin to be disruptive toward the end of the LSI.

- The purpose of interjecting affirmations of the student into every LSI step.

Crisis reveals character... ours!

Chapter 6

THE HELPING PROCESS AND POTENTIAL PITFALLS

Now that you have reviewed the six basic steps of LSI, you are ready to try it. Don't be reluctant to begin. If you are hesitant about your mastery of the background information, keep working on it, but at the same time, begin using LSI. Nothing is as effective for learning as actually doing it. As you begin, remember the basics: the name, purpose, insight, and content of each LSI step. (We know teachers who actually taped the names of the steps to the wall of a conference room while learning the process.) Remember to decode when a student responds with emotionally ladened behavior. And most important, affirm the student as often as you can during every step.

Because LSI is a helping process built on interactions between you and a student, it requires personal attributes and skills as well as understanding. Perhaps there is no other circumstance in which you will find yourself that will illuminate your own personal characteristics and style more vividly than LSI.

The helping process makes huge demands on the adult. In this chapter we take you beyond the basics of LSI to consider the attributes and skills an adult must have to use LSI effectively. We first review some basic qualities needed to be effective in the helping process and provide a list of specific skills. Then we describe ways to provide Emotional First Aid to help students who are so flooded by emotion that rational processes are not working and they are unable to participate in LSI. In the final section of this chapter we describe typical problems that occur when an adult slips away from using these attributes and skills.

ATTRIBUTES AND SKILLS FOR ADULTS IN THE HELPING PROCESS

When working with children and youth in crisis, adults need to acknowledge, accept, understand, believe, and appreciate that emotions are the powerful, personal engines that drive or generate human behavior. Strong emotions are present in every LSI, and they are generated not only by students, but also by us. Our emotions and attitudes may (or may not) be hidden, but they are nevertheless active in our own behavior. Management of these personal feelings and resulting behaviors may become a central force for many adults during LSIs. (Student crisis can evoke a Conflict Cycle in us, unless we guard against it.)

To be helpful to a student in crisis, an adult must have a genuine desire to help the student rather than to express personal reactions to the incident. An adult must simultaneously be direct, realistic, clear, and supportive. When adults are caught up in the stress of an incident, their own emotions are added to the crisis, and their energies go into coping with their personal feelings, behavior, and the reactions of others instead of going into the helping process. You cannot be helpful to a student if you are not able to detach your own needs from the incident, the student's behavior, and the associated stress you may feel. We consider Carl Rogers's (1965) descriptions of trust-promoting characteristics for the helping process extremely important in keeping an adult focused on the job of being a helping person during LSI:

Empathy: Understanding students' feelings and actions

Genuineness: Being consistent, dependable, and real

Positive regard: Conveying caring and interest

Concreteness: Using specific, clear language

Unconditional acceptance: Approving the person but not the behavior

Another resource for adults concerned about being effective as a helping person is Warnock's (1971) four broad guides to universal values. As ''ethics of helping'' we have found them useful. With some license here, we put them in a context as guidelines for the behavior of adults who help students in crisis:

Don't hurt others. Keep your words and actions free from psychological hurt or intrusion. Blame, disgust, sarcasm, or dismay have no place in LSI. Show respect for your students, and don't invade their personal "space."

Be obviously kind. Be abundant in your use of affirmation, praise, and interest. Show kindness, especially during the problem phases of LSI. Convey compassion, acceptance, and respect for the student even while acknowledging the seriousness of the situation and disapproval of the behavior.

Be fair to all, not selective. Let your actions show that you are not selective in your dealings, treating all who are involved in a crisis with the same consideration and fairness. "Fair" should be the adjective students use to describe you as you help them during crisis. Let them know that fairness is a top priority with you.

Don't deceive or distort the truth. Few adults admit to deceiving or distorting reality or "the truth." But it can happen if we avoid painful issues, ignore important concerns of a student, or "fudge" on central issues. Present reality in objective ways. Weigh the significance of the "truth" as the student believes it to be and as others perceive it. Then share this insight with the student.

These general attributes and attitudes reflect a lot about the character of an adult. They are the solid foundation for the helping process. How they are put into action during LSI will influence the eventual success or failure of LSI.

We turn now to some very specific adult skills that contribute to effective LSI. Beyond mastery of the basic six LSI steps, we have observed that effective adults have considerable skills in relationship building and communication. Table 6.1 contains an outline of these essential skills, organized into three categories: relationship, verbal style, and nonverbal body language. We use these three categories as a way to organize the skills into a checklist to guide adults who are learning to use LSI. (This checklist is contained in the back of the book for actual use by adults in

Table 6.1 Checklist of Skills for Adults Using LSI

NONVERBAL BODY LANGUAGE

Convey support and alliance through body posture.
Use eye contact or the opposite as needed to provide "space."
Vary voice quality and volume as needed.
Use physical stance to convey needed adult role and relationship.
Avoid excessive touch.
Maintain physical proximity or distance as needed.
Stand or sit as needed to convey control.
Convey interest, support, or other messages through facial expressions.

VERBAL STYLE

Use concrete words for clarity.
Use imagery to motivate.
Convey ideas clearly.
Maximize student's talk.
Minimize own talk.
Use a time line to help student organize events.
Assist student in clarifying an issue.
Assist student in seeing cause-and-effect relationships.
Use reflection effectively.
Use interpretation effectively.
Decode accurately.
Use third-person form to generalize.
Limit use of references to yourself.
Limit use of negative statements.

RELATIONSHIP

Use active listening.
Communicate respect for student.
Communicate interest.
Communicate calm self-control.
Convey confidence and optimism.
Convey focus and competence.
Avoid intruding into student's private "space."
Avoid value judgments.
Avoid role reversals.
Avoid counteraggression.

training and by their supervisors.) Clearly, these categories over-lap. An effective relationship is built on communication skills, while nonverbal body language is a form of communication and also is a major element in building relationship. The important point is that the adult use as many of these skills as possible during LSI.

You may find it helpful, while interacting with students, to check yourself for these characteristics. Try to identify what you actually do to get across these attitudes about your students. If you cannot find them in yourself, spend some time experiment-ing with ways to communicate them to your students. While monitoring your own words and actions, also keep in mind the four fundamental concerns of most students that we described back in Chapter 1: protection, gratification, relationship, and responsibility. Your students will be looking for these kinds of reassurances from you throughout the LSI.

We encourage readers to also seek supervision while learn-ing LSI. It is not a simple technique but an adventure in teaching, learning, and character building during a time of crisis. It goes far beyond intellectual understanding of the purposes and pro-cedures for the six steps. It requires the adult to have the capacity to relate to the student, to decode behavior, and to accept deeper feelings without becoming defensive, counteraggressive, or caught up in the Conflict Cycle.

EMOTIONAL FIRST AID

Students in crisis may need immediate help and support when their defenses become ineffective and they are over-whelmed by feelings and the demands of crisis. Flooding emo-tions distort reality perceptions and dominate behavior. It is not unusual for a student to cling doggedly to the last emotion or perception experienced in a crisis, and this becomes the reality. When behavior is driven by intense feelings, the student is not responsive to rational thinking or talking. Do not try LSI right away. A student in this condition is not yet ready; Emotional First Aid is needed.

The purpose in providing Emotional First Aid is to reduce emotional intensity to the point that a student can participate in LSI. This process requires a shift from behavioral expression of feelings to rational use of words instead. Emotional First Aid also conveys that a student can count on the adult's support and help in resolving the crisis. Here are several statements teachers have used to convey this reassuring message:

> "It (you) will be all right."
>
> "There is a way to handle every problem."
>
> "I'll bet if we talk about it, we can get a handle on it."
>
> "This situation may not be as serious as it seems."
>
> "I've seen you handle problems like this before; and you did it well."

Redl (1966) describes ways to provide Emotional First Aid, depending upon what is needed to assist a student in putting emotions and behavior under rational management. The following material summarizes these five ways of providing Emotional First Aid.

Drain Off Emotional Intensity

Defusing emotional intensity helps students manage frustration or anger by reducing the amount of emotion injected into a crisis. It also is a useful way to intervene early in a potential Conflict Cycle and avoid spiraling emotions. This can be done by listening sympathetically, identifying and verbalizing a student's feelings of anger, disgust, fear, or panic regarding cruel, frustrating, or disappointing life events. Here are some examples used by adults to reduce students' emotional intensity, maintain communication, and convey understanding:

> "It's really upsetting (unfair, terrible, scary, mean, cruel)."
>
> "It's okay to feel like this; it will get better."
>
> "Other students who have had this same thing happen had those same feelings."
>
> "Sometimes when people feel like this, they think nothing good can ever happen to them again. But there is a way to work it out."

These statements generally assure the student that it is all right to have those feelings and they will not last forever. In the last two examples, the third person form is used to convey that others have had the same feelings (and lived through it), and the student is having a very human experience. Use of the third person form also avoids intruding into a student's personal space. Some students will talk with you about feelings of others when they will not (or cannot) tolerate talk about their own emotions. Here are two teachers' descriptions of how this type of Emotional First Aid was used to reduce emotional intensity.

BILL'S PHYSICAL EDUCATION TEACHER

"The physical education period was just about over when I blew the whistle and asked the group to stop and put away the trampoline. Bill objected, saying that everybody had three turns except him and he wasn't going to be gypped. I sympathized with his situation but explained that our time was up and that the best I could do, if he cooperated, was to see that he would be first next time."

PERRY'S TEACHER

"Perry just completed painting his clay horse and was proudly showing it off to the group when Jim shoved Larry who bumped into Perry causing him to drop his horse which shattered when it hit the floor. Perry was furious and started to attack Larry. When I stopped him, he broke into tears. I said, 'This is a terrible disappointment; after all the work you've done! The other guys feel awful about it, too. They thought it was such a neat horse. Let's stop and think about what should be done about this horse now.'"

The first example, with Bill, illustrates prevention of a blow-up with a simple statement of understanding and concern for fairness; the second, with Perry, conveys understanding and sympathy and offers a rational beginning to problem solving by redirecting Perry away from the boys who accidentally caused the mishap to the problem of what to do about the horse.

Support a Student Engulfed in Intense Emotion

Often an adult steps into a situation that is already out of control. A student's level of panic, anger, fury, guilt, or anxiety is so high that behavioral controls have disappeared. The goal of this type of Emotional First Aid is to protect the student and others from the rage and temporary confusion. Occasionally, this calls for physical intervention until the student's own controls are functioning again. Here are examples of statements adults can make to provide this type of aid:

> "I'll have to protect you until you can help yourself."

> "Teachers are in control here."

> "This is the sort of thing teachers handle."

> "Sometimes kids have to trust adults to take care of problems."

> "There are rules about this situation, and we follow rules here."

Each of these statements provides supportive but firm assurance that external structure, adult authority, and rules guide management of the crisis. The first statement is in first person form, for situations where a student needs to know that this adult is going to be in charge of the student. The next two statements are phrased in the third person to defuse possible personal confrontation between an out-of-control student and the adult. In the last statement, the plural first person form is used to convey teamwork —together things will be all right.

Here are two reports, from a teacher and a coach, illustrating their first aid to students out of control and overwhelmed by emotions:

CARLA'S TEACHER

> "Hearing screams from the corridor, I entered Mrs. A's room and found Vera and Carla in a wild fight. I demanded that they stop, but Carla was too angry to hear. As I separated them, Carla started to hit me, so I had to restrain her physically until she could calm down. I finally talked her down by saying, 'Carla, take a deep breath. I'm going to hold you until your breathing comes easy. Then we can talk about what happened.'"

CHARLIE'S COACH

> "I heard a muffled sound in the boy's bathroom and found Charlie hiding in one of the stalls, crying. I had

no idea why. I put my arm around him and said, 'There's nothing so terrible that we can't work it out together.'"

Maintain Communication When Relationships Are Breaking Down

In an intense crisis, some students withdraw and become uncommunicative or go into a prolonged anger, sulking and refusing to talk. Unless this defense is penetrated, the world of hostile fantasy can be more destructive to them than the world of reality. The purpose of this type of Emotional First Aid is to redirect students away from this reaction by engaging them in any kind of conversation until they feel more comfortable with their thoughts and feelings. Occasionally gadgets, food, or humor are effective ways of thawing a student from a frozen sulk. Here are descriptions of how two teachers used this type of Emotional First Aid to stay in touch with students until they could be reached for a rational LSI.

CRAIG'S TEACHER

"Craig was kept after school to complete his arithmetic assignment. He sat at his desk, folded his arms, looked at me with blazing eyes, and said, 'I'm doing nothing and saying nothing. So leave me alone!' When I tried to talk to him, he covered his ears with his hands. So I sat down at the table near him with the materials I was using to make a board game for reading. The pieces were race cars. Although I was silent, I made certain Craig saw them. It wasn't long before he asked, 'What's that for?' Now we were in a communication mode and ready to begin LSI."

MARTIN'S TEACHER

"Martin kept staring out the window all the time I was talking to him. After many futile attempts to get a response from him, I asked Martin to draw a finger picture on the windowpane and I would try to guess it. Halfheartedly he drew a circle that I said was a purple elephant taking a bath in a white Volkswagen. With great effort Martin tried to hide his smile. Then I said, 'I'll do one now, for you to guess.' Martin nodded that he was interested and guessed my quick design to be an engine falling off a pickup truck and smashing into pieces on an interstate. That imagery made our entry into the first step of LSI.' "

Regulate Social Behavior

Most students quickly learn the rules of a program and are astute in describing conflicting adult standards and contradictions between actions and words, but they often need daily reminders of the rules and regulations if they are to remain in the activity or lesson. The purpose of this type of Emotional First Aid is to provide positive assistance to a student by warning of potential dangers that lie ahead if present behavior is continued. This is an easy form of advice for adults to dispense because it usually is needed before emotions get out of control. Here are several examples:

"I explained to Sally that in order to stay in the group she would have to keep her hands to herself."

"Each day I have to remind Bill that he must finish his assignment before he can play softball when the group goes out to recess."

"When Karen starts complaining in a shrill voice that someone is bothering her, I tell her to use a calm voice and let the person know that it bothers her."

"Roger likes to make smart remarks for attention. I remind him that one of our rules is to do things so others can get their work done. This seems to be enough to drain away his interest in getting someone else upset."

Act as Umpire

Often students try to cast adults in the role of umpire to help settle intense disputes, grievances, or game violations. This form of Emotional First Aid is especially appropriate for students in the preexistential phase of development or going through the

Existential Crisis (see Chapter 4). Many important therapeutic gains can be made when an adult assumes this role as a fair and equitable arbitrator. Most students respect and respond quickly to the adult as umpire. Here are two examples:

RALPH'S RECREATION LEADER

"Ralph came to me complaining about Peter. It seems that every time Peter begins losing in chess he introduces a new rule. The latest one was that his father told him that the king can move two spaces in any direction once both castles are taken. I explained to Peter that although special rules can be worked out at home, we follow standard chess rules here."

DAVEY'S TEACHER

"My class was playing dodge ball. One of the rules we agreed on was that if a ball touches any part of the student in the circle, that student exchanges places with the thrower. In this incident, the ball 'skimmed' Davey's hair. He claimed that he had not been hit but skimmed. When the group demanded that he change places, he turned to me for support only to hear me rule a skim as a hit and that he should have to change places."

After Emotional First Aid

If Emotional First Aid has been used appropriately, adult authority is established, students are supported, emotions are expended, and rational words have replaced behavior as the way to communicate. Most students are sufficiently relieved that they are able to settle down and participate in a full LSI. For those students who are still not able to mobilize themselves sufficiently to problem-solve, you may have to end with a decision that reentry and participation in the ongoing activity are the most that can be obtained at the present. When you make this decision, convey clearly what is expected from the student in behavior and participation. Also be sure the student understands that (a) you have judged it to be the best course of action for the present, (b) the student has qualities that will help make the reentry and appropriate participation successful, and (c) there will be other times to talk again.

POTENTIAL PITFALLS IN USING LSI

Although we ordinarily prefer to use a positive focus when preparing adults to use LSI, it may be helpful to show how problems arise and provide specific suggestions for what can be done about them. We have identified 13 responses adults frequently use that create unnecessary problems. Everyone makes these mistakes at one time or another when using LSI. The important lesson to be learned is the value of self-monitoring so that you recognize why a particular LSI goes astray and what you can do to improve it next time.

Beginning LSI When a Student Is Not Ready

Never go beyond Step 1 until the student has given you some verbal response. Don't even think of it! We have described the importance of reducing emotional flooding before you begin Step 1. We also have suggested the use of "A"s for affirming a resistant student; or the use of gadgets, food, or humor as a type of Emotional First Aid to help a student shift from an emotional to a rational, verbal mode. There will be times when all of these strategies fail to get a student started. If you try all of the above and the student remains noncommunicative, wait. Silence can be a powerful device for mobilizing a student toward verbal interaction. There are several ways to use silence effectively as you wait.

Thinking Silence. Some students need time to get themselves under control, to consider how to express themselves, or to consider the feelings and facts surrounding an event. The body language they use is generally not emotion laden. When you suspect this is going on inside a student's head, respect the silence and sanction it with a statement such as,

> "It's helpful to think about what happened before we talk."

Silence as Confusion. If the student's behavior suggests anxiousness or confusion, it is possible that the student lacks a grasp of what happened or has insufficient verbal ability to get talk started. Sometimes students will pick or twist their fingers, squirm, or kick their feet to convey their confusion or inability to respond. If you suspect this, try tapping into some concrete, key recollection by saying,

> "Tell me what you heard."

> "What was the very first thing that happened?"

Such statements from you may help the student connect to a time line or remember a vivid aspect of the incident that can begin LSI.

Silence as Resistance. Some silence is used by students to test their independence or resist adult authority. If you have attempted all of the above strategies or conclude that a student is deliberately resisting your attempts to start LSI, try one (or two) last attempts to get the student involved with a three-step sequence of reflection, decoding facts, and a question such as these:

> "When you sit there saying nothing you are missing out on what the others are doing—having fun." [Reflection]

> "Sometimes guys don't talk because they don't want to get their friends in trouble." [Decoding fact of not talking]

> "But right now your friends are having fun and what are they saying?" [Question]

A sequence like this usually evokes at least a verbal response of "I don't know!" And you have shifted the student into using a verbal means of interacting with you. If there is no response, go back and try decoding with greater depth, such as:

> "Maybe students don't talk because they don't trust teachers to understand and help."

> "I have the feeling this discussion is upsetting you. It's okay to tell me that you don't want to talk."

> "I see it's difficult to continue. I wonder if your silence is connected to something I said."

In the rare instance when these last statements do not produce some verbal response, the student may be (a) developmentally below the level of comprehending your words, (b) incapable of energizing mental forces (drugs? psychosis? seizures?), or (c) testing you to the limit.

If you suspect that a student is developmentally delayed or having a transient incoherent mental episode, go from Step 1 to Steps 4, 5, and 6 and briefly end the LSI with solutions that psychologically protect the student, while assuring participation in the expected routine and clearly conveying the rules and consequences that follow.

If a student is "digging in" for the ultimate testing of you as the authority (and none of the above strategies has worked), end the LSI but leave an opening for the talk to continue later. At this point, it must be clearly conveyed that nothing goes back to a normal routine until the talk has taken place. There must be talk before the incident is closed. The situation has reached a point of stand-off where authority must be exercised. Usually, a period of time out is necessary in such situations.

Getting Off the Subject

Balancing the extent to which you let a student digress and the extent to which you keep the discussion on the topic is one of the most difficult things to do during LSI Steps 2 and 3. We have found it helpful to keep in mind the purposes of these steps: for the student, to talk, cool down, and use words rationally; for the adult, to begin to shape the real issue for the next step in the LSI.

Sometimes digressions can help the process by giving students an opportunity to communicate on subjects that are not emotionally connected. Digressions also can broaden communication about an incident by indicating information that has significance to a student. We call these "red flags." A red flag appears as an out-of-context comment. It can signal that a topic appearing to be a simple digression actually has some bearing on the incident, has associated feelings, or is part of a greater anxiety that has found expression through the incident. A red flag may change the course of the LSI by bringing out a deeper concern. In other instances, a red flag is used by a student to keep you away from a discussion about the issue. Here are examples of each type of red flag:

EXAMPLE: to bring out a deeper concern

> In the midst of LSI talk about a fight with another boy on the playground, Jeff fires an unexpected question

at the teacher, "Are you adopted?" His teacher suspects it may be a red flag (because Jeff was not adopted and has been setting fires at home), but she is not sure whether it indicates a deeper concern or an attempt to get the teacher off the subject of the fight.

The teacher makes a neutral statement and follows it with a question as a preliminary probe, "Adoption is something a lot of kids think about. Do you think about being adopted?"

There is a long silence, and then Jeff quietly says, "Yes, my real parents were killed in a plane crash when I was a tiny baby."

This led to a dramatic shift in the LSI, to talk about Jeff's fantasy about what happened after the crash and how a judge found ideal parents for him. Jeff told the story with great detail, intensity, and imagination. It was clear that he had told it to himself many times before. The teacher used this red flag and continued the story, reflecting that it is important for tiny babies to have parents that care, and how parents show their care. At the end of the story, the teacher decoded for Jeff.

"Boys think a lot about their parents and have high expectations for them. Sometimes, when parents don't do what they should, their children are angry at them. And that stays on the kids' minds; and then they get into trouble at school."

There was a long silence after that, and then Jeff said, "Yep, that's right."

This was as far as the talk about the deeper concern went for this particular LSI. The teacher then bridged back to the playground incident and continued through the LSI steps. She knew that the topic deeply troubled Jeff and needed to be aired; but enough had been done with it for the first time.

EXAMPLE: to avoid talking about the issue

Rick was brought to the counselor because his math teacher suspected that he had stolen a watch from another student. As soon as he sat down, Rick began,

R: "My math teacher is mean, unfair, and has pets in class. He doesn't know how to do his job. Everyone knows it. It's been going on for years. He should be fired!"

What a temptation for the counselor to react to any one of these accusations and innuendoes! But she recognizes this red flag as an attempt by Rick to keep the LSI away from the real issue of the watch.

T: "This watch you're wearing is the issue."

When he sees that the counselor does not buy into his attempt to get her off the subject, Rick tries several other red flags.

R: "I bought this watch from a guy for $1.00 last night at the ball game!"

T: "Then we need to talk with the person who sold it to you."

R: "I don't know who it was—just a guy."

T: "When a boy comes to school wearing an expensive watch, and another student reports that his watch is missing, what are we to think?"

R: "Are you saying I stole it? Didn't you ever take anything when you were a child? Everybody does it!"

In this last example, Rick tries to reverse roles and control the LSI. When this fails, he tries to divert the focus to the counselor's own childhood. (See LSI Step 2 description in Part Two for more about red flags.)

In either case, don't let a dialogue about a red flag topic continue too long. If it is a real issue from a student's private reality, too much talk may be difficult for a student to handle on a first round. Prolonged discussions on sensitive topics usually result in a student's emotions flooding back again; and you are back to Step 1 unnecessarily. If the red flag was a deliberate digression to keep you away from a topic, don't let the student think you have forgotten about the incident. You do not want a student to control the LSI.

Allowing an Adversarial Climate to Develop

The quality of the relationship between student and adult develops during Steps 2 and 3. Nothing will ice LSI quicker than letting yourself be cast into the role of the judge, the adversary, or the "supreme being." Often this tone creeps into an exchange without the adult being aware of it. Veiled sarcasm, exasperation with the student, disgust, disbelief, or cynicism may convey such attitudes. Judgments about the generally undesirable nature of a student also may be applied by the type of statements an adult inadvertently makes. An experienced juvenile court worker described the phenomenon this way:

> "It's like two players in a tennis game. One makes a shot and the other responds. Sometimes your shot is out of the court and doesn't help you. But sometimes you can get in a smashing shot that is right where it should be. Then you've nailed your opponent!"

You've won the point but lost your opponent. The obvious innuendo in this scenario is adversarial. Somebody wins and somebody loses. During Steps 2 and 3, the potential is there to set up such an adversarial relationship. The student is determined to win, and so are you. Guard against this; it has no place in the helping process!

When you sense an adversarial tone creeping into your LSI, try to change it by deliberately breaking into the stream of discussion and interjecting an "A," affirming some valuable attribute in the student. This action dilutes the potential for conflict. By deliberately aligning yourself with the student and expressing clear conviction that you have confidence in the student's ability to deal with the crisis, you dilute a potential position of being against a student. A posture of "fairness for all," including the student, is another way to avoid adversarial tones in LSI. Remember what we want is a win–win condition.

Reflecting Negatives

As feelings come out during LSI, it often is the adult who initiates the topic by decoding what a student is feeling. When you do this, guard against reflecting negatives without the addition of a positive ending to your statements. If you reflect only negatives, you reinforce a student's perceptions that the student really is impossible (or the situation is hopeless). Here are some examples of negative reflections. We add positive interpretations in parentheses to illustrate how negative statements can be turned around into positive ones by adding a phrase.

"What I see is a boy who is really upset." ("If we talk about it, it will help you feel better.")

"When you do that, it tells me you feel awful about it." ("There are ways to make it better.")

"This just hasn't been a good day, has it?" (". . . but it doesn't have to keep on being this way.")

"When you are out of control like this, it tells me something is really bothering you." ("When you use words, your more mature self is coming out.")

The first parts of these statements are decoded reflections that leave negative images—students with bad feelings and behavior. The intent of the adult was to convey understanding of what the student may be feeling and begin to teach the student that feelings can be put into words. Without the positive addition, the students' unpleasant feelings would be reinforced and their views of self as being full of unmanageable feelings would be expanded. With the addition, the adults have allied themselves with the student and conveyed belief that better things can happen for the students.

Permitting a Student to Seduce You into Counteraggression

Aggressive students have a style of engaging adults in personal struggle by using words and actions that "push the teacher's button." They say things within hearing distance, hoping the adult notices. They choose words and actions that they know the adult will respond to. If the lure is successful and the adult seizes the student's bait, the power struggle is on.

Unfortunately, the battle is one-sided. If the aggressive student wins, aggressive behavior is reinforced. If the student loses, but gets the adult to escalate tension with counteraggression (or open hostility toward the student, or rejection), the aggressive student wins the emotional war by "proving" that adults are hostile toward the student, do not really care, and cannot be trusted.

For the adult caught in this power struggle, the situation seems intolerable. Students' comments, such as shouting "No!"

"So what?" "Forget you!" or "You can't make me do it!" beg for aggressive responses from the adult. What complicates the adult's position is that the angry feelings and actions taken in response seem so justified. It seems logical and natural to put this arrogant, hostile, inconsiderate, hopeless student in his or her place.

The solution for avoiding counteraggression is to recognize the circular nature of the aggressive exchange, and to understand that the more an adult struggles with an aggressive student, the more likely it is that the adult will come out on the short end. For the adult to continue counteraggressive behavior is self-defeating. The adult who acts out personal feelings and does "what comes naturally" is perpetuating the aggressive cycle by mirroring the student's behavior and continuing an aggressive model. As a result, the statement "Aggression elicits aggression" becomes true, as the aggressive student successfully determines the adult's behavior.

What must adults do to stop this destructive cycle? They must recognize and understand the seductive power of the Conflict Cycle they have been caught in. They must acknowledge that the student is actually controlling the adult's behavior. The adult must use self-control to break off the exchange, separating from personal feelings of righteous rage or the wish to subdue the student and win at any cost. And the adult must not seize the student's bait.

Treating Complex Feelings in Trivial Ways

A common mistake is to ask a student, "How does that make you feel?" before the student has the verbal skill to answer the question. Many of the students we work with would fumble around unsuccessfully if asked such a question. Few can describe their complex and deep feelings, even though they are aware that they are having these feelings. To answer this question at all, students with limited verbal skill must come up with a fairly simple set of words to describe overwhelming or complex feelings. When they do this, they often are aware of the superficiality of their attempts to answer the question. By asking the question and accepting the answer, you signal satisfaction with a trivial description of their painful feelings. If a student tries to answer this question at all, the typical answer is "I don't know." A student who responds this way has admitted openly to inadequacy. Equally unproductive is an adult response to a student's answer about feelings that treats the information casually, with marginal compassion, or tries to minimize the student's feelings. For example, a student may be able to come up with, "It makes me feel terrible," in answer to a question about feelings. If the adult responds, "It's not all that bad," or, "Come on now, it's not worth feeling that bad," the student has been put down and has no next move.

There is a time to teach verbal skills for describing feelings, but don't depend on teaching this skill while trying to deal with crisis during LSI. We believe that teaching basic verbal skills to express feelings should begin in noncrisis situations. The sequence for learning increasingly complex expressions of emotion should be taught systematically. The use of multimedia is helpful—through TV, movies, storybooks, fiction, newspaper accounts of other people's crisis, and observations, students learn to talk about the feelings of people they have known in crisis. It can be done within the social science curriculum or during language arts or current events. Such situations put the emotional experience on someone else. It seems much easier for students to learn verbal skills for recognizing and describing the feelings of others than doing this about themselves.

When students can describe the feelings of others, they are ready to describe feelings of their own. Then it is productive for an adult to ask the question during LSI, "How did that make you feel?" When your student begins to show the verbal skills to answer such a question, it often is helpful to phrase and structure discussions in the third person form so that you and the student can talk more freely about other people who have had the same feelings (or about experience that produced the feelings). The most productive gains from such discussions occur when you convey genuine understanding and acceptance of feelings. This begins a process where the student eventually understands what has produced the feelings and how they can be reduced and changed. Such insight requires developmental skills involving multiple perspective-taking that many students do not have until adolescence. However, the readiness skills for talking about feelings and treating them with understanding, respect, and compassion are built long before students get to high school.

Failing to Build LSI Dialogue from a Student's Responses

One of the most commonly heard complaints from adults begining to use LSI is, "I don't know what to say!" Adults whose primary concern is their own performance in managing a crisis sometimes listen only to their own voices and their own agendas, and they seldom hear the message in the student's words and actions. It requires considerable self-discipline and training to become a good listener who decodes and then responds to a student's responses rather then to his or her own responses.

The easiest way to overcome this problem is to use reflective statements (described previously in LSI Step 1) or to use follow-up questions with some of the student's *exact* words or ideas. An elaboration of this technique is to begin a statement with the words used by the student and then tack on your own new idea at the end. Such bridging between you and a student conveys that you are listening and hearing.

Invading a Student's Private Space

Probably the most serious mistake made by adults in LSI is intruding into a student's private psychological space before a student is ready to share it. Unwelcome intrusion usually happens as an adult begins to make interpretations, decoding the student's words and actions. If the interpretation is too private, too bold, or touches a topic too sensitive, a student will close down. Such interpretations often leave a student with the private self exposed, without psychological protection. Saving face, like protecting from painful feelings, is a well-developed defense with almost all students. When your words call up these defenses, it is hard to go further and few, if any, therapeutic gains can be made.

The best approach to a student's private space is an occasional sensitive probe followed by a neutral statement about the topic. A probe is a tentative follow-up question or statement about a topic or idea not touched upon before. Something in a student's words or actions will suggest a probe, although we are seldom certain about its accuracy until we try it. Students' reactions to a probe give you the information you need about its accuracy and their willingness to have that topic included in the LSI.

Fly-fishing makes a nice metaphor to explain probes. In fly-fishing, the lure is lightly cast onto the surface of the stream. The fisherman casts with some knowledge that this is a likely spot to find a fish. However, there is no guarantee that the fly has landed in the right spot. So the fisherman moves up and down the area, casting again and again, until the connection is made. Probes are used in this same way. An interpretation is "cast" in the general direction you believe may be significant. When you get a "strike" you will know it from the student's response (or notable lack of response). In short, you have interpreted accurately, and the new bit of information goes into your decoding network. As the dialogue with a student continues, your probes take on greater accuracy and you will find yourself decoding with increasingly more relevant interpretations about increasingly more sensitive and significant subjects. The student is letting you into that private space. The previous red flag example with Jeff illustrates a student inviting an adult to share that space.

Carefully gauge a student's willingness to let you into the private space. One sure indication that you have gone too far is an increase in the emotional intensity of the student's reactions. If this happens, try an affirmation of the student, to convey confidence that the topic need not be devastating. This usually is sufficient support to continue the topic. If not, back away; go back to talking about the incident again.

Interjecting Personal Comments About Yourself

As the personal, private side of a student unfolds, a caring adult often is caught up in compassion and feeling for the emotional suffering of the student. When this happens, it is natural for adults to want to let students know that they share similar experiences and feelings. This is an attempt to support a student by communicating understanding through mutual sharing. But it is seldom interpreted that way by students. When an adult brings personal experience into a discussion, students often see it as an adult who is more centered on self than on the student ("All she wants to do is talk about herself"). This problem occurs more frequently than imagined, and it is usually not recognized or is vigorously denied by the adult. Here are statements made by adults projecting their own personal experience unnecessarily into LSI:

> "I have felt that way myself."
>
> "When I was your age that happened to me."
>
> "My best friend did the same thing to me."
>
> "People like that make me angry, too."

A better way to show mutual understanding and compassion is by reflecting positively about the strengths of the student, affirming your confidence in the student's ability to resolve the crisis.

Allowing Students to Reverse Roles with You

Students with strong needs to control adults sometimes use the first three steps of LSI to test their ability to control adults. Such students usually are skilled manipulators and are very verbal. In fact, we have seen students who take charge of their LSI, controlling the questions, answers, and the movement through the steps. They also often try to make the decision about when to stop LSI. It could happen to anyone, but don't let it happen to you.

The strategy most frequently used by such students is to ask the adult pleasant personal questions or make observations about the adult's personal characteristics. Here are some typical statements and questions used by students to reverse roles (the student becomes the interviewer):

"Are you married?"

"Do you have children?"

"Are you a real teacher?"

"Your hair (belt, dress, shirt, etc.) is beautiful."

"You have big muscles."

If an adult responds to these harmless questions, the student forms an impression that topics for LSI can be controlled by the student, and an adult who can be controlled is a less reliable adult. If the adult actually answers these innocuous questions, more blatant testing of adult authority invariably follows:

"Why are we cooped up in here?"

"Why didn't you take the other guy out?"

"How long do we have to stay in here?"

"You don't understand the situation!"

"This isn't your business; it's between him and me."

"You really don't want to be fair, you only want to get me!"

If you let yourself respond to such statements, the student is testing for social power, controlling the LSI. Equally devastating is to ignore a student's attempt to control, thereby creating the impression that things get by you or you do not know what to do with the remark. Paradoxically, to show yourself inept or vulnerable to manipulation by a student is frightening to the student. The unspoken message is that you cannot control the situation. Students in crisis have to believe that the adult can handle the crisis in an expert fashion and not abdicate authority.

If an adult fails to communicate expertness in handling the minor attempts of students to reverse roles, they almost always continue to pursue the adult, testing for vulnerability until they find it or become convinced of the adult's dependability in handling the student competently. Typically, the role reversals become more overtly confrontational and personal as these students try to exert power over the adult they perceive to be vulnerable to manipulation. Here are some typical examples of students attempting to control adults with personal attacks (notice the implied attempts to elicit counteraggression):

> ''Your breath stinks!''
>
> ''You don't really know what's going on.''
>
> ''I'm going to get my dad to come down here and tell the principal what a lousy teacher you are.''
>
> ''I can knock your head off whenever I want to.''
>
> ''You're too fat to help anyone.''
>
> ''I'm leaving and you can't stop me!''

By now, your decoding skills should be working, and you recognize that these are really desperate remarks based on concern, anger, or fear. The student in each instance is asking for the adult to show that the student is not going to be left without an expert adult.

There are several ways to respond to students who try role reversal. For the first attempts, when a student uses innocuous questions or statements to see the extent of your vulnerability:

- Turn the question back to the student, rephrasing it to ask a personal question of the student.

- Use an ''A'' to affirm the student's interest in others (careful observations, interest in details, responsive to others, good with words).

- Make a statement such as, "What's important in this conversation is you (or what happened)." Then continue the talk about the incident.

If a student continues the intense personal verbal attack, decoding is essential. The student must know that you understand the feelings behind the attack and that there is another way to help the student out of the desperate situation. Here are several examples of statements teachers have made in response to the previous examples:

"When we're talking about things that are uncomfortable, it doesn't help to try to change the subject."

"I can see how upset you are, but there are ways we can solve this problem."

"It seems to me that you had a choice, and now you regret it."

"What happens to you is what's important, and that will depend on what *you* do!"

Failing to State the Central Issue Simply and Concisely

As you and a student go through the first three steps in LSI, developing the background for understanding a crisis from the student's viewpoint, feelings, and insight, it is easy to bog down. The process often rambles back and forth between Steps 2 and 3 for further information and clarification. The adult continually probes for wider and deeper meaning, and decodes the student's responses. It requires mental organization to analyze a large amount of complex and often disparate information. You may not know exactly what it all means, what is important, and what is simply off the track. Yet it is essential to make a concise and clear statement about the central issue before you leave Step 3.

We find that the best help in this situation is observing the student. Most students' behavior provides indications of what is truly significant. Test your observations by decoding. The student's response will give you an indication of your proximity to the central issue. Sometimes students protest loudly but then show by body language that they are waiting for your next statement. At other times they will be unusually silent, waiting for you to follow up on the idea. When you sense that you are on the right track, ask the student to put the central issue into words with statements like these:

> "What is really important in all of this?"
>
> "We've talked a lot. Some of this is really important if things are going to get better for you. Let's list the important points."
>
> "I think what we are talking about is right to the heart of the problem, do you agree?"

If the student is not able or willing to respond, try to extract the essential focus from the student's viewpoint with lead-ins like these:

> "It seems like what we have been talking about really is . . ."
>
> "What I hear in all of this is . . ."
>
> "It is really hard for kids to have to face . . ."

Once you or the student has made a simple summary of the issue, you are ready to move the LSI into the problem-solving steps. Don't go forward until that concise statement has been made. If you do, you will have no basis for choosing a therapeutic goal and nothing to which you can attach a crisis resolution plan.

Jumping to Solutions Prematurely

Once an issue is out in the open, adults tend to leap to the finish, to "fix it." This same type of reaction also slips into adults' responses when a student describes the incident and the adult says, "Didn't you know that would happen?" Without waiting, the adult often goes ahead and tells the student a better way to handle it. Or the adult reminds the student of the rule that should have been used to govern the behavior. In either case, the adult is providing the solution prematurely. Neither of these responses is appropriate during LSI Step 4, when what is needed at this point is to encourage the student to consider all of the alternative solutions, right and wrong, that might be used to resolve the crisis. Hold off on your opinion or idea about the best solution.

There are two reasons for this. First, you do not want to preempt the student's process of thinking through the choices. To do so would close off the opportunity to consider the results and consequences of each alternative. This is an essential part of accomplishing the LSI goal. Second, if you provide the solution (rule, guideline), it is yours, *not* the student's. You will own it, but the student will not!

Remember that the preferred process for LSI Step 4 is a student-selected solution. Also remember that you should not leave Step 4 until the student states (or restates) the solution that has been chosen.

Failing to Consider Negative Solutions and Consequences

It is not unusual to find adults attempting to lead students away from mentioning inappropriate or negative solutions during LSI Step 5. This desire to skirt any thought of more inappropriate behavior is understandable, but not useful. By sanctioning a review of the inappropriate solutions, an adult actually helps a student rationally anticipate and weigh the consequences. We have found it helpful with some students to role play the various choices and vicariously play out the negative consequences.

Consider the reality of actually implementing a resolution to a crisis that involves new behavior. It is likely that the student may try out the new solution with the best of intentions. But the ineptness of the student or the unpredictable nature of the response of others to the student may produce a failed attempt. To reduce this likelihood, we find it helpful to always rehearse the new behavior with the student during Step 5. Take the role of the potential adversary. When the student tries the new behavior, take the role of someone responding negatively to the student. If the student is overwhelmed, confused, or reacts in anger to your role play, the new alternative has not been sufficiently practiced. You have not prepared the student well enough, or the chosen solution is above the student's ability to accomplish it successfully. You may have to go back to Step 4 and reconsider solutions with the student. If the student handles the practice role play, in which someone responds negatively to the new behavior, you know that you have increased the odds that the student can react to the next real incident with the new behavior. The LSI has been a success.

PRACTICE WITH THE FIVE FORMS OF EMOTIONAL FIRST AID

This chapter has emphasized the challenges to the adult using LSI. The process is one that requires a constant vigil to preserve an attitude of helping, supporting, and confidence. These qualities are demanding—an adult using LSI is in a particularly demanding, and often frustrating, position. We must help ourselves before we can help our students.

Because there is so much material in this chapter, it will require frequent reviews as you actually use LSI. It was difficult for us to select one particular topic to focus on in the final review for the end of this chapter. So, we selected Emotional First Aid, and gave it a new application—applying it to the helping adult. We do this because it is essential to maintain your own emotional strength to be effective with LSI. We also believe that if you can take a concept and apply it to your own needs, it will become easy to use with students in similar situations. This material was originally developed by Carol A. Merritt (1981, pp. 18–19) and is reprinted here with minor changes and permission of the author.

What Emotional First Aid Can a Teacher Use to Get Through the Day?

1. You have just spent several hours developing what you consider to be a wonderful lesson. As you present it to the class, the students begin complaining and saying, ''That's dumb!'' ''We're not doing that!'' You struggle on, but it becomes more and more apparent that you will be unable to use this lesson. Your stomach tightens and your face turns purple. All you can think of is how much time and effort you spent. Which form of Emotional First Aid do you need?

   a. Draining off of emotional intensity
   b. Support when engulfed in intense emotion
   c. Maintenance of communication
   d. Regulation of social behavior
   e. An umpire

2. You are angry with a student who has been uncooperative and disruptive all day. However, you notice he has earned just enough points to have free time. Your disgust with the student makes you want to take away his free time; yet he has

earned it. When you mention your problem to a fellow teacher, his response is, "Hey, he earned his points. Maybe you need to tighten up your behavioral system in the future." Which form of Emotional First Aid do you need?

a. Draining off of emotional intensity
b. Support when engulfed in intense emotion
c. Maintenance of communication
d. Regulation of social behavior
e. An umpire

3. John has been quite verbally abusive and disruptive off and on during the day. It seems that no matter what you try, this behavior continues. As the day ends, he begins once again. You intervene, again, and he spits in your face. Which form of Emotional First Aid do you need immediately?

a. Draining off of emotional intensity
b. Support when engulfed in intense emotion
c. Maintenance of communication
d. Regulation of social behavior
e. An umpire

4. Your students are working independently at their desks on an assignment. Suddenly, it seems that each student requires your individual help, *now*. You try to meet each student's need, but you can feel yourself becoming overwhelmed and wanting to be somewhere else. Consequently, you begin to withdraw slowly. Which form of Emotional First Aid will help you now?

a. Draining off of emotional intensity
b. Support when engulfed in intense emotion
c. Maintenance of communication
d. Regulation of social behavior
e. An umpire

5. All staff members are attending an important meeting. Time is running out and a decision has not been made. One staff member begins on a new tangent, and others become irritated. The director interrupts her and redirects the group to the task, saying, "Remember, we must reach a decision about our new policy today." Which form of Emotional First Aid is the director using?

a. Draining off of emotional intensity
b. Support when engulfed in intense emotion

c. Maintenance of communication
d. Regulation of social behavior
e. An umpire

WHAT WE HAVE LEARNED FOR AVOIDING PITFALLS

- Think of LSI as two-way communication—with a ratio of about 75% student talk to 25% adult talk.

- Maintain positive relationship based on trust.

- Credit a student's good intentions whenever possible.

- Provide simple, genuine affirmations about the student during every LSI step.

- Listen for your words and the tone in your voice to be certain that you are conveying respect and unconditional acceptance even though you cannot approve the behavior.

- Always control the movement from step to step through LSI.

- Stay calm and focused.

- Avoid leading a student into "entrapment." (It is not ethical to manipulate students into "testifying" against themselves.)

- Communicate that change will produce satisfying results for the student.

- Avoid staying "forever" on the topic of the incident and associated problems. Give equal time to alternative solutions.

- With students who review an incident quickly and concisely, it is essential to decode and push them for greater insight.

- Always have students put the chosen solution into their own words and describe the behaviors they plan to use to accomplish it.

- Let the student know that LSI is ongoing ("We will talk again!").

WHAT WE HOPE YOU HAVE LEARNED FROM THIS CHAPTER

- Why LSI makes such heavy demands on the personal attributes and skills of the adult.
- Six basic attributes of an adult that are essential for an adult to be helpful to students in crisis.
- Examples of specific adult skills for LSI, in relationship, verbal style, and nonverbal body language.
- The purpose in providing Emotional First Aid as part of the helping process prior to using LSI.
- Five general ways to provide Emotional First Aid and the specific purpose of each.
- An example of a typical adult statement for each type of Emotional First Aid.
- Four outcomes of effective Emotional First Aid.
- What to do if a student is not ready for LSI after Emotional First Aid.
- A major problem that sometimes occurs with LSI Step 1.
- Three ways to use silence.
- A time when it is useful to get off the subject of the incident during LSI Steps 2 or 3.
- How to determine when it is counterproductive to get off the subject of the incident.
- Ways to avoid developing an adversarial tone in your relationship with a student.
- Why it is important to avoid frequent use of negative reflections or statements, and how to turn them around into positive ones.

- How counteraggression develops and why it is essential to curb it in yourself.

- An example of an ordinary question or statement adults often make that trivializes a student's feelings, without intending to do so.

- How to overcome the problem of not knowing what to say.

- A definition of a probe and how it is used in LSI to get to deeper concerns.

- How you can tell if you have gone too far into a student's private space.

- How to avoid personal comments about yourself, and why this is important.

- What students are trying to do when they attempt to reverse roles with you.

- How to respond to attempts at role reversal.

- How to find the central issues during LSI Step 3 when you are bogged down in so much information from a student that you do not know what is relevant.

- What is lost during LSI Step 4 when an adult tells a student what the rules are or provides a solution to the crisis.

- Why is it important that a student consider negative or inappropriate solutions as well as appropriate ones during LSI Step 5.

PART THREE

APPLYING LSI TO PARTICULAR TYPES OF PROBLEMS

In Part Three we take you beyond a "standard" LSI to illustrate how the process is used with five particular types of problems. During the first three LSI steps, your task is to formulate the central issue which will become the focus for the problem-solving steps. When you have done this, you will find yourself in Step 4 wondering, "Now that the issue is clearly stated, what outcome will benefit this student?" You are asking yourself to formulate a therapeutic goal for the LSI.

A therapeutic goal differs from behavioral, academic, or developmental goals in that it is concerned with helping students gain some new insights about their behavior. Since LSI is a process for talking with students about their perceptions of an incident and helping them to organize these perceptions in ways that allow students to gain new insights, the selection of a therapeutic goal must be specific and central to the individual student.

You may recall from the previous section, in LSI Step 3, the five types of therapeutic goals are outlined: organize reality, confront behavior, build values for self-control, teach new skills, and expose exploitation. A specific therapeutic goal is selected from the particular perception a student has about the incident.

The goals are presented in the form of action statements to highlight the adult's role in shaping the therapeutic focus. We present the original perception a student might hold about an incident, which gives you the information you need to choose a particular therapeutic goal. The goal you select will shape the remainder of the LSI. It will guide what you do and say, and will structure the way the crisis is resolved. If effective, the LSI outcome changes the insight a student has about the incident, feelings and behavior, and others' reactions.

The next five chapters illustrate LSIs used to accomplish these therapeutic goals. You will see that the basic step-by-step process is not modified. Neither are the general purposes and contents of the LSI steps changed. What changes is the direction of the solution and the outcome for the student. Then, in the final chapter, we include an abbreviated form of LSI for very young or developmentally delayed students.

Chapter 7

THE GOAL: TO ORGANIZE REALITY
WHEN A STUDENT DOESN'T SEE AN EVENT
AS OTHERS DO ("REALITY RUB-IN")

Student's Perception: "I have been treated unfairly and gypped, and now they are blaming me for the problem."

Uses: With students unable to interpret or perceive events accurately; those unaware of their own behavior or the reactions of others; or with mentally disorganized students, with thought disorders or with limited ability for abstract thought.

Goal: To interpret event accurately, correcting distortions, social myopia, or misperceptions about the incident.

Focus: Organizes students' mental perceptions and sequence of time and events; developmentally the most rudimentary of the therapeutic goals.

Student's New Insight: "I only saw and remembered part of the problem. Now I see the part I played and what I need to do to correct this problem."

This therapeutic goal uses LSI to help students reorganize and clarify reality by discussing their blurred, distorted perceptions of the incident, identifying what is real versus what is imagined or constructed to avoid reality. These students have been described as having "social blindness" or "tunnel vision"—seeing only the part of an event that is personally threatening. They forget the sequence of events, distort reality, separate feelings from behavior, and intensely remember the last hostile action (remark, punch, hit, words) inflicted by a peer. The common characteristic among all students for whom this type of LSI is appropriate is inability to connect cause and effect in the series of interpersonal exchanges culminating in an incident.

Originally called a "Reality Rub-In" (Redl, 1966), this type of LSI has its foundation in the characteristic way a student thinks and reacts to an incident. Redl describes the rationale for this therapeutic goal:

> The trouble with some of our youngsters, among other things, is that they are socially nearsighted. They can't read the meaning of an event in which they get involved, unless we use huge script for them and underline it all in glaring colors besides. Others are caught in such a well-woven system of near delusional misinterpretation of life that even glaring contradictions in actual fact are glided over by their eyes unless their view is arrested and focused on them from time to time. More fascinating even, are the youngsters whose preconscious perception of the full reality is all right but who have such well-oiled ego skills in alibiing to their own consciences that the picture of a situation that can be discussed with them is already hopelessly repainted by the time we get there. It is perhaps not necessary to add how important it is, strategically speaking, that such children have some of this "reality rub-in" interviewing done right then and there, and preferably by persons who themselves were on the scene or at least known to be thoroughly familiar with it. (p. 44)

THE ADULT'S TASK

With this type of LSI, your goal is to correct a student's blurred and distorted perceptions of the incident. To accomplish this goal, the task is to help the student clarify reality by discussing the student's perceptions of the incident and sorting out distortions about what occurred. In the process, you help a student reconstruct the details of the incident, consider the resulting consequences, and organize the new insight into a behavioral plan for future use.

A time line is essential for this type of LSI. The adult helps the student reconstruct the incident and the feelings involved.

Together, they trace through the sequence of events, identifying who, where, when, and what happened. In doing this, students are helped to organize perceptions and learn that behavior evokes behavior from others. They learn the likely consequences of similar actions, and see the possibility of altering future events by changing their actions.

In a perceptive discussion about the use of this type of LSI with disturbed adolescents, Robert Bloom (1981) summarizes the importance of maintaining a helping relationship during such LSIs.

> Challenges to youngsters' perceptions of reality often are emotionally painful to the youths. Much comfort is found in being a victim of events; recognition of their contribution to their pain and misfortune can be vigorously avoided. And, in the typically conflicted adolescent-adult push-pull relationship, an adult on whom all this misfortune and pain can be projected is readily available. (p. 23)

Bloom suggests four guidelines to keep on track: (a) Maintain an ongoing relationship, (b) avoid creating a sense of urgency, (c) keep your own hostile feelings under control, and (d) be calm and focused.

EXAMPLES OF STUDENTS WHO CAN BENEFIT

Of the five therapeutic goals for LSIs, the most frequently needed is to organize (and reorganize) students' perceptions about the reality of an incident. Here are brief vignettes of students needing this type of LSI, where the focus will be to help them see reality in a clearer way, their own role in it, and the way others see the same events.

STEVE   After a fight, all Steve can remember is that Bill hit him and called him a "mother name." There is no initial recall of his teasing, provocative behavior toward Bill, which triggered the fight. The LSI focuses on the details of the event and Steve's feelings to help him recall the events realistically and see his own role, as well as Bill's, in the crisis.

DEIDRA

Everyone in the school knows about Deidra and her violent rages. Teachers and students alike try to avoid incidents with her because they fear her terrible wrath. When in this condition, she is violent, vicious in her physical attacks of anyone in her way, and almost impossible to control. She is a physical person, limited in her language ability and abstract thought but responsive to concrete imagery and adult relationship, when she cools down. One teacher is able to "talk her down" to the point where she can participate in a basic LSI that helps her focus on each step in the incident leading to the crisis and her role in it in a very concrete and vivid way. Simultaneously providing Deidra with affirmations of each tiny positive behavior she exhibited is essential for helping her to see herself in a different light, as a person who can control herself.

MARTIN

After a lot of annoying behavior from the class, Martin's teacher tells the class that no one can leave his or her seat without permission for the rest of the period. Almost immediately, Martin gets up, walks across the room, and asks for a pencil. When he is told that he is using the pencil as an excuse to break the rule, he protests that he needs the pencil to do his work. Then he kicks over a chair and screams at the teacher, "You always pick on me! You want me to fail my work!" LSI with Martin begins with this basic therapeutic goal, correcting his misperceptions of the incident. (Decoding is essential in this incident, and may lead to a different central issue and change in focus to another type of therapeutic insight. (See Chapter 8.)

TONY

During the transition from one activity to another, the students are moving their chairs to another part of the room. As they come together, Tony deftly pulls the chair away from under Bob, just as he starts to sit. The inevitable happens: Bob lunges for Tony and Tony fights back. It all happens in a matter of seconds, and then the teacher steps between them. "He's killing me!" shouts Tony. "Don't let him near me or he'll be sorry!'' Tony is a bundle of outrage, fear, and anger. The teacher takes Tony to the conference room for LSI. Her first strategy will be to consider LSI for organizing reality, where she uses a time line to help Tony see the cause-and-effect relationships in this crisis.

SALLY

Sally looked very depressed after the math test was returned. This perplexes her teacher because Sally received a score of 95% correct. When the teacher looks up a few minutes later, he sees that Sally is crying. As they begin the LSI, it comes out that Sally thought she had disappointed the teacher and he was mad with her. It seems that yesterday he told her to study her homework and perhaps she would get all the problems correct on the exam. The teacher chooses to help Sally review in detail his behavior and hers in each step that built to the incident. His goal is to teach her ways to understand the behavior of others realistically and with greater objectivity.

DARYL

In the conference room, the teacher is trying to drain off some of Daryl's rage before beginning LSI about an incident on the bus. Daryl paces from wall to wall kicking vehemently and saying, ''I hate you, you old fool! I hate this dumb class. They do baby stuff.'' When the teacher does not respond (trying silence), Daryl approaches her and says, ''You'd better pray. I'm going to hit you.'' When she says, ''I'm sorry you feel so bad,'' Daryl turns away, falls on the floor breathing heavily, and then starts to cry. The teacher uses several supportive comments and suggests that he will feel better by breathing deeply, like an athlete before a sports event. When the emotional flooding passes, the teacher knows there is so much feeling and confusion in Daryl that he first needs LSI that will help him organize reality.

SHERRY

Because she shows many behavioral characteristics associated with a diagnosis of schizophrenia, it is a struggle for 11-year-old Sherry to sustain participation in group activities. After a successful field trip to the park, Sherry became almost frantic as the group began boarding the bus. She lashed out at the recreation therapist who had moved close to her for proximity support. When she did this, she knocked a cup of tadpoles from the therapist's hand. When she saw what had happened, she looked bewildered and started mumbling words out of context. The therapist recognized an opportunity for LSI that would help Sherry anticipate when future situations would evoke similar feelings of confusion and disorganization. He organized the LSI around the sequence of events preceding the tadpole incident.

These vignettes illustrate some of the problems that are suited to this first type of therapeutic goal—organizing reality to broaden students' perceptions and insights. We suspect that you have found some of your own students among these descriptions.

EXAMPLE OF LSI USED TO ORGANIZE REALITY

In the following example, we illustrate a full LSI used with the therapeutic goal of organizing reality. We chose this LSI, with Al, because it contains many typical problems you will face in conducting LSIs for students like these. You will notice the confusion and distortion early in this LSI; it is typical. You also will see how quickly Al responds to affirmations and support from the counselor; this also is typical. Another effective technique in this LSI is the counselor's use of open-end sentences which Al responds to easily. This technique allows a student who has not often used words to deal with crisis to participate in interactive verbal communication successfully.

Teacher as the Dumping Ground

Nine-year-old Al learned early that expressing anger or hostile feelings is a dangerous thing to do. When he does so, the adults in his life retaliate harshly. Therefore, he has developed passive-aggressive strategies to cope with the world around him and to provide an outlet for the anger within. Al is a master of the art of argument, confusion, and misdirection. When an adult makes him angry, he defensively denies responsibility and transfers the blame to others, pointing out what he has done correctly. He defends what he had done as right, and he usually responds to adults' requests in halfhearted or unacceptable ways.

Al has been under a lot of stress prior to this particular incident, some of which was known to his teachers and some of which was not. This day, while Al's teacher is assigned to administrative duties and is out of the classroom, a substitute teacher is filling in. It is stressful to Al and to the substitute. For the past 2 days, Al has been particularly disruptive in his classroom. Finally, his acting out escalates to such a degree that he can no longer be maintained in the room by the substitute teacher, and he is sent to the counselor. Al (A) enters the counselor's (C) office without any difficulty and is able to sit in a chair. He seems angry and restless, but controlled.

STEP 1: FOCUS ON THE INCIDENT	C: Al, I need to be able to understand what has been happening in the classroom today. What's been going on?
	A: Nothing. I wasn't doing nothing wrong; and she sent me in here.
	C: You weren't doing anything wrong?
	A: All I was doing was playing with my toy cars. What's wrong with that?
	C: What were the directions given to your class at that time?
	A: We were supposed to be at the table with our math books.
	C: And then what happened?
	A: She made me come in here.
	C: Let me understand this. You were playing with some toys when the directions were to be at the table with your math book. Is that right?
	A: That's right.
	C: And then the teacher told you to come in here?
	A: She's not my teacher!

C: The substitute teacher?

A: Yeah, and I wasn't hardly doing nothing!

AI has described the reason for the LSI in this basic exchange.

STEP 2:
STUDENTS
IN CRISIS
NEED TO
TALK

C: I hear that you think it wasn't fair. Did she ask you to put the toys away before she told you to come here?

A: Yeah.

C: How many times?

A: But I wasn't hardly doing nothing!

C: I understand that's the way you feel. The question I asked was, "How many times did the teacher ask you to put away the toys?"

Counselor begins to construct time line.

A: A lot.

C: Help me understand what "a lot" means to you. Was it three times, five times . . . ?

A: She didn't have any business sending me away!

C: Well, let's see if we can get the facts organized. First, everyone was supposed to be . . . where?

A: At the table.

C: Where were you?

A: At the table!

C: Good! So, everything so far is going well for you.

A: (Sits up a little more confidently.)

C: Now, what was everyone supposed to do at the table?

A: Math.

C: Okay, and what were you doing?

A: (Long silence.) Playing with the toys.

C: Okay, so this substitute teacher told you to put away the toys and . . . ?

A: (Another long silence.) I don't know.

C: Think about what you told me a few minutes ago.

A: I guess I didn't put them away.

C: And then, the teacher . . .

A: She got mad at me.

C: Did she tell you again?

A: Yeah.

C: Could she be thinking that you were ignoring her?

A: Maybe.

C: And is there a rule about obeying teachers?

A: Yes. (Shuffles feet and looks out the window.)

C: What is that rule?

A: Do what teachers tell you to do.

C: That's exactly right! And so, when you didn't do that, what did she do?

A: She sent me out of the room, to here.

Counselor has established a basic time line with Al but is still uncertain about whether or not the incident is the issue.

STEP 3: FIND THE CENTRAL ISSUE AND SELECT A THERA- PEUTIC GOAL

C: It seems like the teacher was trying real hard to give you directions to help you stay in the program, and you weren't able to listen to her or follow her direction. Is that a fair interpretation?

A: I don't know. (Looks out the window again.)

C: When a teacher tells a student the rule, and the student knows the rule, and still doesn't follow the rule . . . I wonder how come? Do you have any ideas about this?

> Counselor begins probing for Al's motivation to change and perception of the incident.

A: She made me mad!

C: And what do you do when someone makes you mad?

A: Sometimes I don't follow directions, and I talk back, and play around.

> Counselor recognizes that Al's perceptions can be expanded and there may be sufficient motivation for changes in his behavior.

C: You have a good understanding about yourself, Al! It sounds to me like you were trying to make this teacher mad.

A: I was!

C: But from what we've said so far, it doesn't sound like this teacher did anything to you to be mad about. And yet you're acting like she is mistreating you. I wonder if there is something else you were mad about?

A: Yeah. (Hops out of chair and paces around the room.)

C: You seem to be pretty upset.

A: Yeah . . . I got a D on my report card for conduct.

C: I can see how that would upset you. Did it upset your mom, too?

A: Yeah . . . she beat me up for it.

C: That must have made you feel even more upset.

A: Yeah . . . and my teacher, *not* the substitute, she's gone somewhere. She said she had to do something else.

C: So, here you are, without your regular teacher, and all this pressure. It seems like you've been carrying around an awful lot of feelings lately. And what's been happening to you?

Counselor selects therapeutic goal—organizing reality to expand Al's insight about feelings that are influencing his behavior.

STEP 4: CHOOSE A SOLUTION BASED ON VALUES

A: I'm in trouble. I don't follow directions, and I talk and play around.

C: You have a lot of personal problems.

A: Yeah! (Slumps in seat.)

C: But these can be worked out. Think about what you just said.

A: (Looks up with new interest.)

C: You gave a very thoughtful and accurate description of your problems. And what did you do with those problems?

A: (Looks interested but confused.)

C: You were upset because of your personal problems and then dumped it on the substitute teacher. I think you may have convinced yourself that the teacher was mean and unfair to you.

A: Yeah. (Nods in agreement.)

C: Al, I'm impressed that you figured this out! Most students your age don't understand how feelings at home can cause problems at school.

A: Right. (Conveys interest in this idea.)

C: So, when you feel upset what is the thing to do?

A: Talk to you about it instead of disobeying that substitute.

Al states solution.

STEP 5:
PLAN FOR
SUCCESS

C: That's one solution; keep it in mind. I'll be here if you need to talk. What else could you do, if I'm not around?

A: (Long pause.) Talk to my regular teacher?

C: That sounds like it might work. Do you think she understands about how feelings can mess up your behavior?

A: I don't know.

C: Do you think if we talked together with her about this, she would understand how to help you when you get upset again?

A: I think so.

C: Shall I tell her that you are trying to keep from letting personal worries make you break rules and ignore teachers? Or do you want to tell her yourself?

A: You tell her.

C: Okay, that is something we can do together. I believe she will understand and help you.

A: Let's tell her now.

C: So, we had a problem and now you have learned something very important to keep this from happening in the future. What's the new strategy? Words or actions?

A: Words.

Al has stated the solution in a form he is capable of handling successfully.

| STEP 6: GET READY TO RESUME THE ACTIVITY | C: (Humorously.) By golly, you're right! When you use words, did lightning come out of the sky? |

STEP 6: GET READY TO RESUME THE ACTIVITY

C: (Humorously.) By golly, you're right! When you use words, did lightning come out of the sky?

A: (Grins sheepishly.) No.

C: Did fire come out of my nose?

A: (Grinning.) No.

C: Did you fall into a thousand pieces?

A: (Laughs.) Naw!

C: Talking about things always helps. And so does understanding what is really behind what you do.

A: (Smiles broadly.)

C: We'll have to wait until after school to talk to your teacher about this. But what can we say now, to your substitute teacher when I take you back to class?

A: Maybe you should tell her that everything is all right now.

C: And what about your behavior in the classroom?

A: Tell her I know to obey the teacher and listen to what she tells me to do.

C: Okay. Let's go and talk with your teachers!

(Our thanks to Gregory Malone for describing this LSI for us.)

Initially, this LSI focused on the here and now, when establishing the time line and course of events. Al's confusion and myopic perceptions were evident. Once the time line was established, the counselor used decoding to probe for the issue. Al distorted reality to displace his need to express angry and hostile feelings toward his regular teacher (who had left temporarily) and his mother (who had punished him). The counselor recognized a basic developmental anxiety in this situation—abandonment (see Chapter 3 for a review of this anxiety).

The first objective was to get Al to see that the issues about which he was angry were not caused by the people against whom he was acting out. The second objective was to help him with the insight that direct, verbal expression of hostile feelings does not necessarily lead to retaliation. There are many circumstances and places under which this type of verbal interaction can be helpful and rewarding. We would like Al to think that school is one of those places. Helping Al feel comfortable in using words to deal with stress and emotions will be a long and laborious process. It is our belief that this LSI was one step in that direction.

WHAT WE HOPE YOU HAVE LEARNED FROM THIS CHAPTER

- The primary purpose in using the therapeutic goal discussed in this chapter.

- The common characteristic shared by all students who can benefit from this therapeutic goal.

- Characteristics of students for whom this form of LSI might not be applicable.

- What essential process is used by adults to provide the structure for achieving this goal.

- The importance of using small details in this form of LSI.

- Why a helping relationship must be maintained by an adult to do this form of LSI effectively.

- Why abundant affirmations are essential for students with whom you use this form of LSI.

- Four guidelines to aid adults in being effective with this form of LSI.

- Why this form of LSI is most frequently used with students when they first begin a program.

- How a student's insight changes if this therapeutic goal is successful.

- What indications suggest that different therapeutic goals should be used with a student.

Chapter 8

THE GOAL: TO BENIGNLY CONFRONT UNACCEPTABLE BEHAVIOR WHEN A STUDENT DOESN'T WANT TO CHANGE ("SYMPTOM ESTRANGEMENT")

Student's Perceptions: "I do what I have to do even if it hurts others." "I have a right to be free." "I have a reputation to maintain." "I have no need to change."

Uses: With students who are too comfortable with their deviant behavior, receiving too much gratification; those who rely excessively on aggression, passive aggression, manipulation, or exploitation of others.

Goal: To make a particular behavior uncomfortable, by confronting the rationalizations and decoding the self-serving narcissism and distorted pleasure the student receives from the unacceptable behavior.

Focus: Uses external adult authority, power, and judgment to help student obtain new insight into deviancy and the inevitability of dire consequences.

Student's New Insight: "Maybe I'm not as smart as I tell myself." "Maybe I've been cruel." "Maybe I've been tricking myself."

Unlike other types of LSI, this therapeutic goal is used to confront behavior by increasing students' anxiety about what they are doing and saying. This goal is needed with students who typically have a highly skilled way of protecting themselves from feelings of guilt by tossing responsibility and fault to another student, the program, or adults.

These students generally receive so much reinforcement from their peers that they do not believe anything is wrong with the way they behave. Therefore, they have little motivation to change or give up their power position in the peer group or with adults they can control. They use their verbal skills to avoid talking about an incident and the feelings evoked by it. They try distractors, role reversals, and "red flags" to divert the adult from the issue (see LSI Step 2 in Chapter 5 and "Getting Off the Subject" in Chapter 6). Often, students switch from one tactic to another until they are successful in confusing and frustrating the adult. A singularly consistent strategy among such students is their attempts to control adults and the focus of the LSI through alibis and rationalizations. Here is a list of the most common rationalizations we have documented in our own LSI experiences:

"He did it to me first."

"He called me a dirty name."

"Everybody else did it too."

"I was only kidding."

"She had it coming to her."

"I had to do it."

"I gave him a warning."

"Didn't you ever do anything like this when you were little?"

"I'll tell the truth if you promise not to call my parents."

"I was upset today because my mother hit me this morning."

"I'm not talking because you won't believe me anyway."

"It was an accident."

"My dad told me to hit anybody who did that to me."

"I have a bad temper, so he shouldn't make me mad."

Unlike other types of LSI, the therapeutic goal here is to confront this behavior, increasing students' anxiety about what they

are doing and saying to the point that they have some understanding and motivation to change their behavior. Fritz Redl
(1966) described this type of LSI as a "Symptom Estrangement"
interview. Here is his description of what is needed by these
students and why:

> Our children's egos have, in part, become subser-
> vient to their pathological mechanisms they have
> developed. They have learned to benefit from
> their symptoms (behaviors) through secondary
> gains, and therefore are in no way inclined to
> accept an [adult's] idea that something is wrong
> with them, or they need help. . . . We use many
> of their life situations to try to pile up evidence
> that their pathology [behavior] really does not
> pay or that they pay too heavily for what meager
> secondary gains they draw from it. It is not a sim-
> ple matter of arguing these children into letting
> go of their symptoms [behaviors]. We must enlist
> part of their "insight" into wanting the change.
> . . . And, our actions [adults' daily behavior and
> values] have to be well attuned to our words in
> this interview more than others. (pp. 44–45)

THE ADULT'S TASK

The first task of the adult is to keep the focus on students'
inappropriate behavior and to avoid being led astray by students'
verbal barrages or attempts to control or camouflage their
behavior. The next task is to develop changes in the students'
insight by decoding their alibis and behavior to show them how
they justify and enjoy their "righteous aggression."

If successful with this type of therapeutic goal, students learn
that they are not fooling themselves or others by their behavior
and that it is not bringing about the benefits they desire. When
this level of insight has been achieved, these students usually
need follow-up LSIs with a change in therapeutic focus to learn
substitute social skills for achieving what they need from others
in more appropriate ways (see Chapter 10). It also is essential that
the environment and the adult's own behavior do not continue
to reinforce the student's unacceptable behavior. If aggressive or
manipulative behavior is not acceptable, adults must provide

alternative open, acceptable ways to convey feelings, protect against stress, and obtain gratifications. This open process seldom can be learned quickly by these students. It will take considerable perseverance from adults to "stay on top" of these students' ever-ready attempts to control. It also takes many, many LSIs, but the message will get through!

EXAMPLES OF STUDENTS WHO CAN BENEFIT

The following vignettes illustrate typical problems of students who can benefit from LSI that confronts their behavior and disallows its continuation by revealing the student's inner feelings of "enjoying" the reactions, discomfort, and confusion of others.

PETER

A model race car brought to school by Jim for a group project has disappeared. After discussion with the group and a search of the room, the model still cannot be found. As the students get ready to leave for the day, the teacher notices a sizable lump in Peter's coat which turns out to be the model car. As the subsequent LSI begins, Peter shows no remorse or guilt about taking the model from his friend. He calmly defends himself by saying, "I was planning to return it." When the teacher continues to probe for the details in LSI Step 2, Peter changes his response to, "Jim shouldn't have brought it to school in the first place. Everything gets ripped off around here." During Step 3, as the teacher begins decoding Peter's behavior, his responses change again: "Someone stole my model about 3 weeks ago."

Peter's teacher sees that the necessary therapeutic goal requires confrontation for Peter to grow in insight and understanding about himself. The teacher will support Peter while rejecting his unacceptable behavior and decoding the feelings that accompany it.

DICK, JIM, AND BOB

In the hall, near the boys' restroom, Bob flipped Dick's pen out of his pocket. Dick demands that Bob pick it up and return it to him. As Bob starts to do so, Dick grabs the pen out of his hand; and

Bob retaliates by calling Dick "a gay queer." With this insult, Dick starts chasing Bob. As they run, Bob bumps into Jim who is watching from the sidelines. When this happens, Jim grabs Bob's hand and twists it until Bob drops to the floor in pain. When Bob is on the floor, Dick comes over and kicks him in the ear. The teacher hears the commotion and finds Bob rolling on the floor, crying and rubbing his ear. When asked what happened, Jim and Dick reply that Bob has been teasing them and that he finally got what was coming to him. They show no concern about their cruel tactics but actually seem pleased about it.

Their teacher takes all three boys to the conference room for a group LSI that focuses on the event and the inappropriate behavior of each boy. The therapeutic goal needed for this LSI is to confront their behaviors as being thoughtless and unkind, getting them all in trouble. The needed insight is that such behavior is not acceptable and brings no benefits.

ANTHONY   During social science class, a new boy in the group, Anthony, is working with Wayne on a map project. Wayne raises his hand for help. As the teacher starts over to Wayne, Anthony says, "You'll help that white boy when you won't help me." The teacher ignores the remark but moves toward Anthony, sensing the need for proximity control. Before she gets there, Wayne jumps up and hits Anthony in the face. Anthony slugs him in return. Wayne doubles over moaning and screaming that his stomach has been "damaged."

These overt acts of aggression do not follow the typical profile of manipulation by students needing to control events and others. However, both the aggression and the racial comment need direct confrontation in the LSI the teacher conducts with both boys. The therapeutic goal in this situation has two parts: First, aggression needs to be confronted as an unacceptable way to deal with people who upset you. Second, Anthony's and Wayne's feelings behind the racial remark need to be decoded. For Anthony, the remark signals his uncertainty about his new situation; for Wayne, the remark triggers confusion about how to respond. Once the decoding has been accomplished, the LSI can move forward into new alternatives for dealing with such feelings. This is another LSI that will switch therapeutic goals to teaching new social skills after the student gains insight into why these behaviors are unacceptable and counterproductive (not worthy of the students and of no benefit to them).

ELISHA

During math class, Elisha is unable to do a math problem and when she reaches for her eraser in her backpack, it is missing. In frustration she throws her pencil across the room, accidentally hitting Daniel in the face, causing him to cry. The teacher says, "Elisha, come pick up the pencil and then apologize to Daniel." Angrily Elisha retorts, "No! Let the crybaby pick it up himself." Then she sticks out her tongue at Daniel. The teacher, now angry herself, sternly orders Elisha to "pick up the pencil!" To this, Elisha coyly looks at the teacher and says, "I have to use the bathroom," and starts to get out of her seat. By now the teacher is fully aroused and walks over to Elisha, "Young lady, you will sit right here until you apologize." The teacher does not believe that Elisha has to go to the bathroom. At this point the student's anger and stubbornness are in direct conflict with the teacher's unwillingness to let herself be manipulated. A Conflict Cycle between teacher and student is well under way. Elisha says in a matter-of-fact way, "I guess I'll have to go right here," and she starts to wiggle in her chair. The teacher has had it! She yanks Elisha out of her chair and yells in an exasperated voice, "You don't speak to a teacher like that; we're going to the principal's office."

The LSI conducted by the principal relies on a carefully reconstructed time line to begin to get to the sequence of events and the feelings behind them. The therapeutic goal is to confront Elisha about her manipulative behavior by decoding it to expand her insight about how her frustration with the math problem and the missing eraser transferred over to Daniel and the teacher. (Maybe the principal should conduct LSI with the teacher as well!)

BILL
AND JOE

Bill was in a rotten mood and seemed to enjoy it. His friend Joe came by and flipped him a "bird." Like a chemical reaction, Bill jumped up and punched Joe in the ribs. Suddenly the two were in a violent fight filled with screams and kicks. The teacher yelled for them to stop and grabbed Joe while another teacher held Bill. Ten minutes later they were in the conference room, controlled enough to talk about the incident. Bill's interpretation was that Joe was asking for a beating. He had taken Bill's prize pen, cheated in Monopoly, tripped him in the gym, and called his mother a name. He deserved to be punched out, and he was lucky the teachers broke it up or Joe would be a bloody stump by now. Joe's perceptions were entirely different. He had an explanation for each event. "It's true, I gave him the 'bird,' but what's the big deal? We do it all the time. It's like saying hello. I don't know what got into Bill today. We usually get along, but today he acted like a crazy man." Bill's angry response was, "For a week you've been picking on me, and I should beat you up four times instead of one time!" After some review of the sequence of events, Joe seems to understand Bill's feelings but Bill continues to stick to his story—"Joe got what he deserved!" The teacher sends Joe back to the room and begins LSI with Bill to confront his unacceptable behavior.

(This LSI is described in full by James Tompkins, 1981, pp. 26–28, in an insightful article about this type of LSI. It is summarized here with permission of the author.)

These examples show how difficult confronting unacceptable behavior can be. When conducting this type of LSI, adults have to keep focused, maintain communication with the student, control their own emotions and reactions, and avoid moralizing, while clearly conveying disapproval of the behavior. We must be quick thinking and verbally facile. The aim is to decode the behavior, revealing feelings students are protecting or gratifying. To this end, we are exposing students' defenses to themselves. There always is the risk that they cannot hear it and will close down or defend more vigorously. When this happens it requires secondary interpretations of the denial (see Chapter 3 on decoding). The other task is to illustrate how these students pay too heavily for the small gains they get from their behavior. There are more gratifying ways to obtain what they want (and to feel better).

EXAMPLE OF LSI USED TO CONFRONT UNACCEPTABLE BEHAVIOR

This LSI with Scott illustrates many of the strategies we have found to be effective with students who need to be confronted by an adult about their inappropriate or unacceptable behavior. Scott is a prototype of many such students with whom we have worked. The important theme that emerges in this LSI is that Scott really wants to be seen as an admired person who has values beyond self-serving; he has the capacity to gain new insight into the importance of changing his own behavior. (Our thanks to Amanda W. Seeff for conducting this LSI.)

It All Depends on You

Scott is a very bright, elflike, manipulative, and sometimes passive-aggressive 10-year-old. He is quite skilled and articulate in turning situations around so that problems created by him seem to be exaggerations of the adult's imagination, and any consequences earned by Scott's behavior seem to be mean and unfair impositions by the adult. Scott also is a practiced master at recreating his own stress in others, including his teachers, mother, and even his therapist.

After recess one day, students are taking turns going to the water fountain to get a drink. During this time, Scott gets up and walks casually from his desk to the teacher's desk, looks at the point chart, plays with pencils, and picks up a book from her desk. Carrying it back to his seat, Scott flips noisily through the pages. After all of the students have had a drink, except Scott, the teacher begins the next lesson. At this point, Scott cries out, "I didn't get my drink of water!" The teacher reminds Scott that he has chosen not to get a drink. Scott is unable to accept this natural consequence. He folds his arms, puts his head down on the desk, and bursts into tears, crying, "You're not fair to me! You're jumping down my throat!" The teacher quietly reminds Scott of his behavior. Scott loudly cries out again, "I should have stayed home, like yesterday, sick! It's my first day back and look how you're treating me!" Scott throws his pencil at the wall and then kicks over his desk. His teacher responds, "I can see how upset you are, but I cannot accept this way of showing it in the classroom. We need to go out for a talk about this." Here is the LSI between Scott (S) and his teacher (T) that follows.

STEP 1: FOCUS ON THE INCIDENT	**T:** Let's review what happened, step by step, Scott.
	S: (Smiling sweetly.) Nothing happened.
	T: We wouldn't be here if something had not happened.
	S: (Sits down with an air of confidence.) Nothing happened.
	T: Let's talk. Something happened!
	S: (Knits brow and shouts.) You're not fair to me! You gypped me out of my drink.

Scott has described the incident in the simplest of terms, from his perspective.

STEP 2: STUDENTS IN CRISIS NEED TO TALK	**T:** We need to figure out what exactly happened.
	S: You know what happened!
	T: People often have two views of the same event.
	S: I was minding my own business and you were playing favorites.
	T: Right after recess, we came back to the classroom. What did I ask the students to do?

Teacher begins to construct a time line.

S: You said, "Sit down." I was sitting!

T: That's true. At first, you were sitting. Then what happened next?

S: Nothing. (Smiling sweetly at the teacher again.)

T: Then, I called the students to . . .

S: Get a drink of water!

T: That's right! Now, try to remember. Who did I call?

S: I don't know. Don't ask me. I'm not them.

Scott starts playing with a small toy car. The teacher waits, giving him a chance to answer.

T: Well, I'll tell you what I saw. David was sitting at his desk quietly. Mike was quiet; Tom was sitting at his desk, too, and you were standing by my desk.

S: (Cocks head to one side, looking quizzically at teacher and smiling.)

T: (Decoding.) I see you're smiling. That tells me that you *do* remember these events and going to the desk. Okay, now when you went back to your desk, what did you do?

S: My science teacher told me to find out why both the North Pole and the South Pole are cold.

T: That's true. You were using a science book. Whose book is it?

S: It was on your desk. But I was using it carefully!

T: During the time you were looking through the book, what was I doing?

S: Calling kids to get a drink—except me.

T: There must have been some reason why I didn't call you.

S: You don't want me to learn about the North and South Poles. I said I was looking through the book appropriately!

T: But was this the right time for reading? It seems to me that you had a choice to get water or not to get water. I think you made a choice.

S: (Smiling, in a singsong voice imitating the teacher.) I know! "It all depends on you! It all depends on you! It all depends on you!"

STEP 3: FIND THE CENTRAL ISSUE AND SELECT A THERA- PEUTIC GOAL

T: That's right! In this class there is one way to get good things for yourself; you have to choose that way.

S: (Looks away, swinging feet.)

T: What is that way, Scott?

S: (Smiles furtively.) By doing bad deeds.

T: (Decoding.) Your smile tells me that you don't really believe that. I'm sure that in the past you've been able to get your way with grown-ups. But now you're learning that sometimes you have to do what adults say. And what happens to you is based on what you do.

Teacher finds the central issue—the developmental anxiety of conflict between independence and dependence.

S: (Tries to avoid issue by racing his toy car around the chair.) Vroooom, vroooom!

T: Nice car, Scott. Where did you get it?

S: School store.

T: Really? You must have had to work hard for it.

S: Pretty hard.

T: So, there was a time when you could choose behavior that earns points and gets a car. That's great! You must have made good choices to get that neat car.

S: (Looks pleased.)

T: That sort of reminds me of what happened in class today; except that it turned out differently. Instead of getting a car, you could have gotten a drink of water. Now, Scott, this is the big question, "What did you choose?"

S: (Again using a singsong voice.) "It all depends on you."

T: Exactly right! What does "all" mean?

S: Everything you can get.

T: So, how would you get good things?

S: Sit in my seat and not get up.

T: Instead of . . .

S: (Loudly.) Playing around your desk!

> Although Scott has stated the expected behavior, it carries little conviction. The teacher chooses the therapeutic goal to confront his unacceptable behavior.

T: I'm just wondering, Scott, whether you've had this kind of problem before—making a choice to get good things or not. Have there been other times when you've chosen to not get good things for yourself?

S: I don't know.

T: I remember one time when you talked with me about you and Keith choosing to throw the beanbag on the bus and losing your chance to ride the bus. And another time when you chose not to put away the Lincoln Logs and then lost your chance to play with them.

S: It's time to go back to class.

T: Think about what I said.

S: Yeah, I know.

STEP 4: CHOOSE A SOLUTION BASED ON VALUES

T: I have a thought I want to share with you. I know you are very smart, so when you make a poor choice it must have a hidden meaning to you. I'm beginning to believe that you enjoy your behavior even though it gets you into trouble!

S: (Smiles.) Maybe.

T: You know what the penalty is for throwing materials and kicking over furniture?

S: Yeah, out-of-the-room!

T: That's right! I'm glad you remember the rule. Now, that means 10 minutes out here. This is your chance to show me that you can choose the right behavior. You pick the chair you want and sit on it. I'll be back in 10 minutes. If you make a good choice by following this direction, then it will show me you are ready to return to class.

Teacher selects solution based on Scott's values—his own needs.

STEP 5:
PLAN FOR
SUCCESS

S: Okay. I want to go back to class.

T: I'm glad to hear that. Now, review the plan.

S: I'll sit in this chair and then I'll go back to class.

T: When . . . ?

S: When you come back.

T: And see . . . ?

S: And see me sitting in this chair.

T: Good, I hope you can make a good choice for yourself now.

Teacher leaves hoping that Scott is able to connect his behavior to its consequences. But it is never simple with Scott. When she returns, he is not in the same chair. He has moved directly across the room.

S: (Looking at teacher and smiling sweetly.) Is my time up now?

T: I see you made a poor choice.

S: (Defensively.) I'm sitting in a chair, just like you said to do!

T: Wrong choice, Scott! I guess it has to be done again so you can show me how to follow directions.

> Teacher uses indirect "practice" to teach the natural consequences of behavior.

S: (Angrily.) You aren't fair!

T: I'm glad you value fairness. Fairness means going by the rules. And now you have another chance to show me that you can follow directions and go by the rules.

S: (Throws himself into another chair and breaks down into heartrending sobs.)

T: We'll try again. I'll be back in 10 minutes. Maybe you can make a better choice this time. It all depends on you!

> Teacher returns in 10 minutes and Scott is sitting half upside down in the chair.

STEP 6: GET READY TO RESUME THE ACTIVITY

T: I see you've made a better choice this time. It shows me you are beginning to show responsibility. You figured out that it works to your benefit. Things should go better for you this way!

S: I want to go back to the room.

T: What will happen when you walk back in?

S: I don't know and I don't care!

T: Your friends are interested in what has happened to you.

S: I'll tell them . . . (starts to make another defiant remark but stops himself). I'll say, "It's all up to you."

T: Good way to handle it, Scott. It's the way things really are!

Scott's teacher will have many more opportunities to teach him that there is a cause-and-effect relationship between his behavior and what happens to him. Over time, this LSI process and firm, consistent limits helped Scott accept responsibility for his own behavior, rather than blaming it on adults. As he began to show some regulating of his behavior choices, Scott's teacher shifted to other therapeutic goals in subsequent LSIs. In particular, she began challenging him to use values at a higher level, to strengthen his self-control (see the next chapter for this type of LSI and Chapter 4 for a review of values).

The effectiveness of this LSI with Scott depends to a great extent on the teacher affirming some genuine attribute in Scott while holding the limits and allowing natural consequences of his unacceptable behavior to provide motivation for change. Her focus on his earning the race car made this point for Scott because it was tangible evidence that he *could* make choices that worked for him, instead of against him. Unless an affirming, caring dimension is preserved during this type of LSI, adversarial tones creep in and the relationship will deteriorate to the point where a student will not change, at any cost. See Chapter 3 for a review of the developmental anxiety of conflict between independence (controlling) and dependence (being controlled).

WHAT WE HOPE YOU HAVE LEARNED FROM THIS CHAPTER

- The primary purpose in using the therapeutic goal discussed in this chapter.

- The common characteristic shared by all students who can benefit from this therapeutic goal.

- What is done deliberately in this form of LSI that is not done in any other LSIs.

- What process is used frequently by adults to change students' insight during this LSI.

- What new insight a student should have as a result.

- One follow-up problem that must be avoided if the results are to be lasting.

- The essential message that should be conveyed to both overtly aggressive and passive-aggressive students, when using this form of LSI.

- How a carefully detailed time line helps this form of LSI.

- Why this form of LSI can be particularly difficult for adults.

- Why this form of LSI also can be particularly difficult for a student.

THE GOAL: TO BUILD VALUES THAT STRENGTHEN SELF-CONTROL ("MASSAGING NUMB VALUES")

> **Student's Perception:** "Even when I'm upset, a part of me is saying, 'Control! Stop yourself' . . . but I don't."
>
> **Uses:** With students who, after acting out, are burdened by remorse, shame, inadequacy, or guilt about their own failures or unworthiness; those with a destructive self-image; and those who have a negative social role.
>
> **Goal:** To emphasize student's positive qualities; strengthen self-control and self-confidence as an able and valued person with qualities like fairness, kindness, friendship, or leadership potential.
>
> **Focus:** Extends student's self-control and confidence by abundant affirmations and reflections about existing socially desirable attributes and potential for future acclaim by peers; developmentally this goal requires a shift in source of responsibility—from adult to student.
>
> **Student's New Insight:** "Even under tempting situations or group pressure, I have the capacity to control myself."

This type of LSI is used to help students who are burdened by anxiety about guilt or inadequacy, yet do not use controls at the right time. These students experience anxiety at the wrong time and in the wrong proportions. Their remorse or guilt usually is felt after the fact, which is emotionally destructive and not helpful to them in controlling their own behavior. Their remorse is often so intense that they seek punishment (set themselves up to be caught and punished). It is ironic that students who have been physically or sexually abused often develop this self-abusive response and make degrading comments such as:

"I'm no good!"

"It must have been my fault."

"I wish I was dead."

"Go ahead, hit me!"

"I guess I'm just a loser."

"I can't do anything right."

"It was a stupid picture." (after tearing up a drawing)

Initially, the adult is sympathetic toward such students, since they are so distressed about the results of their own behavior. They promise to be good, never do it again, write confessions, make new resolutions, and shed abundant tears. However, like a tempting bag of popcorn, their intentions do not last long. It is not that they deliberately defy or ignore rules and regulations. The problem is that they are easily stimulated, lack adequate self-control, are sensitive to what is happening around them, and then become unproductively guilty when they are caught in the Conflict Cycle.

The therapeutic goal for these students is to strengthen their self-control by building up their self-esteem. If they have sufficient belief in their own capabilities and attributes, they will be more likely to use control and put on brakes *before* they react to stress with impulsive, unacceptable behavior. Adults often mistakenly believe that these students can control themselves if they want to do so. It is essential to understand that these are students who do not like their own behavior but because of weak controls cannot use the constructive values they have internalized. Fritz Redl (1966) called this form of LSI "Massaging Numb Values" to convey the point that the values are there, within the student, but not in use. Unless they are reinforced and expanded, such values lie dormant or atrophy.

> No matter how close to psychopathic our children may sometimes look, we haven't found one of them yet who didn't have lots of potential areas of value appeal lying within him. . . . Admitting

value sensitivity, just like admitting hunger for love, is quite face-losing for our youngsters. There are, however, in most youngsters some value areas that are more exempt from peer group shame than others. For instance, even at a time when our youngsters would rather be seen dead than overconforming and sweet, the appeal to certain codes of "fairness" within their fight-provocation ritual is quite acceptable by them. Thus, in order to ready the ground for "value arguments" altogether, the pulling out of issues of fairness or similar values from the debris of their daily life events may pay off handsomely in the end. (p. 45)

THE ADULT'S TASK

The first task for adults is to see through the smoke screen of aggressive behavior and provide affirmations of the student, reflecting on positive qualities and leading the student into also acknowledging them. Use of a time line in LSI Step 2 is helpful in focusing the student on details that bring out examples of positive attributes, no matter how small these may be. (Any strategy that creates more inadequacy, shame, remorse, or guilt in these students only complicates the problem.)

The method is to highlight and magnify the flickering signs of control. When examples of constructive behavior and positive qualities become part of the LSI discussion, decode to help students gain insight into why they need to succumb to behavior they know is not acceptable. Then connect this insight to the underlying positive values they represent (see Chapter 3 for decoding strategies and Chapter 4 for a review of developing values). In Step 4, shape the central issue around the positive behaviors that have been observed and the values they represent. Finally, in Step 5, use these qualities as part of the plan to resolve the crisis and reinforce values to guide behavior choices in the future.

Anthony Werner (1981) offers a major caution for adults using this type of LSI.

Responding to this pupil's acting out behavior in punitive or moralistic ways does not help the pupil learn and grow but only creates additional problems. . . . He will either feel that he has "paid the price" and lose contact with his feelings of guilt and the desirable values behind them, or he will feel increased guilt and will act out more intensely. (p. 31)

Werner goes on to describe the particular qualities needed by adults to make this type LSI effective.

The effective interviewer is one who is able to be supportive, compassionate, and empathetic. With these qualities and skill in LSI, this interviewer can become an advocate for the child. He [or she] can help turn a specific problem incident into a valuable experience for the child. (p. 31)

EXAMPLES OF STUDENTS WHO CAN BENEFIT

The following vignettes illustrate three types of problems that respond to LSIs that strengthen students' self-control by recognizing and emphasizing existing values the students reveal in their behavior. When selecting this therapeutic goal, it is important to avoid using it in a way that will reinforce unacceptable behavior. This is less likely to occur if you base your own responses on (a) a student's level of development about who should take responsibility for control of behavior (the existential phases discussed in Chapter 4); (b) the type of developmental anxiety with which the student is struggling (discussed in Chapter 3); and (c) the general level of values the student currently uses and the possibility to challenge the student to the next level (discussed in Chapter 4). In the following illustrations, we show how an adult adjusts responses to avoid reinforcement of unacceptable behavior, and we summarize the information used to choose this particular form of LSI.

DOUG
(age 12)

During free time, Tom talks Doug into helping him let air out of a car tire in the staff parking lot at the Middle School. Doug is very upset when they are caught and spontaneously says, "It was a dumb thing to do."

In the LSI, they discuss exactly how the incident developed and who had the idea. Doug says that he refused to go along with Tom twice before he agreed to help him. With this information, the teacher is able to support Doug by saying, "When you were able to say no twice, it shows me that a part of you is trying to stop doing something you know is wrong."

Doug is surprised but pleased that the teacher supports his flickering attempt at control rather than punishing him for his impulsive behavior.

The teacher selects this approach to Doug because she recognizes that his behavior in this incident was not up to his level of conscience development. He showed traces of self-control and a value system that should have been operating before the incident occurred. Doug seems to have the potential to regulate his own behavior at a level of "my better self."

PETE
(age 10)

During recreation, Pete seems fine as the students line up to pick teams. After the first student is picked, Pete begins to "boil over." His expression changes and he throws down his cap saying, "I don't want to be on that team; they're losers!"

The teacher walks him away from the group and talks to him about team work. This seems to satisfy Pete, and he rejoins the group. He is picked for a team and play begins. A little later, it is Pete's turn to bat. He strikes out! Immediately, he turns toward the umpire in rage, threatening him with the bat and screaming, "That was not a strike!"

The teacher intervenes quickly and takes Pete away from the group for LSI. In answer to her first question, "Why are we back here talking again?" Pete blurts out, "That was *not* a strike! My team needed a hit. I know the umpire's supposed to always be right, but he wasn't watching."

The teacher believes that the most beneficial therapeutic goal for Pete in this situation will be to focus on strengthening his self-control by emphasizing a positive quality. She builds on Pete's comments about the umpire as the authority and the value of being a good team member. The teacher affirms for Pete that he has shown that he can be a team player. She has seen the sign of it when he went back to join the game the first time. The teacher knows that Pete's value system has developed to the level where group relationships are beginning to be important (a winning team member). She also knows that he is beginning the post-existential phase of development, where students begin to handle their problems themselves (however ineptly). And she knows that Pete is struggling with the developmental anxiety of conflict typical for his age, where the drive for independence is intermittent with a resurgence of need to be controlled by an adult. While there is some risk that this choice may not work, the teacher concludes that it is better to take the risk because Pete seems to already have a clear view about events and perceptions of others, and he has conveyed that his value system can play a role in his choices of behavior. She is confident that, little by little, Pete's desire to be a a valued team member can be used to strengthen his own self-control.

JACKIE (age 11)	During a writing lesson, Jackie suddenly rips his paper into shreds. It is a task that he had previously done successfully and shown pleasure in doing. As his teacher takes him aside for LSI, he has to be physically restrained to help him gain control. He is gasping and sobbing. In a strange, deep voice, he says, "I'm a bad boy! I'm a bad boy! I'm a baby! I'm a baby! . . . Let me go; I'll kill you!"

When the emotional flooding is over and Jackie breathes more calmly, his teacher sees that he is ready for LSI. She has tentatively decoded the paper-destroying act as frustration and anger at himself for performance below what he had hoped for. She knows Jackie is still in the preexistential phase of development, where adults are viewed as the source of authority, even though he is 11 years old; and she also has seen considerable evidence that Jackie is struggling with two basic developmental anxieties —abandonment and inadequacy. When she considers Jackie's value system, she sees a parallel—meeting his own needs but also a desire for adult approval. For these reasons, Jackie's teacher chooses the therapeutic goal of building self-esteem, recalling Jackie's previous achievement, affirming the value of his work and her approval of its quality.

Although they are at varying developmental levels and having quite different problems, these students have one common characteristic: Their values are so fragile or so deeply buried that they are not used to regulate behavior until *after* an incident. By using the type of LSI that strengthens self-control, a teacher is able to be proactive, moving such students developmentally forward by activating their latent or ineffective values and expanding them into positive forces for students to use themselves—the beginnings of inner control of behavior.

EXAMPLE OF LSI USED TO BUILD VALUES THAT STRENGTHEN SELF-CONTROL

The following example illustrates a full LSI with the therapeutic goal of reinforcing and fortifying latent values to strengthen self-control. We chose this incident with a high school student, Sidney, because it illustrates how LSI can be used to relieve the burden of guilt and at the same time support a student who must face the serious consequences of completely unacceptable behavior. You will see how important it was for the coach (C) doing this LSI to identify the student's underlying developmental anxiety and level of values before choosing the direction the LSI would take during the resolution steps. This also is a good example of decoding and framing the central issue. It would have been easy for the coach to go off on a tangent of homosexuality, mentioned by the student. He chooses not to let this particular LSI digress into that topic because the reality of the incident has to be addressed first. Sidney's outpouring of concern about himself and girls gives the coach ample understanding about his underlying anxieties and values.

I Don't Want Them to Think I'm Weird, or Anything

This high school incident has everyone talking. It seems that Jay came screaming to the girls' physical education coach with her blouse torn, buttons off, and tears pouring down her cheeks. When she finally quieted down, the coach was able to learn from her that she had been "attacked." She had agreed to meet Sidney out in the parking lot for a smoke. When she got there, she said, "He jumped on me! He ripped my blouse and said all sorts of terrible things. He raped me! He's a maniac!" With that, Jay dissolves into sobs and tears again. The coach calls the guidance counselor to assist Jay. Then he goes to get Sidney from history class.

Sidney quietly follows the coach out of the room and down to his office. Sidney is a serious, very bright 16-year-old who has no close friends and always seems to be on the fringe of every group. He offers no resistance and seems to anticipate that something terrible is going to happen. He sits down like a prisoner about to receive a death sentence.

STEP 1: FOCUS ON THE INCIDENT

C: You know why I came to get you?

S: Well, yeah. I guess you mean about Jay.

C: Right. She has made a serious charge against you. She is really upset.

S: (Cracks knuckles and looks around the room.)

C: Let's talk about what happened.

S: Nothin' happened! (Looks up briefly then looks away.)

C: This situation is not going to go away. I know it is difficult to talk about, but I need to hear what happened from your viewpoint.

S: Nothing! I *said*, nothing! She's making all that stuff up.

C: What do you think she has said?

S: I don't know, but it's not true. I was just trying to be friends and she acts peculiar.

C: Let me see if I understand. From your point of view you were trying to be friendly toward Jay, and then she reacted in a way you hadn't expected.

S: That's about it.

STEP 2: STUDENTS IN CRISIS NEED TO TALK

C: So, where were you when all of this began?

S: I saw her in the hall and asked if she'd like to meet me outside.

C: Has she ever met you outside before?

S: Naw, I don't know her very well. But I'd like to.

C: She's attractive to you?

S: Yeah, I guess so.

C: What did she say?

S: Nothing. Just nothing! She sort of smiled and kept on walking.

C: So you didn't know whether or not she heard you?

S: Oh, she heard me! She showed up by my car in the parking lot!

C: What did she say when she first got there?

S: Nothin'. Just nothin'. She just stood there and smiled sort of funny.

C: What happened next?

S: I don't remember. Anyway, I didn't do anything to her!

C: So, what happened next.

S: I don't remember. (Looks away and begins to crack his knuckles again.)

C: She says that you attacked her. She called it "rape." Is that what you would call it?

S: *No!* That's crazy. I just wanted to be friendly with her. You know, like getting to know her sort of close like we were friends and all that sort of stuff. (Looks down and then begins biting his nails.)

C: Let me see if I understand your view of this. You find Jay attractive and want to be friends with her in a special way?

S: Well, yes.

C: And now she is all upset and says you messed with her.

S: *It's not true!*

C: Then what happened to make her say that?

S: I just tried to touch her, to be friendly.

C: Do you think she liked being touched?

S: She made out like she didn't want me to touch her.

C: Where did you touch her?

S: Just a little, under her blouse.

C: How do you think she felt when you did that?

S: She said mean things and jerked away. That's when *she* tore the blouse.

C: Do you remember what she said, exactly?

S: She said something like, "Keep your hands off me . . . you fag!"

C: You were surprised by her reaction?

S: (Silence.)

C: Was she saying, you went too far and to back off?

S: (Nods in agreement and cracks knuckles.)

C: Let's see if I have this straight. You had one set of intentions and she had another. And now she's saying you raped her. And she may bring a complaint against you. Does that about sum it up?

S: Yeah, that's basically it.

STEP 3: FIND THE CENTRAL ISSUE AND SELECT A THERA-PEUTIC GOAL

C: After you touched her, what did she do?

S: She pretended she didn't like me to touch her, but I know she really did.

C: What gave you that idea?

S: Everyone knows girls really like guys who touch them, but they don't want to show it.

C: How do you know that?

S: Some of my friends told me.

C: Most of your friends have girlfriends?

S: Yeah. (Becoming very anxious and awkward in his speech.) They're always talking about girls and what they do to them. I try to stay out of it, but they're always goofing out and joking about you know, male-female relations, all that kind of stuff. It's hard to keep your mind on your work with all that talk going on.

C: It puts a lot of pressure on you?

Coach begins to probe for level of insight about feelings and the central issue.

S: Yeah. (Gets up and walks around the room.) They're nice guys but they have a problem you know, of getting everything they can. You know, so basically they sort of tease me about, you know, about not just getting what I can with every female I can. You know, they say I'm too slow.

C: What thoughts go through your head when they say you're not as swift as they are?

S: Well . . . yeah, to an extent, but . . . they're my friends and all. I don't want them to think I'm weird or anything. Or, you know, out of touch. Or above them; or, God forbid, a homosexual.

C: Have they brought up the topic of your being homosexual?

S: No. They always bring up the subject of girls. They're always putting pressure on me to perform, about sex and everything. I used to be interested in this one girl and I found out she was interested in me. In my way I guess I'm what you would call slow; I'm very slow. I take time. I want to get to know a person and just don't want it to be just . . . (pops his hands together) . . . *bang*, hop into bed, you know. I still believe in being friends and stuff like that. So I went to visit her and show I'm interested, but I guess I'm kinda shy. I didn't make any roads, and the guys would all the time hassle me, "What's your problem? Why is it taking so long? What are you gonna do? Come on, you need to do something or you're gonna be gone." They have this saying, "Good guys finish last," and they're always telling me that. I don't know what it means, and if they see an article in a magazine or somewhere, "Good guys finish last," they show it to me.

C: They really know how to get to your feelings.

S: Yeah, they always say, "Hey, when's the last time you've been with a girl?" and all that kind of trash. I mean, in public. I don't think it's too cool.

C: You have a sense about what is acceptable and unacceptable talk.

S: Yeah, even talking about that kinda stuff is trash.

C: Sounds like it's confusing to you, about yourself and what is really the right thing to do. And the guys are putting so much pressure on you that you forget to act according to your beliefs.

> Coach states the central issue.

S: That's it! To mention that you have a lack of . . . or either getting too much, I mean that's not cool. I mean, like it's my business if I take too long, and if a girl doesn't want to wait, then it was never meant to be. I think things take time . . . not to rush!

C: That sounds like a guideline to live by. How did it work today with Jay?

> Coach begins probes for motivation to change level of insight and behavior.

S: Girls are strange. They say they want that kind of guy. That's what kinda pisses me off.

They say they want a gentleman. They say they want someone to treat them nice, take 'em out, but actually they're . . . it almost seems that good guys do finish last. I've seen some nice girls, and they go out with these guys that just totally dog them out and they play on 'em and go out with other girls and go to bed with other girls. They're the ones getting all the girls. I mean it's . . . and here I am by myself.

C: And in some kind of trouble!

S: Yeah. It's so messed up. It's so weird. They say they want a gentleman, but they really don't. Then I hear them explain to each other, "He never takes me out. He doesn't spend money on me. He doesn't do this. He doesn't do that."

C: Sounds like some of the girls have mixed feelings about what they want out of a relationship.

S: Yeah, I guess so. (Pause.) I'm different from a lot of guys.

C: You're referring to your attitude toward girls and sex?

S: I don't know why they can't see it the way I think. One of my friends—his parents have been separated and he was raised by his mother, like me, but it's not the same. He doesn't act the same. Even the little things I do bother him, even about how . . . protective . . . I guess I mean. Like my body. I'm not into exposing, it's just like my chest. I can't walk around without a shirt on. He says it's prudish, but it's me. I don't feel comfortable. My mother always said, "Only unbutton your shirt to the first button." So that's the way I've always done it.

C: And when things don't go the way you expect them to, it upsets you?

S: Yeah, the things they say upset me. Like I'm an outcast. I guess I'm more my mother's son and she raised me to respect women. You're not supposed to do this to women. You're supposed to treat a woman like a lady all the time.

C: And now you're interested in Jay, and the guys are putting pressure on you again?

S: Yeah, they've been saying, "When are you going to make your move?" They're telling me, "If you don't do it now, forget it!"

C: And sex is an important, special thing—where you want to wait for that special person?

S: I think so. You have to have some feeling for them. And you just don't jump into bed. I worry about getting someone pregnant and things like that, and all these sex diseases. You just don't know what's going to happen.

Coach recognizes that Sidney is struggling with intense developmental anxiety, with guilt and confusion about sexuality and with his identity as a person. Sidney wants others to like and respect him. Sidney's level of insight about the incident seems to be that he knows what is acceptable behavior and that what happened was wrong, but he is not exercising acceptable behavior in keeping with what he knows or believes.

C: Is Jay like the other girls?

S: She's different, but she says those things, too. She told me she was totally bored.

C: So, you thought she was telling you that she likes guys that move in on her?

S: Something like that.

C: So you did?

S: Well, I wasn't going to hurt her. God forbid! I really like her. I wanted to give her what she wants.

C: And you thought she wanted to be physically handled?

S: Yeah, something like that. But she acted strange when I touched her.

C: We seem to have two issues here. You wanted Jay to be that special person. And you thought she wanted you to come on with her, until you touched her.

S: And I really didn't want to scare her. I'm not that kind of person.

> Coach chooses to shape the remainder of the LSI around the therapeutic goal of building Sidney's self-control by strengthening existing values. The coach believes that, with support, Sidney can learn to use his insight and values *before* he acts.

STEP 4:
CHOOSE A
SOLUTION
BASED ON
VALUES

C: How is Jay going to know that you are not like the others? Her feelings have been hurt, her blouse is torn, and she probably doesn't trust you anymore.

S: I guess I need to let her know I'm not like that!

C: Do you think she'll talk to you?

> Coach begins review of alternative solutions.

S: Probably not. She'll say I touched her and tore her blouse.

C: Those seem to be the facts.

S: Well, sort of . . . I did touch her, just a little, under her blouse.

C: What sort of reaction would you have, if someone touched your private parts?

S: Probably the same.

C: So you can truthfully say to her that you know how she might feel?

S: I think so. I really am sorry. I'm not that kind of guy.

> Coach sees that a solution can be developed from Sidney's value system, at the level of "my better self."

C: Maybe if she heard that from you, she might believe you.

S: Maybe.

C: And what about her blouse?

S: She thinks I did it. But it happened when she jerked away!

C: Is that her responsibility, or yours?

S: I guess it's mine.

C: How will you handle the blouse problem?

S: I can tell her I'm sorry: "It was a real pretty blouse, and I'll buy you another one."

C: That's a sign of maturity, to face responsibility. What if she still accuses you of attacking her?

S: She's probably like all those other girls who don't really know a good guy when they see one.

C: You'd like her to think of you as a person who respects others?

S: Yes! I do respect others.

C: How can you convince her?

S: I'll tell her!

C: What are two points you would like her to know about you?

S: (Looks out window and pauses.) I'm not like the others.

C: You are a person who wants to be kind and considerate of others' feelings.

S: (Nods in agreement, then looks away and shuffles his feet.)

C: And . . . what is the second point?

S: I'm sorry about touching her and tearing her blouse. It was a mistake and I want to pay for a new one.

Sidney restates the solution.

STEP 5:
PLAN FOR
SUCCESS

C: Now what's the next step?

S: Talk, I guess.

C: You're right! Jay is in the guidance coun-selor's office. Do you think she will be calm, and listen to you?

S: I hope so. I don't want her to take out a warrant against me.

C: So what will you say?

Coach begins rehearsal of new behaviors and anticipation of responses from others.

S: I'll walk in and smile at her.

C: How could she interpret that?

S: I like her!

C: How else might she interpret your smile, especially if she doesn't know you like her?

S: (Long pause.) I don't know. Maybe that I'm laughing at her.

C: That's right! What would be an alternative strategy for you?

S: To look like I'm sorry?

C: That probably will work better. And who should talk first?

S: I guess I should.

C: What if she's still so angry with you that she wants to dump it on you?

S: Then, maybe I should let her talk about it first?

C: Good idea! People who are angry usually can't hear what others have to say. Then she can get it off her mind before you apologize.

S: (Long silence.)

C: Talking with Jay is not going to be easy.

S: I know.

C: What if she starts screaming at you or crying?

S: I'll be cool. I'll tell her I'm really sorry. I just didn't know what she wanted, and I'll pay for her blouse.

C: And what if she says you're lying?

S: It's the truth!

C: How can you convince her, when she was there?

S: I'll remind her about what really happened.

C: That may work, unless she tells about it from another point of view.

S: (Long silence.)

C: Can you just sit and listen to Jay, even if you don't agree?

S: I think I can.

C: I think so, too. You are a person who values your good reputation, and it's really on the line today.

S: Yeah, I know.

STEP 6: GET READY TO RESUME THE ACTIVITY

C: Let's hope Jay has her feelings under control and is ready to talk with you.

S: I'm ready.

C: When we go into the counselor's office, Jay will be there. What will you do?

S: I'll just say "Hello" and sit down.

C: Maybe you should wait for the counselor to invite you to sit down.

S: Okay.

C: And the counselor will begin. Maybe you will be able to get your two points across after the counselor talks.

S: Okay.

C: Like you said, this won't be easy. She still might want to take out a warrant against you. We don't really know how she feels now about what happened. Whatever comes, be cool!

S: I can do that. I know I don't want a warrant. My reputation is at stake.

C: What will you say to the guys after you finish in the counselor's office?

S: I'll just tell them, "It's a private matter!"

C: Great! I believe you are mature enough to face this and stay in control—the whole distance!

This LSI was followed by a second one that included both students and the guidance counselor. By then, Jay had received Emotional First Aid (see Chapter 6), during which it became clear that she had not been raped; but her feelings had been roughly handled. When she calmed down, the counselor used LSI to organize the reality of the episode for her (see Chapter 7). The therapeutic goal of the joint LSI with Jay and Sidney was to teach acceptable social skills for both students to communicate their interest and expectations toward each other (see Chapter 10).

WHAT WE HOPE YOU HAVE LEARNED FROM THIS CHAPTER

- When an LSI therapeutic goal should focus on a student's values.

- Characteristics of students who benefit from this form of LSI.

- The type of insight such students have about their own behavior.

- When, and how, these students typically show their values.

- How you can improve a student's self-regulation of behavior by reinforcing latent values.

- The four sequential tasks for adults using this form of LSI.

- Two strategies to avoid in this form of LSI, and what happens if they are used.

- Three types of information you need to avoid reinforcing unacceptable behavior while using this form of LSI.

- The common characteristic shared by all students who benefit from this form of LSI.

- How this therapeutic goal can be used with older students to help them face, in a responsible way, the serious consequences of their unacceptable behavior.

THE GOAL: TO TEACH NEW SOCIAL SKILLS ("NEW TOOLS SALESMANSHIP")

> **Student's Perception:** "I want to do the right thing, but it always comes out wrong."
>
> **Uses:** With students who have the right attitudes or values toward adults, peers, and learning but who lack the appropriate social behaviors to be successful.
>
> **Goal:** To teach new social behaviors that student can use for immediate positive gain.
>
> **Focus:** Instructs in how to do specific behaviors that will have immediate payback in desired responses from others; developmentally reflects emerging independence and responsibility.
>
> **Student's New Insight:** "I have the right intention, but I need help to learn the skills that will help me make friends, achieve, and get along with adults."

Teaching new skills to students who lack successful social behavior is a major way for adults to help students manage and change socially inappropriate behavior. Most troubled students use inappropriate or dysfunctional patterns of interpersonal social behavior. Yet many of these same students have appropriate intentions and feelings.

These students want to belong to a peer group, have friends, achieve, be acclaimed as a person others admire, and have a close relationship with someone. Although they have such positive feelings and good intentions, they lack the acceptable social behaviors that could result in greater fulfillment of their desires. The behaviors these students use often produce just the opposite reactions in others. They want to be friends, so they communicate this by shoving, hitting, bragging, touching, or making negative remarks toward others. They want to be approved by adults, but they try to achieve this by overreacting, contradicting, directing adults, being silly, boasting, covering up for inadequacy or failure by disavowing interest, posturing as know-it-all, or clowning. Adolescent students often take on negative, alienated postures toward adults and peers as defensive covers for their social and interpersonal ineptness and their failure to achieve approval or affirmation.

Using this form of LSI, an adult can support a student's good intentions and desires and teach new social behaviors. The therapeutic goal is to change a student's ineffective, unacceptable behaviors by teaching substitute behaviors that bring about desired reactions from others. To accomplish this goal, the adult helps students realize that their current way of expressing feelings and needs does not always bring the results they hope for. Then a connection must be made between the student's desire for acceptance and behaviors that will bring about these desired results.

Fritz Redl (1966) describes this form of LSI as widening students' "adaptational skills." In crisis and under stress, students' defensive behaviors often break down or change form. Redl sees this as a significant time to make therapeutic gains.

> Use many of their life experiences to help them draw visions of much wider ranges of potential reactions to the same messes. . . . Even the seemingly simple recognition that seeking out an adult to talk it over with is so much more reasonable than lashing out at nothing in wild fury may need to be worked at hard for a long stretch of time with some of the children I have in mind. (p. 46)

THE ADULT'S TASK

A trusting relationship with the helping adult is essential to the effectiveness of this type of LSI. For this reason, it is rarely used during the initial stages of students' programs when they typically spend their time defending against adult help and support. Once beginning trust develops and a student wants to learn and have friends, the adult can serve as a model for the student and use a range of techniques to teach prosocial behavior.

Perhaps the hardest part of this type of LSI is to initially identify a student's genuine feelings and desire for socially rewarding results. This is the task during LSI Steps 2 and 3. With the help of decoding, you can validate these intentions, if a student is not able to convey them directly (see Chapter 2). Then it is important to support and affirm the student by reinforcing the idea that what the student wants from others is desirable—and achievable. Throughout the process, providing empathy, warmth, and real concern for the student's feelings is essential.

The next task is in LSI Step 4 where you and the student review alternatives for achieving desired results. There are numerous ways to evoke changes in others' reactions (breaking the Conflict Cycle, described in Chapter 2). From the alternatives discussed, help the student select new behaviors that are realistic and achievable. Keep the choices simple and in small, sequential steps. Then, in Step 5 let the student rehearse the new behavior in role play with you. Be sure to include simulations of situations where uncomfortable feelings may reoccur when others say or do things that fail to acknowledge the student's "new look."

While these tasks require a fairly straightforward effort in teaching new social skills, Leonard Sanders (1981) offers an important caution when a student's behavior is so outrageous or frustrating that it is difficult to attribute any "good intention" to the student.

> Take care not to destroy the good intentions and growing positive attitudes of the pupil. The goal is to hook up the pupil's desire for acceptance with acceptable behavior. (p. 32)

This is an important point: New skills are taught on the premise that the student has shown good intentions and wants a different result. Clearly, if an adult belittles or fails to recognize the student's intentions (however obscured they may be by unacceptable, negative social behaviors), the foundation for building solid new behaviors may be lost.

When a change to new behavior is agreed upon by student and adult, behavior can begin to change. But don't be discouraged if it does not occur rapidly. Change will be gradual. It is difficult to quickly change old coping strategies that typically have developed from defensive maneuvers to protect from feelings of rejection, alienation, or inadequacy.

EXAMPLES OF STUDENTS WHO CAN BENEFIT

Almost all students can benefit from this form of LSI because they need more effective social behaviors. However, we believe that this therapeutic goal is particularly suited for students who give some indication by their words or actions that they are not happy with the outcome of an incident. This level of insight often develops after the other forms of LSI (described in the three previous chapters) have been used successfully. This sequence of therapeutic goals makes sense: The first goal helps a student organize perceptions of the incident. The second goal benignly confronts a student's unacceptable behavior to create awareness that the behavior only makes matters worse. The third goal builds a student's value system into active use in making behavior choices. And then, this fourth therapeutic goal follows, for the student who has come through these other levels of insight and now (a) is aware of what happened and why, (b) acknowledges that unacceptable behavior has no payoff, and (c) accepts the idea that the "better self" (values) can be a guide to behavior. This fourth therapeutic goal is for the student who has accomplished the previous insights and now wants to learn new ways to bring about better results. Identifying a student's intention is a key for choosing to teach new skills as the therapeutic goal. Here are several examples of students who meet these criteria and will benefit from this form of LSI:

AL   Whenever Al wants help with his math assignment, he taps his pencil on his desk. If the teacher ignores him, he calls out disruptively, "Come over here! . . . I said, 'Get over here!' "

Versions of Al's irritating and somewhat disrespectful behavior toward the teacher are fairly commonplace, wherever adults make assignments for students to do work independently. What is Al's intention? His intention is to accomplish the task successfully, but it is obscured in his defensive attempts to capture the teacher's attention and reassurance without appearing to need help. LSI that builds on this intention will be successful in motivating Al to change the way he tries to enlist aid from the teacher.

JOHNNIE   Although he is only 9 years old, Johnnie has an
AND   independent, streetwise style of bullying his
ROBERT   peers. He has seen daily violence between the
                      adults in his family, and he usually catches the

spillover himself. He has learned to cope with devastating feelings of inadequacy by covering up with toughness. Today, Johnnie is having trouble with his reading, and Robert starts calling him "stupid" and "first grader." Johnnie yells back, "Shut up, you bastard!" Robert only grins and retorts, "Retard!" By the time the teacher gets to them, Johnnie has jumped on Robert, who continues to laugh at Johnnie and taunt him.

What is the hidden, "good intention" in this episode? In the resulting LSI with both boys, the teacher recognizes that buried beneath Johnnie's unacceptable behavior are enormous inadequacy and the counterintention—to be known as someone who can do well. It is time to change how he deals with his friends and replace physical forms of social interaction with words instead. For Johnnie, the new words will focus on responding to Robert's taunts by disinterest, emphasizing instead Johnnie's interest in doing top quality work.

Does Robert have any buried "good intention"? The teacher thinks so as she recognizes Robert's concern and dislike for those who represent failure and the need for expunging such miserable failure in others. Robert also may bear hard feelings toward Johnnie for past injustices and is using this opportunity to get back at him. For Robert, the teacher chooses to decode his unacceptable, taunting behavior as an attempt to retaliate against Johnnie's previous attacks. This interpretation works well, as it conveys understanding of his feelings. It also works as evidence to show Johnnie that tough behavior toward peers brings the wrong reaction. Instead of being admired, he may be disliked. The teacher steers the LSI into the topic of keeping out of other people's business, allowing every person the "right" to a quiet and safe learning environment. She develops the theme that Johnnie's work is of good quality because he works hard at it. She decodes Robert's behavior as awareness that others sometimes have a hard time, and then she helps the boys develop alternative behaviors that show consideration and fairness while staying out of others' troubles.

KATHY   A seventh-grade class is engrossed in a project designing a group game about environmental issues. Kathy is asked to join a goup of four other students in which Matthew, a student she likes, is a member. Matthew smiles at her and she says, "It's not my idea to join this group." Matthew reacts and responds, "Looks like 'Acid Rain Kathy' is here!" The group laughs and Kathy says, "I'm *not* joining this group! Nobody tells me what to do, and I'll damn well do what I please!" With that she stalks away.

In the LSI that follows, the teacher chooses to teach Kathy new, more acceptable ways to convey her feelings about her friends and her resentment toward perceived adult authority. This therapeutic goal is chosen because the teacher knows that Kathy's intention is to be an accepted group member and be liked by others. She learned aggressive behavior as the primary way of solving problems from models in her home. Emotionally and socially immature for her 13 years, Kathy often gets into situations that are too demanding for her limited intellectual and emotional capacity. But when Kathy feels good about herself, the teacher has observed that she is kind and protective toward younger or smaller students in the class. The teacher uses these past observations to build on Kathy's "good intentions" to be a helpful, valued member of the group. The teacher supports Kathy's feelings that sometimes the group may not make a person feel good and isn't a comfortable place to be. Then, together they select alternative ways to respond when peers hurt her feelings and it looks as if she may not be successful in the group. The teacher reaffirms belief in Kathy's attributes and ability to use different behaviors to convince others that she has something to offer the group. They practice these new behaviors, and Kathy finishes the LSI with new confidence that she is a person her peers will want to include.

JOSH   Josh and Lee are attempting to organize a relay race with the others in their group during recess. None of the other students seems invested, and they are not being cooperative. Before the boys get the race under way, the teacher calls; it is time to return to the classroom. Josh storms away and starts to cry in a fit of temper.

As the teacher takes Josh away for LSI, she views this incident in light of Josh's past behavior. He is an impulsive 10-year-old with almost no frustration tolerance. He has daily episodes where he has outrageous responses to situations of disappointment. His response to any sort of disappointment has almost always been out of proportion to the event. He usually jumps into people's faces and tries to punch them. He does not seem able to put brakes on his own behavior, even when reminded to do so. And so his behavior snowballs into crisis after crisis. The fact that this time he walked away and had a temper tantrum was an improvement from his physical attacks of the past. As a result, the teacher chooses to focus the LSI on teaching new skills for dealing with his disappointment. We include excerpts of Steps 3, 4, and 5 from the actual LSI because they illustrate an effective way this teacher (T) handles the slow but gradual changes in Josh's (J) behavior as new skills are slowly learned.

T: The Josh I knew when he first came to my class would have gotten into a fight about this incident. Today, you ignored the others who weren't cooperating and walked away instead. This is a big accomplishment for you!

J: I really wanted to punch them out!

T: I know you did, but you didn't! You have really shown that you have grown. Instead of fighting, what did you do?

J: I got mad . . . and hollered . . . and cried.

T: I noticed that. It was disappointing that the relay didn't work. But you handled your disappointment in a different way today.

J: (Smiles at teacher.)

T: And I've noticed that there are three different sides to Josh.

J: (Looks up with interest.)

T: Josh Number One pushes people around and jumps into their space.

J: But I don't do that anymore.

T: I know, so we have Josh Number Two. When he gets disappointed he cries and throws things. Today, you didn't throw things. Now I think we can work for Josh Number Three. You are ready to make that change. This new Josh will be able to say, ''I'm angry and upset, but I'm not going to have a tantrum over it. It's not worth it!''

J: (Looks at floor and kicks his foot back and forth.)

T: What do you think about a third Josh; is this the time?

J: Okay, okay, okay. Can I go now?

T: In a few minutes, when we solve this problem.

J: Okay. (Looks directly at teacher.)

T: Let's take a look at the situation again. Let's pretend we had Josh Number One. What would he have done when the other guys wouldn't cooperate?

J: Hit 'em!

T: That's right! And what would Josh Number Two do?

J: Cry and scream and have a problem so that he ends up here. (Grins at teacher.)

T: Right again! And how would Josh Number Three handle this situation?

J: Just say, "Okay, the period is over and maybe we can do it tomorrow."

T: That's great! Let's try it on a new situation. Let's suppose that it is lunchtime and you and Roger finish so you can start a game. You are 5 minutes into the game and I say, "It's time to clean up." What are you going to do?

J: Not throw the game around and just clean up and get ready for class.

T: You got it! You are showing that you understand a better way to handle things when they bother you. That's real progress.

J: I won't get into trouble.

T: When you have a big disappointment you will have to work hard to keep Josh Number One and Josh Number Two out of it, so Josh Number Three can show the mature way to handle disappointing situations.

(Our thanks to the teacher, Jody Tepper, for sharing this LSI.)

EXAMPLE OF LSI USED TO TEACH
NEW SOCIAL BEHAVIORS

The following example of a full LSI illustrates the use of this same therapeutic goal—to teach new, acceptable social skills. In this incident, between Denise and John, the LSI is used to build on their good intentions—to interact with each other. This example also illustrates how two students who are both central to the incident receive individual LSIs before a joint LSI, but with different strategies on the adult's part. You will see that the adult adjusts her strategies to the students' different values, different views of adult authority, and different types of anxiety. Yet for both students, the therapeutic goal enhances their self-esteem by recognizing their good intentions and shows them alternative behaviors to accomplish what they seek—successful peer relationships.

That's Disgusting

Denise is an 11-year-old with a learning disability who has been in a special program for several years. She is somewhat overweight, and this causes her problems with some of her classmates. Often, she sets herself up for ridicule without realizing what she is doing. Also, her weight causes her to be teased and picked on frequently. She looks at this as attention and acceptance from others, not realizing that it is negative.

John is a 10-year-old boy in Denise's class. He is constantly picking on someone or making fun of others, which gain him attention and acceptance from other students. John constantly makes unnecessary and rude comments to peers and adults, not realizing he is deprecating them. He has a tendency to attack others verbally and never takes others' feelings into consideration. He is an honest boy and sometimes feels guilt when he does not tell the truth. His weight problem is the opposite from Denise's: He is extremely thin for his age.

In the lunchroom, while the teacher is out of the room, the class is in an uproar. Denise and John are attacking each other verbally while the rest of the students in their group are coaching John and laughing profusely. The teacher returns and removes Denise and John to the hall to find out what happened.

Denise and John do not need to drain off any physical anger. At this point, all their anger is verbal. John tries to continue baiting

Denise, and she yells back at him. The teacher interrupts, and both students begin to calm down. The teacher recognizes that there is need for LSI with both students. Because this teacher (T) is not comfortable doing LSI with both students together, she chooses to begin with John (J), telling Denise (D) to return to the lunchroom to talk after she finishes her talk with John. (This LSI could have been done together; however, it is up to the adult to decide whether to do individual or group LSI first.)

JOHN'S LSI

STEP 1:
FOCUS
ON THE
INCIDENT

T: I see that you are very angry at Denise. Take some time to calm yourself down so we can talk about what happened.

J: Denise is so disgusting! I can't stand sitting next to her at lunchtime—or any time!

T: Why don't you tell me what you think happened, and we will try to work it out together. I need to understand exactly what happened if we are going to understand Denise.

J: Every day at lunchtime, Denise is always eating her lunch right in front of my face and chewing with her mouth open. She does this on purpose because she knows it gets me angry!

The reason for the LSI has been brought out.

STEP 2:
STUDENTS
IN CRISIS
NEED TO
TALK

T: So, in other words, when Denise eats her lunch, she chews with her mouth open and that gets you angry?

J: Yes, she does it right in front of me every day. It's disgusting! When she eats, everything is all over her mouth, and I can't stand looking at her.

T: What do you do when Denise does this?

J: I tell her that she is a fat pig and to turn around because I can't stand looking at her.

T: Is this what you said to her today?

J: No, I told her that if she doesn't turn around and eat her lunch, I was going to throw it in the garbage.

T: I see. You gave her a warning that what she was doing was annoying you?

J: Yes, and she continued to eat that way. So I got up and threw her lunch in the garbage.

T: Let me see if I've got this right. Denise was sitting and eating her lunch. She turned around with food on her face and chewed her food right in front of you. Then you warned her, and she didn't listen. So you threw her lunch in the garbage. Is that it?

> **Teacher has sufficient understanding of the incident to begin to focus the issue.**

STEP 3:
FIND THE
CENTRAL
ISSUE AND
SELECT A
THERA-
PEUTIC
GOAL

J: That's what I did, but then Denise turned around and *hit me* on the arm, for no reason. I warned her what I would do!

T: I see. She hit you after you threw her lunch out. Would you rate it as an easy, average, or hard hit?

J: It wasn't too hard; but she had no right to hit me!

T: You're right. Hitting doesn't solve problems.

J: (Nods in agreement.)

T: Well, John, Denise was very angry that you threw her lunch away.

J: I didn't want to fight with her; I only wanted her to stop being so disgusting!

T: Let me see if I've got this straight. You were trying to tell Denise that she wasn't eating properly, and when she kept on you threw her food away, and then she hit you?

> Teacher sees that the incident is the issue in this crisis and restates what the student has told. She also hears John's "good intention" trying to help Denise improve her eating skills. This information suggests that the therapeutic goal for this LSI with John should be to teach him new ways to behave when another student is inappropriate.

STEP 4: CHOOSE A SOLUTION BASED ON VALUES

J: Yes. I was trying to help her.

T: Is there anything else you could have done to handle that situation better, instead of throwing her food away?

> Teacher begins review of alternatives.

J: No . . . well . . . I guess I could have waited for you to come back and tell you.

T: That would have been a better choice; but what else could you have done?

J: I don't know!

T: I believe that in your own way you were trying to tell Denise that she wasn't eating very properly. That was a good idea, but is there another way you could have been helpful so she wouldn't become so angry?

J: I guess so.

T: What other way could you let her know that you don't like the way she eats in public?

J: I guess, tell her. But she won't pay any attention.

T: Maybe there is a way for you to say the same thing without hurting her feelings?

J: I guess so.

T: What could you say?

J: "Take my advice, Denise. Don't eat like a pig!"

T: Try to tone that down so she will really listen to you.

J: Okay. "Denise, don't eat with your mouth open. I don't like it."

T: That's good. Now, could you improve on that statement so she will know it's just not you that doesn't like what she does?

J: I could tell her that nobody wants to see inside anybody's mouth when it's full of food.

Student states solution.

STEP 5:
PLAN FOR
SUCCESS

T: That sounds like a solution that will work! Denise won't think you are trying to hurt her feelings or pick on her. But if she says, "I don't care whether you like it or not," what will you say?

J: She's dumb! She doesn't know how that puts people off!

T: You believe that Denise really wants others to like her?

J: Yes. Can I go now?

T: First we need to practice how to help Denise understand the message without getting mad. I'll be Denise, and you tell me the message.

J: Denise, do you want people to be disgusted when they see you eat?

T: Can you make it a little kinder?

J: Denise, it puts people off when they can see into your mouth full of food.

T: That is much better! I think Denise could listen to that without getting her feelings hurt or getting angry. Can you do this the next time it happens?

J: I guess so. I'll try.

STEP 6: GET
READY TO
RESUME THE
ACTIVITY

T: All I could ask is that you give it your best try. Let's go back to the lunchroom now. You have about 10 minutes to finish your own lunch. And while you're doing that, I'm going to talk with Denise.

DENISE'S LSI

STEP 1:
FOCUS
ON THE
INCIDENT

T: Denise, I saw how upset you were at John. You seemed very angry. Why don't we talk about it and see if we can work this out?

D: (With intensity.) John is always picking on me, every day . . . all day long. He never leaves me alone!

T: It sounds like an ongoing problem with John, but let's discuss what happened today at lunchtime.

D: All I was doing was eating my lunch, and the next thing I knew I turned around and John threw my lunch in the garbage!

T: I see. John threw your lunch in the garbage for no reason at all?

D: Yes! I was eating my lunch. John was making fun of me. He said I was a "pig" and to turn around!

Denise has described the incident.

STEP 2:
STUDENTS
IN CRISIS
NEED TO
TALK

T: Why do you think he was calling you a pig?

D: I don't know! He is always calling me names.

T: No one likes to be called names. It's upsetting.

D: It hurt my feelings!

T: I can understand how you felt. Do you think you gave him any reason for saying something like that?

D: No, he was just being mean. He threw my lunch out for nothing!

T: Did he say that he was going to throw it out before he did it?

> Teacher begins organizing events into a time line.

D: Yes, he told me that he was going to throw my lunch away. He had no right to do that!

T: Did he say anything else?

D: He told me to turn around and stop looking at him.

T: I see. Were you doing anything while you were looking at him?

D: Well, I was eating my lunch, minding my own business, and talking to Ben.

T: Let me see if I understand this correctly. You were chewing your food at the same time you were talking to Ben. Then, you turned around and looked at John, and he threw the rest of your lunch in the trash?

D: That's right!

> The time line has been established.

STEP 3: FIND THE CENTRAL ISSUE AND SELECT A THERA-PEUTIC GOAL

T: Did John say anything to you before you started talking to Ben?

D: Yes, he told me that if I looked at him like that again he would throw my lunch away.

T: Why do you think he said that?

D: He said that the way I eat makes him sick.

T: What do you think he meant by that?

D: I don't know.

T: Do you think it could be possible that it may not look very nice, talking and chewing food at the same time?

> Teacher identifies the issue and states it tentatively.

D: That's disgusting! My brother does that at home all the time.

T: Do you think that maybe that was what you were doing, without realizing it?

D: (Long pause.) I guess that's possible. But he doesn't have to throw my food away!

Denise verifies the teacher's statement of the issue.

T: No one likes to be threatened and have someone take away their food.

D: (Smiles at teacher.)

T: What happened after he threw away your lunch?

D: I hit him, of course!

T: I see. You hit him. How did hitting John help solve the problem?

D: It didn't. But I didn't know what else to do!

T: It sounds as if you wanted to get along with him.

D: I want to be friends . . . but not if he's going to throw away my lunch!

Teacher recognizes Denise's desire to be accepted and have friends. She chooses to teach Denise new social skills, as the therapeutic goal for this LSI.

STEP 4: CHOOSE A SOLUTION BASED ON VALUES

T: Now that you have a clearer picture of what happened between you and John, how do you feel about it?

D: I don't think I handled it very well.

T: What could you have done instead of hitting John?

Teacher begins Denise's review of alternatives.

D: I could have told you, but you weren't around.

T: So, if I wasn't around, is there anything else you could have done?

D: I didn't know what was bothering him. He just threw out my lunch.

T: Maybe we have two different things to handle here. If you knew what was bothering him, what could you have done?

D: I could have apologized. He didn't give me a chance.

T: And what about chewing?

D: I guess I should pay attention.

T: What is a general rule that helps with this eating problem?

D: Not to talk and chew at the same time.

T: That's a helpful rule! It makes others want to be around us while we're eating.

Denise has stated the solutions.

STEP 5:
PLAN FOR
SUCCESS

T: Let's practice. I'll be you and you be John. And I've forgotten that rule about not chewing and talking at the same time. (Teacher pantomimes.)

D: That's disgusting!

T: I'm sorry, John. I wasn't paying attention to what I was doing. I didn't mean to make you feel disgusted.

D: That's okay, Denise. Just don't let it happen again.

T: I'll try to remember. Let me know if I'm doing it. But please don't throw away my lunch.

D: (Smiles confidently.)

STEP 6: GET READY TO RESUME THE ACTIVITY

T: Do you think you can try this out at lunch tomorrow?

D: Well, I'll try, but you may have to remind me once in a while if I forget.

T: I'll help you to remind yourself, as long as you give it your best. Try and be aware of how it looks to others when you're eating lunch.

D: I can do it.

T: And you'll like the results. The kids will not mind sitting at the table with you. It's a good step for making friends.

D: (Nods in agreement.)

T: You have a new plan for this situation, and I have talked to John about his behavior. I believe you are ready to go back to the class-room now. The lunch period is about over; and the class should be on its way to our room by now. What will happen when you go back into the room?

D: I'll just go back in like everything is all right.

T: And if the others start looking at you and asking, "What happened?"

D: I'll ignore them.

T: That would be okay, but what else can you say?

D: Maybe I'll tell them that John and I had a problem and we settled it!

T: Good idea! That will stop their wondering about what happened, and lets them know you and John are handling it.

Denise will need continuing LSIs for social skills and appearance. Her weight and manners while eating show that she needs a lot of assistance. John needs assistance in learning self-control of his rude comments to others, especially Denise. He needs to concentrate on taking other people's feelings into consideration before he says something. This will demand a great deal of work on John's part and a number of LSIs focusing on social communication skills.

WHAT WE HOPE YOU HAVE LEARNED
FROM THIS CHAPTER

- The primary purpose in using the therapeutic goal discussed in this chapter.

- The common characteristics shared by all students who benefit from this therapeutic goal.

- The developmental process that is emerging in students who successfully accomplish this goal.

- The new insight students should gain as a result.

- Why this form of LSI is rarely used when a student first begins a program.

- Why it is often difficult to identify a student's good intentions.

- The importance of being accurate in your assessment of what a student values in accomplishing this goal.

- Three characteristics students have that indicate readiness for this form of LSI.

- Three major processes adults use to accomplish this goal.

- One potential problem an adult can cause, if caution is not exercised.

- Where this particular type of therapeutic goal fits in the logical sequence of goals for LSIs, as a student's insight expands and grows.

THE GOAL: TO EXPOSE EXPLOITATION OF STUDENTS WHO ARE ABUSED, ISOLATED, OR EXPLOITED BY OTHERS ("MANIPULATION OF THE BOUNDARIES OF THE SELF")

Student's Perception: "It's important to have a friend even if the friend gets me into trouble."

Uses: With students who are neglected, abused, scapegoated, isolated, or who seek out destructive friendships by acting out for them.

Goal: To help a student see that another student (or adult) is manipulating events in a way that is working against the student's best interest.

Focus: Provides insight into the behavior of others; views social interactions from the perspective of motivations and behaviors of others; developmentally, this goal requires considerable maturity on the student's part, as the student learns to understand how others think, feel, and behave.

Student's New Insight: "A friend is someone who helps you solve problems and feel good rather than someone who gets you into trouble."

At issue in this form of LSI is the use of a student by others to gratify their own needs. Students who allow themselves to be manipulated and used by others are vulnerable because of their own emotional needs and inept interpersonal skills which frequently lead to peer teasing, scapegoating, rejection, or isolation. For these students, rejected and socially isolated, the desire for acceptance and the pain of developing social skills make them particularly easy targets for a more sophisticated student. They become victims of their classmates and "friends." Sometimes these students set themselves up to be victims. For others, the role is imposed on them by circumstances or associations in which they are helpless to protect themselves psychologically or physically. Their need for friendship is so great that they seem willing to pay any price to obtain it.

Friendship is a complicated relationship, based on mutual trust and respect. Neither individual consciously exploits the feelings or resources of the other. Friendship is essential to one's self-esteem and feeling of belonging. It is an emotional bond creating interpersonal freedom and comfort. Friendship means closeness, sharing feelings and secrets, and providing support. Friendship is a source of psychological support for students over 9 years of age (see Chapter 4 for a discussion of the existential phases). For adolescents, making and keeping friends is a major activity, providing a major quality to life and a significant topic in many LSIs.

Troubled students find friendships difficult to make and maintain. Yet their need for friendship is great. This need for friendship makes them vulnerable to being socially exploited by the skills of a more sophisticated student. They are "invaded" psychologically by their "friends" who influence their thoughts and behavior. The instigator manages to be deviant but keep out of trouble, while the victim carries out the behavior and becomes the unlucky fall guy. An instigator usually knows what to say at the right time in the right way to seduce the victim to blow up and act crazy so that adults will control and punish this disturbing student. These are intriguing dynamics. They show how the defenses of one student are used to the detriment of another.

Fritz Redl (1966, p. 46) vividly described this type of problem:

> From time to time one invariably runs into a child who combines with the rest of his explosive acting-out type of borderline aggressive pathology, a peculiar helplessness toward a process we like to refer to as group psychological suction. Quite vulnerable to even mild contagion sparks, he is often discovered by an exceptionally brilliant manipulator . . . and then he easily drifts into the pathetic role of the perennial "sucker."

A poignant parallel is drawn when these same students are exploited by adults, particularly parents or other adults in a "helping" role. The most blatant form of this is abuse—sexual, physical, and psychological. The theme is the same as with misguided friendship: Someone else is using the student to gratify his or her own needs. Students allow this to happen because of their own emotional needs for an adult who will nurture, support, and care for them. A caring, close relationship is a fundamental human need. These students seek it so desperately that they are willing to be blind to the truth and allow themselves to be targets for adults in need of self-gratification. These students pay a terrible price for adult attention—subjugation and exploitation. They are true victims, psychologically helpless and physically vulnerable.

As with peers, an adult instigator often is a sophisticated manipulator who plays on a student's fears of abandonment, fears of rejection, guilt, and an enormous sense of inadequacy. The result often is that the adult instigator makes the student feel such intense anger or so hopelessly guilty and inadequate that emotional development stops and the student fails to develop a sense of independence, self-esteem, and identity. This further binds the passive victim to the needs of the adult.

The therapeutic goal for LSI with such students is to expose to them the self-serving control of the student by another (peer or adult), showing how the instigator gets power and pleasure from what has been done while the victim gets nothing but trouble (the perceived "friendship" or "caring adult" does not really exist). To accomplish this therapeutic goal, it is necessary to help a student consider how caring relationships or friendships are expressed in the behavior of others. This is difficult territory because, as you highlight behavior that conveys what is genuine, the contrast with the behavior of the manipulator becomes vividly clear. Betrayal is revealed, and the deepest and most painful parts of a student's private side are touched in the process.

THE ADULT'S TASK

Foundations for a successful LSI with this therapeutic goal depend upon a carefully constructed time line in Step 2 based on what happened, in detail. With this information, you should be able to find specific descriptions of the behavior of the instigator that can be identified as "unfriendly" toward the student. A student who has been victimized this way will almost always defend the "friend" or adult, so it is not useful to openly cast the instigator in a negative light. Wait until your student has described the incident in such detail that the conclusion is obvious. Your student's "friend" was looking after himself or herself, and the exchange was not an act of genuine friendship or caring. This is a difficult insight for most students to develop. It requires a fairly high level of interpersonal perspective-taking to see an action from another's viewpoint. To understand that another person benefited deliberately from your student's distress is even more difficult for the student to see and accept.

Once the facts of misuse and self-service have been established, be sensitive to the sense of loss (bereavement) your student will be experiencing. A significant adult or friend, no matter that it was an abusive relationship or a false friendship, has been lost. This is your second task—affirmation and support of the student's positive qualities to bolster sagging self-esteem and feelings of being left a victim without support or friends. During LSI Step 4, this can be done as you search out the student's values. When you tie your affirmations and support to what the student really values, you infuse the situation with new hope. This shift moves the student away from total emotional dependence on others by reaffirming the student's own self-esteem and integrity as a valued person.

The third task, also started during LSI Step 4, is to provide some hope for alternative behaviors the student can use to obtain support and relationship from different adults or friendship and acceptance by peers. Here you assist students in reviewing alternative behaviors. Focus the discussion on other adults or peers and their behavior that may indicate better opportunity for genuine relationship and friendship. Because these students have so few social skills with peers, it is often necessary to emphasize that "to have a friend, you must be a friend." This theme leads to a review of specific behaviors that build friendships. If the problem is centered around adult exploitation, the theme should include the "right to your own space (privacy, respect, or fair treatment)" and ways to break out of the dependency for emotional support from that adult. You may find yourself combining therapeutic goals at this point to teach new social behaviors while helping your student maintain the new insight that all "friends" and adults are not friendly.

Your final task is to conduct a second LSI, this time with the student who did the manipulating and exploiting. You probably recognize that the most fitting therapeutic goal for use with students who are instigators would be to confront them with their unacceptable behavior and work for insight that such behavior will not benefit them; it only backfires when others understand what is being attempted. (Chapter 8 describes this type of LSI.)

EXAMPLES OF STUDENTS WHO CAN BENEFIT

The name given by Fritz Redl to this form of LSI ("Manipulation of the Boundaries of the Self") highlights the characteristic problem common to all of the students who can benefit from this therapeutic goal: The psychological boundaries of their personalities are so vulnerable to influence and manipulation by others that they are unable to form a confident or self-assertive presence. They have such a marginal sense of self-esteem and personal worthiness that they allow someone else to "pull their strings." It is easy to see how such students fail to develop responsible, independently directed acceptable behavior. Their lack of self-esteem is so enormous that they give every indication of being hopelessly immature and completely devoid of capacity for responsible, self-directed behavior. Here are two examples of students who are typical of those who can benefit from LSI that helps them see that they are being victimized.

BOB

George is extremely clever in getting Bob to blow up in the classroom. He knows Bob is very sensitive and is having trouble completing an assignment. After a few minutes, George says in a loud voice, "I'm all done! The teacher gave *me* the easy job!" (smiling and looking over to see Bob's reaction). This is enough to trigger an explosion from Bob. He leaps out of his desk, throws his chair in George's direction, and shouts, "You're a dumb ass!" As the teacher moves in to avert further disaster, George says in her direction, "Bob's really out of control; how can we work with all this noise?"

In the LSI that follows, Bob is able to recount the sequence of events. He also is able to describe alternative behaviors that are more acceptable ways to handle frustration. On several occasions previously he has lost control after George had fueled a situation with his incendiary remarks. The teacher recognizes that Bob is not aware of how he is being manipulated by George. So the teacher chooses the therapeutic goal of exposing George's manipulation and focuses the LSI on helping Bob consider how his behavior resulted in this talk while George was enjoying seeing Bob in trouble. This was a new idea for Bob! From this point, the teacher helped Bob consider why George might want to see

Bob lose control. The insight Bob developed was that maybe George liked to see others in trouble doing things that he wanted to do but wasn't willing to be caught doing. ("He wanted someone else to do his dirty work!")

MARY

Mary's reputation as an uncooperative, huge (145 pounds), unwashed 14-year-old makes her the target of many nasty comments about the size of her breasts and her odor from boys in her special education class. Her precocious sexual development does not help her situation. The level of poverty, deprivation, and abuse she experiences at home seems to be a major cause of her social and emotional problems. There have been repeated reports of concern by her teachers and school social workers that incest also is occurring. But even with home visits and attempts to encourage Mary to talk about what happens at home, they have failed to verify the suspected sexual abuse by both father and brothers. To build up her nonexistent self-esteem, she is seeing one teacher every day for an hour of private "girl talk." The focus is on health, grooming, bathing, eating, fashion, and related subjects. The results have been unbelievable. Mary has begun to talk spontaneously, showing glimmers of humor and reporting that she has washed her clothes herself (in a bowl—no washing machine and no money for a laundromat). She begins reaching out to the teacher in a relaxed and comfortable relationship. One day, while going through a teen fashion magazine, Mary (M) begins a dialogue with her teacher (T) with this comment:

M: People should let people alone.

T: Yes, every person has a right to privacy.

M: (Silence.)

T: People need privacy at school sometimes, don't you agree?

M: Yes! I like privacy . . . like here. No one yells or butts in here.

T: So we can talk about anything we want because we have absolute privacy!

M: (Silence.)

T: People like privacy at home, too. You know, like a private place where no one butts in on them.

M: (Turning pages rapidly.) I don't have any private place at home. I can't even go to the bathroom without someone butting in on me.

T: A bathroom is one place where everyone needs privacy. Is there a lock on the door?

M: We did, but my brother ripped it out of the wall once. (Pages turned even more rapidly.)

T: Have you told your mother you needed privacy?

M: I did, but she didn't do nothin' or say nothin'.

T: She leaves you on your own?

M: Yeah, I guess she can't be bothered. She's sick all the time.

M: Have you mentioned to your dad that you need privacy?

M: (Long silence. Then she changes the subject.) Do you like this dress? I wish I could make one like that.

This student has slightly opened the door to her private world, one of torment and terrible abuse. She is clearly being exploited psychologically, physically, and sexually by several adults. Her opening comment is the invitation to her teacher to talk about forbidden topics. The teacher conveys respect for Mary's privacy by talking around the problem during this initial discussion. Abbreviated LSIs follow every day, as Mary opens the door a bit farther. The teacher allows her to close it down when discussions come too close to the real issue—invasion of her body almost daily before she comes to school. In addition to using the LSI goal of exposing exploitation, the teacher also varies the goals through a series of LSIs that build self-esteem by reinforcing latent values held by Mary and teaching new social skills that help change Mary's ideas about herself. She gradually begins to see herself as a valued person who can protect herself from exploitation and abuse.

EXAMPLE OF LSI USED TO EXPOSE EXPLOITATION

While the teacher is out of the room, Alex shoots off a cap gun belonging to Leon. As the teacher returns, he hears the noise and walks in to see the smoke. When the teacher confronts Alex, he simply says, "It's not mine. It's Leon's."

Knowing that Leon is a smart, sophisticated streetwise sixth grader who has the reputation of being clever and verbal, the teacher wonders why he gave the gun to Alex, a lonely, insecure student who acts impulsively at times and seems to always be at the fringe of the group. He chooses to do LSI with the two of them together because it involved Leon's cap gun. Whenever there is a crisis, Leon seems to be at the center of it but always manages to escape responsibility. The teacher suspects that Leon may have instigated this latest event. We selected this LSI to illustrate how the teacher (T) managed the topic of exploitation with both victim (A) and manipulator (L) present.

With a Friend Like You, Who Needs an Enemy?

STEP 1:
FOCUS
ON THE
INCIDENT

T: This is the most interesting gun I have ever seen. It has no handle or barrel . . . just a trigger and a firing cap. Whose gun is it?

L: It's mine!

T: Where would you get a gun like this? It's so unusual!

L: I found it in an alley when I was riding my bike.

T: You must have sharp eyes to see such a small gun. I bet you were happy to find it.

L: Yeah!

T: Are you always this lucky?

L: Sometimes.

T: Did you have to clean it up to make it work?

L: Well, I took it home and got it working.

T: Did you have any caps at home or did you have to buy them?

L: I had to buy them.

T: Did you get the roll kind or the stick kind?

L: The stickum kind.

T: It seems you got the right kind. When did you first bring this special gun to school?

> Teacher begins constructing a time line with the students.

STEP 2:
STUDENTS
IN CRISIS
NEED TO
TALK

L: Today.

T: Where did you keep it?

L: In my pocket. (Looking relaxed and enjoying the discussion.)

T: Did you have it in your pocket this morning?

L: Yeah, I had it all day.

T: Who did you show it to before Alex got it?

L: No one.

T: What! (Showing surprise.) You mean you had this neat gun in your pocket all day and you did not show it to any of your friends? Is that true?

L: Yeah.

T: That's amazing. I'm very surprised. You must have excellent self-control! If I had brought that gun to school, I would have shown it to everyone. I could not keep it in my pocket all day.

L: (Begins to look uncomfortable.)

T: Alex, now you tell me how you got to use Leon's gun.

A: I sit in front of Leon and while we were cleaning up to leave, Leon tapped me on the shoulder and showed me this gun.

T: I'm sure you were surprised to see it. So, what happened next?

A: Leon said I could hold it.

T: I assume then that you are friends. Tell me, are you two (a) real good friends, (b) good friends, or (c) just friends?

A: Well, we are good friends.

T: Is that true for you, Leon?

L: Yeah, we're friends.

T: I see. So, as a friend, you gave Alex your prize gun.

L: That's right.

T: Then what happened?

A: Well, I was playing with it, and Leon said you were out of the classroom so it was okay if I wanted to shoot it.

T: That must have been very tempting for you. Here a friend gives you his gun and then tells you the teacher is out of the room.

A: I told Leon I better not shoot it.

T: What did Leon say?

A: He said if I didn't want to, Bill would shoot it.

T: Let me guess, Alex. You looked around the room, didn't see the teacher, and you shot the gun. When I came back what did you do with the gun?

A: I put it in my desk.

STEP 3:
FIND THE
CENTRAL
ISSUE AND
SELECT A
THERA-
PEUTIC
GOAL

T: How did I know you had done it?

A: (Looks very nervous and starts swinging feet.)

T: Looking at you, Alex, you seem very upset and worried about this problem.

A: A little. (Clearly upset.)

T: What I don't understand, Leon, is why you didn't shoot the gun.

L: Well, Alex is my friend so I let him go first.

T: I see, but help me understand why you told Alex to shoot the gun or you would give it to Bill. Is that right?

L: Well, Bill said he would do it, and I didn't think Alex would do it.

T: Oh, you didn't think Alex would do it! So you were kidding him or teasing him?

L: Yeah, I was just having fun.

T: That's right. You were having fun because you are a very smart person! You knew that the guy who shot the gun would be in trouble. Since you didn't shoot the gun, then we can all agree that you are very smart. Alex, if Leon is very smart what does that make you?

Teacher tests for Alex's insight.

A: (Deep silence.) Dumb!

T: That's right, your friend Leon gives you the gun, tells you the teacher is out of the room, calls you a chicken, and then tells you he was only kidding and he didn't think you'd be dumb enough to pull the trigger . . . (Long silence.) . . . Alex, let me ask you an important question. Is a friend someone who helps you or hurts you? Is he someone who gets you into trouble or someone who helps you stay out of trouble? Think about it. Remember, Leon is a very smart guy. He has excellent self-control. He can keep his gun in his pocket all day; but when he takes it out, who does he get to fire it off in the classroom?

| | Teacher selects therapeutic goal, to expose the exploitation. |

A: (Looks uncertainly at the teacher while Leon is looking anxious and angry.)

STEP 4: CHOOSE A SOLUTION BASED ON VALUES

A: I'm not sure . . .

T: Alex, it is important to have friends, and I want you to have friends. But you need friends you can trust. Not friends who take advantage of you. For example, if you fell out of a boat, would your friend give you a helping hand and save you, or would he push you away from the boat so he wouldn't fall out? (Silence.)

T: I know what kind of friends I want. Maybe it is different with you. Maybe you enjoy being used for Leon's pleasure and fun. If so, you are free to continue, but don't trick yourself into believing you have a trusting friend. It's your choice.

A: Leon does get me into trouble. Even my mother doesn't want me to play with him.

T: You have a difficult situation to think about, but it sounds as if you have a new idea about friends.

A: Yeah. Leon can get me into trouble.

Alex states a solution based on values of "adult approval." The teacher begins the next step but continues to challenge Alex to a solution at a higher stage (see Chapter 4), where he has shown some awareness of "my better self."

STEP 5: PLAN FOR SUCCESS

T: You've got a good insight there. How should this change what you do around Leon?

A: Well . . . maybe I ought to stay away from him.

L: That's dumb!

T: Just listen, Leon. You will have your turn to talk. Alex has some ideas about choosing friends.

A: Yeah. I like Leon and I want him to be my friend. But I don't want him to get me into trouble. That's not what friends do.

T: Right! So, you have this decision, how to stay friends with Leon and still stay out of trouble.

L: That's B.S.! I don't want to see him in trouble.

T: So, I hear Leon saying that he wants to be your friend, Alex. What about that?

A: Maybe we could be friends some of the time. Like, when we're not going to get into trouble?

L: I don't want no trouble!

T: I believe, Alex, you have a workable solution to this problem. Maybe you can plan to ignore Leon when he is trying to get you to do something that you don't want to do. What about that?

A: Yeah! Then, me and Leon can be friends some of the time, but not all of the time.

T: Do you think it also will help if I change your seat, so that you sit on the other side of the room?

A: Yeah, that would be good. Then I won't get into trouble.

T: Where you sit may help, but it's up to you whether or not Leon gets you into his game.

Let's review, what is the most important thing you can do about this situation?

A: Ignore him when he's doing something I don't want to do.

Student restates the solution, showing some responsibility for independently managing his own choice of friends.

T: That should work well for you. And maybe you can make other friends who know how to be true friends.

STEP 6: GET READY TO RESUME THE ACTIVITY

T: You have a difficult situation, but it sounds like you have the right idea! Tomorrow we can talk about it some more, but you can leave for now.

L: What about me?

T: We have much more to talk about.

L: It's unfair; I didn't do nothing!

T: I hear you, but we still need to talk about how you treated Alex.

You probably recognized the need now for a second, follow-up LSI with Leon, with a different therapeutic goal—to confront his unacceptable behavior (see Chapter 9). This was done around the theme that Leon thought he was getting pleasure and power from exploiting Alex. But in the future, as others catch on to what he is doing, he will no longer have friends and others will avoid him.

You also may have noticed that the teacher selected Alex's exploitation and need for friendship as the central issue rather than the incident itself. In this LSI, Alex was able to understand the dynamics of his relationship with Leon and how he is being influenced and used by this false friendship. Alex will need additional assistance in follow-up LSIs to learn new social skills for making and keeping friends. Social communication instruction along with other social skills curriculum will all be helpful to Alex. In addition, he probably will need reinforcement of his values from time to time (see Chapter 9) to strengthen his self-control.

The success of this LSI was in large part due to the interest the teacher showed in the gun and in the care with which each detail of the incident was brought out. The initial focus on the gun reduced Leon's defensiveness and made it possible to move the students freely into a discussion about the sequence of events. By bringing out the small details as he worked through this LSI, the teacher was able to show Alex the facts. The time line developed in Step 2, with questions of "Where," "What," and "Who," provided these facts without the teacher having to make accusations which would have heightened the students' defensiveness and made this LSI more difficult.

WHAT WE HOPE YOU HAVE LEARNED FROM THIS CHAPTER

- When the LSI therapeutic goal should focus on exposing a student's exploitation by others.

- The common characteristic of students who benefit from this form of LSI.

- The role friendship plays in the events that lead to exploitation.

- A common characteristic of students or adults who exploit others.

- Why some students allow themselves to be abused by adults.

- Why you shift the central issue from the incident to the exploitation during this form of LSI.

- What painful feeling results after a student realizes that he or she has been exploited by a "friend" or adult.

- Three tasks for adults using this form of LSI.

- Why a carefully constructed time line during Step 2 is essential.

- When it might be appropriate to move slowly and avoid excessive details while doing this form of LSI.

- Another, inevitable, LSI that must follow when you use this form of LSI.

"First, we talk about it.
Then we fix it.
Then I smile!"
(from Sam)

LSI FOR YOUNG OR DEVELOPMENTALLY DELAYED STUDENTS

This problem-solving "recipe" was contributed by a 5-year-old boy named Sam. In its simplicity there is the essence of adapting the LSI process when working with young children. Sam's recipe begins with "we," signaling recognition that crisis resolution is the combined effort of two people—the child and the adult. The adult provides support, security, understanding, and guidance. The child must go beyond the stress of the moment to actively participate in the crisis resolution by using words, not emotions. "Talk" is the way Sam describes this process of converting emotional expressions from unacceptable or destructive behavior into a manageable, rational form. Without talk, we have no way to transform a crisis into a solution and better feelings.

Sam's next ingredient, to "fix" the problem, is a statement of naive faith in adult omnipotence. We know there is no quick fix. Yet young children have a need to believe that when something is wrong, somebody (an adult) can fix it. This is an indication of the preexistential phase of development described in Chapter 4, when children cling to the belief that it is the responsibility of adults to take care of problems. In abbreviated LSI, the point of "fixing it" is to communicate that adults can help children deal with crisis in ways that are within their capacity to use. Knowledge of developmental skills and developmental anxieties (see Chapter 3) becomes essential for an adult attempting to determine whether to use a full LSI or to abbreviate it to meet the needs of young children or developmentally delayed older students.

The "smile" in Sam's recipe touches every child's need for optimism and confidence for the future. Things must get better! While this may appear forced to some adults, it is essential that young children be sensitized to what they are feeling; and the more positive the feelings, the happier the child. Bruno Bettelheim (1977, p. 125) said, "If a child is for some reason unable to imagine his future optimistically, arrest of development sets in." This need for hope is a powerful determinant in motivation (or its lack) for changing behavior. Lasting change happens when a student genuinely believes that better things are to come. Our responsibility is to provide the process whereby this happens and negative feelings are replaced with feelings of success, pleasure, self-confidence, security, and trust. For young children, the time immediately following the selection of a new behavior holds promise for successfully sensitizing them to their own feelings of relief at crisis resolution and pleasure in the results of a new, successful behavior. This learning provides the foundation for future motivation, teaching that changes in behavior can produce desired responses from others. This is what Sam's "smile" ingredient is all about. At the end of this chapter we include the abbreviated LSI used with Sam when this recipe emerged.

ABBREVIATED LSI

To resolve crisis, a young child must struggle through a series of difficult understandings that are the same essential ones included in a full LSI with older students, but in embryonic form. The same six LSI steps we have described throughout this book are distilled into three general steps for the very young: (a) talking about the incident and the issue, (b) developing a solution by selecting new behavior, and (c) enhancing self-esteem. Sam surely caught the essence of these steps in his recipe.

Here are the three parts of the abbreviated LSI we use with very young children and the learning processes that are tapped during each part.

"TALK" PHASE (Includes Steps 1, 2, and 3)

- Draining off intense feelings so that perceptions of reality dominate behavior.

- Using words (instead of behavior) to describe these events and actions of self and others.

- Ordering the events into a sequence that conveys what came first, next, and last.

- Discriminating between relevant information and elements that confuse or cloud the incident.

"FIX IT" PHASE (Includes Steps 4 and 5)

- Understanding that something can be done by oneself to make the situation better.
- Learning a new way to make it better.
- Stating the new behavior in words.
- Wanting to try the new way of behaving.
- Rehearsing the new behavior.

"SMILE" PHASE (Step 6)

- Acknowledging that better feelings have resulted from the crisis resolution.
- Practicing the expression of these new feelings to carry them over into the next activity and for the future.

Even in this abbreviated form, the essential LSI elements are there. The process remains a strategy for support and teaching, providing a structure to move a child developmentally forward from marginal ability to manage behavior toward full participation in talk about a crisis, its resolution, and the child's responsibility.

HOW TO TELL IF A STUDENT NEEDS AN ABBREVIATED LSI

Young children in crisis (or those of any age who are developmentally delayed) usually do not understand the source of their psychological pain, but they feel it *acutely*. They do not possess sufficient impulse control to curtail the expression of these feelings through regressed or unacceptable behavior. They do not understand the Conflict Cycle (see Chapter 2), and fail to connect their behavior to the responses of others. Their own needs are paramount (see the values section of Chapter 4). When thwarted in their attempts to gratify their own needs, these children can only express outrage. In a child 3 or 4 years of age, this behavior is normal (although still unpleasant for adults), but the lack of skills is anticipated as adults try to teach and intervene in constructive ways. In a school-age child, expectations are increased, and such primitive behavior is no longer tolerated. Yet these are the students who often have failed developmentally and have few skills for higher functioning behavior.

These students, developmentally unprepared to benefit from a full LSI, may have difficulty relating bits and pieces of an incident to a central issue. Their sequencing of events may be faulty, tending to focus on one particular aspect of the incident (usually the part that has evoked the greatest emotional response). Their expressive verbal skills are limited, and often they fail to comprehend the meanings of words and ideas adults use. They cannot communicate simple feelings verbally. Nor can they connect their feelings to their behavior. Memory is fragmented, and they may need assistance to reconstruct the simplest sequence of events.

In their social and emotional development, such students are all-important to themselves, and concern about others is seldom present. Fairness is viewed as something that benefits themselves only. They want adults and other children to agree with them, and they will go to extremes for adult attention and support.

When you recognize a student like this, who is developmentally unprepared for a full LSI, consider using the abbreviated form to teach the necessary prerequisite skills while dealing with the crisis at hand. You may recall these prerequisite skills from Table 1.1. They can be grouped roughly into five fundamental requirements:

- Attention span for listening and retaining what has been said.

- Minimal verbal skills to use language spontaneously and with sequential thought.

- Sufficient comprehension to understand the meanings of the adult's words.

- Mental reasoning to understand the essence of the incident and the problem it produced.

- Trust in the adult.

If a student has these minimum skills but you are in doubt about whether to use a full LSI or adapt your LSI to the simple three-step model we describe in this chapter, make an informal assessment of your student using the list of prerequisite skills included in Chapter 1. If you find that your student lacks some of these skills, the abbreviated LSI can be used to teach him or her needed skills for participation in a full LSI later.

Here is an example of a student, Donna, who has a few of the prerequisite skills listed above for a full LSI but can benefit from an abbreviated LSI.

DONNA
During a music activity, 6-year-old Donna is laughing, giggling, kicking her feet, and squirming. The teacher's attempts to redirect her back to participating fail. Donna throws her tambourine and tries to snatch rhythm band instruments from the other children. All the while, Donna is grinning from ear to ear and the other children are showing increased anxiousness. When the teacher attempts to take Donna away from the group, the child kicks and hits the teacher. As the teacher forcibly removes her from the room, Donna makes a last attempt, between kicks and screams, to bite the teacher. In the conference room, Donna is alternately volatile, hyperactive, threatening, and playful. By holding her in a gentle but firm manner, the teacher is able to help Donna calm down. Now what? Donna behaves in this manner frequently and seems to enjoy the excitement and contact with the teacher. She does not want to talk. Should the teacher leave her alone? Go back to the classroom with her? Ignore her? Or begin an abbreviated LSI, talking to build a foundation for future self-control? What should the teacher focus on? The child's laughing? Kicking? Throwing instruments? Or that she was angry when she was physically taken from the room? Or that she was disrupting the others? What can Donna learn from this incident (and what should the teacher try to teach her)?

All of these questions lead to possible approaches with Donna. The teacher chooses to do an abbreviated LSI because Donna has shown that she has the basic requirements for LSI: (a) sufficient attention span, (b) minimal verbal skills for social communication, (c) comprehension of adults' words, (d) genuine trust in the teacher, and (e) some awareness of expected behavior. Yet Donna's teacher also recognizes that emotions distort Donna's perceptions of events and drive her behavior. Her teacher knows, from previous activities, that Donna also lacks skills to put events in the sequence of occurrence. She understands that Donna needs to work on several developmental skills while the current crisis is also addressed. This can be done as part of the abbreviated LSI.

DEVELOPMENTAL OBJECTIVES FOR ABBREVIATED LSI

Many developmental objectives can be worked on while doing an abbreviated LSI (Wood, 1986a). Here are examples of such objectives that we have used with young children during abbreviated LSIs (Wood, 1979, reproduced here with permission of the publisher).

BEHAVIOR OBJECTIVES

- To recall routine.

- To show awareness of expected behavior through actions.

- To tell about expected behavior using words.

- To refrain from inappropriate behavior when other children behave inappropriately.

- To explain why an event occurred (cause and effect).

- To tell about different ways to behave.

SOCIAL COMMUNICATION OBJECTIVES

- To use simple words to answer an adult's questions.

- To use words to describe events.

- To use words to describe characteristics and behavior of self and others.

- To show feelings appropriately through words and actions.

- To make positive statements about self.

- To participate in talk with others.

SOCIALIZATION OBJECTIVES

- To seek adults spontaneously.

- To initiate appropriate behavior toward another child.

- To interact appropriately with other children in play.

- To cooperate with other children in group activities.

- To share and take turns.

- To imitate appropriate behavior of other children.

- To label events with simple values.

PREACADEMIC OBJECTIVES

- To identify details in objects and people.

- To tell a story sequence.

- To use concepts of "same" and "different."

- To discriminate opposites.

- To categorize things that have the same characteristics.

- To give reasons "why."

- To listen to a story with comprehension of what happened.

- To learn rules of play and participation.

If you already use a developmental approach to working with young children, you will have recognized these objectives as major milestones of development. They are essential skills for students to master in the process of normal social and emotional growth. When you use the abbreviated LSI with a child who lacks these developmental skills, the process becomes a strategy for teaching specific readiness skills in preparation for future participation in full LSI.

You also may have recognized the connection between these developmental skills and the specific therapeutic goals of LSI described in the preceding chapters. As you teach specific developmental skills in the abbreviated LSI, you also are able to provide for several of these therapeutic goals: to organize reality (see Chapter 7), to build values (see Chapter 9), and to teach new social skills (see Chapter 10).

WHAT HAPPENS IN AN ABBREVIATED LSI

Young children (and older students who are developmentally delayed) typically respond to crisis with emotions flooding their behavior. When this occurs, they generally display noticeably regressed behavior. You may see this in temper tantrums, screaming, crying, sobbing, kicking, spitting, biting, thumb-sucking, rolling around on the floor, waves of profanity or other "shock" words, or running away. When such signs of regression are evident, Emotional First Aid (described in Chapter 6) is necessary before attempting to use LSI. We look for indications that the child has calmed down physically before attempting any talk. Many well-meaning adults choose to ignore a child in this condition, using strategies such as, "I'm waiting for you to get it together," or "time out." Neither approach is particularly productive if children lack developmental skills to calm themselves without help. To be left alone may increase confusion in children who have a fragile grasp of reality, or it may increase resentment and anger in children who struggle with anxiety about abandonment (see Chapter 3).

It is important for a child to settle down before you begin LSI, but ignoring or extended periods of waiting are not usually the best ways to get their attention and readiness to begin. This is a time to show understanding, support, and your availability to help solve the problem. It is essential to convey to children flooded with emotion that nothing can happen until their behavior shows that they are ready. Be specific and concrete in your suggestions for ways to calm down.

"I can tell you're ready to talk when you breathe slowly."

"When you shout, I can't hear you; talk softly and we can take care of this problem."

"It's going to be all right. When you can sit (up, quietly, in the chair), I'll know you are ready."

"Talk" Phase

When a child shows any change from regressed behavior, try to shift the communication mode to words and away from emotionally driven behavior. A sympathetic question that requires only a yes or no often helps to get the process started.

"Are you feeling better now?"

"Did someone hurt you (your feelings)?"

"Do you want to solve this problem?"

"Something has upset you; can we talk about it?"

"What happened isn't nice, is it?"

When asked, "What happened?" (Step 1), children typically respond with,

"I don't know."

"It's his fault!"

"I didn't do it!"

"I don't care!"

It is not unusual for young children to begin to talk and then cycle back to emotional flooding again. When this occurs, just repeat the expectation that they can show you they are ready to talk by a specific behavior. Caution: Don't use a time reference such as counting one, two, three, or demand a specific quiet behavior for a sustained period. These expectations can become ends in themselves, and young children may lose the drift of the incident and the issue as they concentrate on the performance you demand.

Simple, direct questions are suitable for young children who have some verbal ability to respond, even if it is with limited words or phrases. If they can coherently describe simple experiences at times when there is no crisis, they should be able to benefit from answering your questions. They may emphasize one aspect of the event to exclusion of others. However, the important point is that the child is sharing perceptions with you. The session may be short, and a full investigation may not be appropriate. But the exchange teaches the child that you listen and value what the child has to say.

For children who are not ready to share information openly with an adult, putting the "Talk Time" into the form of a story narration that parallels the incident usually is effective. This is a useful alternative when a child cannot tolerate or accept painful reality because of too much mental or emotional confusion, guilt, difficulty comprehending the incident, or poor reality testing. With this strategy, the adult reconstructs a parallel incident with make-believe characters in a make-believe situation (Wood, 1981).

Begin with a standard story opener like, "This reminds me of a story about . . ." Ask questions to obtain the child's input and use this to construct the story, touching on the important points of the real incident. Join your words to the child's perceptions, however limited or strange the perception may be. By dealing in a fantasy mode, some children are able to "hear" and accept the problem and alternative behaviors required to resolve the crisis. Bring closure during the make-believe part of the story, weaving the story line around the hero or heroine going to a private place to talk about solving the problem ("fixing it") with an adult. End the story with a return to reality and the real situation. When this strategy is used, children are able to transfer the make-believe solution to the real incident and generally can rephrase the change needed in their own behavior.

Another variation of this approach is role play. Introduce the crisis as a more distant, third-person event that can be role played. Begin the role play as the narrator and then take the child's role in portraying the crisis. Usually, the child can take the adult role at this point. Here is an example of reversed role play, using Donna's crisis described previously.

> **T:** We are sitting on the rug for music. (Teacher sits on floor and indicates for Donna to sit also.)
>
> **T:** Donna says, "I don't want no tambourine!" (Teacher mimics this action of throwing the tambourine.)
>
> **T:** Now, Donna, you are the teacher. What do you say to this student?

D: Stop that!

T: Why, Teacher?

D: 'Cause, I said to stop that!

T: I don't want to play this dumb thing!

D: Well, why didn't you tell me?

T: I don't know.

D: (Long pause.) You need to tell the teacher when you don't like what you have.

T: Okay, Teacher. I'll remember that. I'll say, "Teacher, let me have the drum." Is that okay, Teacher?

D: That's good!

By participating in role play, a child verifies the events of the incident and begins the process of learning to sequence events. It also allows for some independent effort, to come up with a workable solution. When this happens, the child "owns" the solution by putting it into words or actions.

Creative art materials provide excellent alternatives for communicating during the abbreviated LSI. Drawing materials, chalk, music and rhythms, puppets, dolls, animals, and tape recorders stimulate certain children who have no use for other avenues of communication. The adult must structure the activity so that the child is given sufficient content and direction to explore the incident. The material should already be familiar and preferred by the child. Many young children will communicate through creative media, reliving or expanding on a crisis. This is an excellent opportunity for teaching such children to symbolize a critical event and then to use words to describe what they have created and experienced. These media also provide opportunity for the adult to structure an appropriate resolution using verbal and nonverbal channels, such as solutions through creative play, pictures, pantomime, or role play.

"Fix It" Phase

In the material above, the emphasis is on communicating about the crisis and the behaviors and feelings connected with it. You probably noticed how easily the abbreviated LSI flows from the "Talk" phase into the "Fix It" phase. With young children, talking itself is part of the process of crisis resolution, because they are learning to use words to solve problems instead of acting it out through unacceptable behavior. The actual "Fix It" phase may be fairly brief, with the solution almost always being generated by the adult. This should not be surprising to you, if you recall from Chapter 4 that young children are almost always in a preexistential phase of development where they look to adults for authority and solutions to problems. Remember, it is not until about age 8 or 9 years that children enter the Existential Crisis phase of development where they begin to see themselves as having responsibility for problem solving.

With any of the strategies described above in the "Talk" phase, the adult structures the resolution and reentry process. Children feel relief when they are assisted in finding a solution to the crisis. At their stage of development, young children must rely on adults for resolution and new alternative behaviors. If you tend to think that children are old enough to solve a problem of their own making, remind yourself that if they knew how to handle things in a better way they probably would do so. Almost all children want good things to happen to themselves.

Most abbreviated LSIs focus on the therapeutic goals of organizing reality or teaching new skills. Keep these goals in mind as you move to close down an abbreviated LSI. These goals, in the "Fix It" phase, form the ideas you and a child will use to solve the crisis, change behavior, and relieve destructive feelings. If a child is primarily disorganized or in fragile touch with reality, clearly the therapeutic goal is to organize reality, and constructing a sequential time line of events that occurred may be the first achievement. Solutions should be in keeping with ways to assist the child in staying organized by learning to observe what happens with others, staying in touch with events, and reporting them to the teacher.

With young children who are in touch with reality, are aware of what happened, and have the desire for new, more successful ways to behave, the therapeutic goal of teaching new social behaviors is appropriate. This goal might include such behavior choices as learning to signal the teacher when problems seem to be creeping into an event or actively seeking the teacher for help. Other solutions include planning ways, with the child, for the teacher to restructure or redesign an activity so that the child has the opportunity to demonstrate more successful behaviors.

It is essential in an abbreviated LSI to include an attempt by the child to state (or restate from your model) the solution to the crisis. Children like Donna usually can summarize a simple behavioral alternative such as,

> "Sit down."
>
> "Tell teacher."
>
> "Listen to the teacher."
>
> "Do what the teacher says."
>
> "Share."
>
> "Don't hit; ignore him (or tell him . . .)."
>
> "Tell him to stop it."
>
> "Wait for a turn."
>
> "Find another toy."
>
> "Use words to let teacher know."
>
> "Be friends by . . ."

Learning to do and say these sorts of things seldom comes spontaneously. Because of their stage of development, young children need help from adults to achieve the alternative behavior. They seldom can successfully implement new alternatives without initial massive support from an adult.

Structure the resolution and the child's reentry into the classroom or ongoing activity carefully with the child so that it is clearly understood how and when the child will need to use the new behavior. Children must be reassured that adults will help with the resolution and reentry. Here are several statements used by adults to convey this idea of ongoing support:

> "I'll be right beside you when you . . ."
>
> "I'll be there, and I know you will do this really well."
>
> "When you go back, you can remember just how well you did it here with me. It will work again."
>
> "Each time this trouble comes up again, remember that you know how to handle it now!"

Find numerous ways to recognize the new behavior as the child tries it out. Smiles, approving looks, pats, and positive words convey genuine interest in the child's new behavior and provide social reinforcement in a warm and meaningful manner. Remember that an abbreviated LSI is only as effective as the quality of the follow-up the child experiences.

A crisis is resolved, "fixed," when children are sufficiently aware of the behavior required of them that they can (a) state it in some form, (b) demonstrate it with an adult, and (c) indicate confidence that everything will be all right when they do what is required.

"Smile" Phase

The smile itself is not the requirement for this last phase that ends the modified LSI; it only serves as a symbol for feelings of pleasure and accomplishment. Each child who goes through an abbreviated LSI with you should have a sense of relief and accomplishment at the end, with the knowledge that he or she had a problem and has solved it!

Helping children to have a sense of accomplishment and to be aware of their own feelings of relief and pleasure in resolving the crisis is a fundamental premise for every LSI. It is based on the idea that self-regulation of behavior must include the ability to control and direct emotions that fuel behavior. No child is too young to begin to learn this lesson, but the form this instruction takes must be suited to the age and developmental stage of the student. For young children, traditional preschool education instructs in the meaning of "feeling words" such as "smiling faces" and "happy," "sad," or "mad" activities. In this way, preschool teachers begin to teach words for these feelings and their behavioral expressions. But long before young children are ready to use these basic feeling words, they have been decoding their parents' facial expressions. Infant researchers tell us that an infant's smile and the parent's response are the foundations for the human infant's first interpersonal relationship. Because of its built-in availability as a powerful interpersonal tool, we chose the smile for use at the end of Sam's crisis, to tap his new feeling of elation that he has mastered a new skill—how to solve a problem. We want him to be aware of the sense of pleasure he is actually experiencing over this new mastery. We selected the smile because Sam actually grinned as he marched back to his classroom to tell the others about how to solve a problem. If you watch carefully, almost every young student with whom you use LSI successfully will show some smile, distinctly or fleetingly, as he or she completes the LSI and returns to the activity.

There are many other ways to complete this part of the abbreviated LSI where the child is assisted in getting in touch with positive, pleasurable feelings. Some teachers use an interactive model, such as "Let's shake!" And as they do so, they communicate a smile themselves. Children automatically respond with similar behavior. Other teachers follow up with some sort of food or snack to nurture and comfort. They accompany this activity with words that reflect the well-being and pleasure for the satisfactory end of the crisis. Still others provide proximity and a sense of psychological closeness which conveys support and caring, accompanied by words of positive feelings. In all of these strategies, the intent is to obtain similar verbal statements of well-being and pleasure from the child.

You may be asking, "Why go to all this time and effort after you have resolved the crisis and developed a new, alternative behavior for the child?" This emphasis on the resulting feelings as a last phase of the abbreviated LSI is essential for enduring behavior change and social emotional growth. In Chapter 2, we introduced the idea of every child's private reality. This is the emotional memory bank where thoughts, feelings, and anxieties are stored. Recognizing that this storehouse of emotion drives behavior and molds self-esteem, we believe that changes must be made there if lasting change is to occur in a child's self-concept, self-esteem, and in the behaviors that express these attitudes toward self. Because young children often are unaware that the quality of their feelings has changed during LSI, adults must provide a way to sensitize them to these positive changes in their own feelings.

EXAMPLE OF AN ABBREVIATED LSI

Saga of the Yellow Towel

Five-year-old Sam walks into the day-care center sobbing his heart out. His mother leaves after a quick explanation that Sam wants a yellow towel for the swimming trip today, but she has no yellow towel to give him. Sam's teacher tries to console him and redirect his attention to his beautiful blue towel. The strategy doesn't work. Sam's sobs increase and tears flood down his cheeks. He is in an emotional spiral that quickly turns into a tantrum. Between gasps, sobs, and tears, Sam blurts out, "But . . . I want . . . a *yellow* towel!" He breaks down again in racking sobs. Sympathy and reassurance have no effect on him. "I . . . want . . . a *yellow* towel! I *have* to have it!" Sam declares. More sobs and another outburst follow. The teacher tries another strategy, reminding Sam of the fun they will have at the pool and reflecting on Sam's previous accomplishments in swimming. "We're going to have a really good time at the pool today, Sam. Last time you were the one to jump in and swim to the other side. Remember that? Everyone clapped when you got to the other side." Sam responds with another flood of tears interspersed with, "I can't go without a yellow towel!" Sam's teacher then tries the "reality" approach. "Well, Sam, I guess you'll just have to use a blue towel if you want to go swimming today."

With that, Sam disintegrates into uncontrollable sobbing, "I don't want to go swimming!" By now, Sam's emotional state is having a contagious effect on the other children. They stop their play, staring at Sam and the teacher. Thumbs go into mouths; tears well in eyes. One child pushes another. A blanket of unhappiness sweeps through the room. It is time to reverse the rapid disintegration of the emotional climate in the classroom. A break is needed in the Conflict Cycle occurring with the interactions between Sam and his teacher. The teacher, finally recognizing this need to divert Sam from emotionally driven behavior to a rational, problem-solving mode, chooses to use abbreviated LSI with Sam. As the teacher (T) begins, she knows that the first step is to get Sam (S) to talk without breaking down into emotional flooding.

"TALK" PHASE

T: Do you know, we can do something about this problem?

S: (Shakes head to indicate no, but sobbing slows and he glances at the teacher.)

T: Sam, let's do something about this problem. Let's talk about a yellow towel, with words not tears.

S: (Looks again at teacher, with some interest; then breaks down again, sobbing.)

T: We can fix things if we can talk about it . . . with *words* not tears.

S: (Sobs continue but less intensely.)

T: I can tell your words are wanting to come out and are waiting for you to put away your tears.

S: (Sobs end.)

T: Good work! Now we can figure out a way to fix things so you will feel better.

S: (Sobs begin again.)

T: Tell me in words, Sam, not tears.

S: (Haltingly.) I *have* to have a yellow towel to go swimming! (Tears begin again but teacher interrupts.)

T: Use *words*, Sam, to tell me about it.

S: I need a yellow towel.

T: Where do children get yellow towels?

Teacher begins helping Sam recall events and organize reality.

S: (Silence.)

T: Did you pack your towel in your bag before you went to bed last night?

S: Yes, but it wasn't a yellow towel. I didn't like it!

T: What color was the towel you packed?

S: Blue, but I don't like blue.

T: You like yellow. I can understand that. It is your favorite color, right?

S: Yes! And I'm not going swimming.

T: Did you take a bath before you went to bed last night?

S: Yes.

T: Think about last night. Can you remember the color of the towel you used?

S: No . . . but it wasn't yellow. I hated it!

T: Could it have been this blue towel?

S: No! It was white. I hate that blue towel!

T: So, let's see what you did when you got up this morning. Your mom brought you to school?

S: Yes.

T: And did she fix breakfast for you before you left?

S: Yes.

T: What did you have?

S: Cereal and raisins.

T: Sounds delicious. Your mother really takes care of you, giving you a good breakfast like that.

S: (Nods and looks directly at teacher.)

Teacher notes Sam's responsiveness to the idea of being cared for.

T: So, your mom took care of you this morning. She fixed you a delicious breakfast and then helped pack your bag for our swimming trip today?

S: (Looking guarded now.) Yes, but I *told* her I didn't want that towel.

Teacher observes that Sam has demonstrated all of the processes at work in the "Talk" phase, so she begins to look for resolutions.

T: Where do children get towels?

S: (Silence, but looks at teacher hopefully.)

T: To get a towel, we have to know where towels are kept.

S: (Nods agreement.)

T: Now, what I hear you saying is that you like yellow towels and you do not have one today. Is that right?

S: Yes! I want a *yellow* towel.

"FIX IT" PHASE

T: To get any kind of a towel, we have to first find the place where towels are kept.

S: Right! (Leans toward teacher and conveys interest in the idea.)

T: Do you know, we keep towels here at school for children who sometimes forget their towels?

Teacher knows there are no yellow towels at school; but she also knows that Sam is not yet emotionally ready to hear the full facts.

S: (Shakes head.) I don't know where the towels are.

T: Can you think where they may be, in this building?

S: (Now very interested.) Maybe in the office?

T: If not there, where else should we look?

S: Maybe in the supply room? Maybe in the bathroom?

T: Good ideas on how to fix this problem, Sam!

S: (Beaming.) Let's go look!

Sam and teacher go to the places he has suggested. The last stop is the supply room where the school keeps a few clean towels on hand for emergencies.

T: Sam, look what we've found!

S: The towels . . . we found the towels!

T: Right! Let's look at them.

> They go through the stack of towels naming the colors as they go, "Blue, green, pink . . ."

T: What funny color is this, Sam?

> Teacher singles out a grayed-yellow towel that has once been white. Sam studies it dubiously.

S: I don't know.

T: Some people may call it white, but I wouldn't, would you?

S: No, that's not white.

T: Then, what color is it?

S: I don't know.

T: Some people call this color "cream" or "beige."

S: (Listening but looking disappointed.)

T: Do you know how people make this color called "beige"?

S: (Shakes head but looks interested.)

T: You mix white and yellow to make "beige." So we have found a towel that is made of white and yellow.

S: But I want a yellow towel.

T: Now we have found a towel made with *yellow* and white; so you have fixed the problem of no yellow towel! Now you can go swimming with us!

> Sam's love of swimming overcomes his resistance. He reluctantly takes the towel, clutches it to his chest, and starts back to the classroom.

T: When your friends tell you what color towels they have, what will you tell them about your towel?

S: (Thinks about question, looking confused again.)

T: How many colors are in your towel?

S: Two! My towel is two colors!

T: That's right. Your towel is a special color made of two colors. That's something they will be interested in. Before we came down here to hunt for towels, we had a problem. Do you remember?

S: (Nods.)

T: What did we do about it?

S: We found the towels.

T: And before that?

S: We talked about it.

T: That's right! We talked about it, and then we fixed the problem! Now there is something else we need to do, so you will feel good.

S: (Clutching towel; impatient now to rejoin the group for swimming.) I want to go swimming!

T: Before we go back, let's see how you can tell the other children about fixing a problem. When you have a problem, what do you do first?

S: *Talk!*

T: Right. Then what do you do?

S: Find the towels.

T: That's it! How did you *fix* the problem?

S: My towel's white and yellow. (Grinning happily.)

Sam has accomplished the five processes involved in the "Fix It" phase and has stated the solution in his own limited way.

"SMILE" **T:** And after children fix a problem, they feel
PHASE better! I can see that you really feel better now
because you are smiling.

> Teacher begins final phase, sensitizing Sam to a new feeling and
> giving him practice that will carry over into the next activity.

S: (Thinks about it and gives teacher another faint smile.)

T: What will you tell the others about how they will feel when they fix a problem?

S: *Smile!*

T: Sam, you have learned something important today that the other children in the group will want to hear about—how to take care of a problem. Everyone has problems, but not everyone knows how to fix them.

> As Sam and the teacher return to the room, the other children come
> over to him with great interest. They also had been caught up in
> Sam's emotional outburst.

T: (To other children.) What do you do if there is a problem? Sam had a problem and now he knows what to do. Sam, tell them the three steps you use to take care of a problem.

S: (Beaming with pride and eagerness.) First you talk about it. Then you fix it. Then, *smile!*

SUMMARY

Children and youth of every age can be caught in a Conflict Cycle resulting in crisis. We have used LSI with children as young as 5 years of age. LSI can be a helpful strategy for solving or resolving their crisis, if they have the prerequisite skills to participate. They must be able to listen and attend mentally, have some verbal skills for sharing information, understand meanings in words, understand that there has been an incident, and trust that the adult is a helping person.

But solving one crisis is not the end in itself! Life presents one crisis after another; and the solution to one crisis does not necessarily free the student from having another, and another, and another. The important point is to teach students ways to deal with crises that are within their capacity to use. When they try new behaviors with successful and pleasurable results, the motivation is laid for retaining the new behavior to avoid future crises by breaking into Conflict Cycles that bring unpleasant reactions from others.

It is hard to imagine that there is one process useful in working with students of varying ages. Yet the basic LSI process is a road map for dealing constructively with any crisis, on any subject, with any person. Focus is on the students in LSI—their self-image, behaviors, feelings and anxieties, and reaction of others to their behavior. The full LSI process should be abbreviated when a student does not demonstrate all of the prerequisite skills. Most very young children and those with developmental delay will be in this category. They will need to learn some basic communicating skills while solving a crisis. To do this, we suggest an abbreviated LSI in which the six steps of the full LSI are collapsed into three phases that focus on talking, fixing the problem, and smiling as better feelings result.

Having worked your way through to the end of this book, we hope you have developed a perspective about crisis as occurring when students are not able to manage their own feelings and behavior in ways that bring about the results they desire from others. We hope you have an understanding of LSI, and its abbreviated form, as a process you can use for supporting and teaching children and youth in crisis.

We also hope that you view your roles and responsibilities during LSI as changing to adjust to the particular crisis, behaviors, anxieties, and developmental needs of each student. The roles and messages you will be required to deliver are complex. The variations are endless. Sometimes you will have to be an adult authority who enforces rules, order, and justice. At other times you will be the benevolent nurturer, the teacher, the sympathetic supporter. Sometimes you will protect a student from situations too overwhelming to handle; at other times, you will back off from a central role to allow experience ("the great teacher") to teach the hard way. In all of this, the adult remains the source of support, affirming belief in the student—whatever age, behavior, or problem arises to produce crisis.

WHAT WE HOPE YOU HAVE LEARNED
FROM THIS CHAPTER

- Why it is necessary to use an abbreviated LSI form with some young children and those who are developmentally delayed.

- What single process in an abbreviated LSI is particularly beneficial for young or developmentally delayed children.

- The qualities an adult must provide while using the abbreviated LSI form.

- How the steps of a full LSI are condensed into three general phases in abbreviated LSI.

- Four learning processes included during the "Talk" phase.

- Five learning processes included in the "Fix It" phase.

- Two learning processes included in the "Smile" phase.

- Six behavioral characteristics of students who can benefit from the abbreviated form.

- How to determine if a student needs a full or abbreviated LSI.

- Why developmental objectives are well suited for inclusion in abbreviated LSI.

- Two types of therapeutic goals that are usually appropriate for very young children and those who are developmentally delayed.

- What needs to occur in the child before beginning an abbreviated LSI.

- Three different strategies for conducting the "Talk" phase with students who are limited in their verbal ability.

- What structures the crisis resolution during the "Fix It" phase, and the developmental reason for this.

- Why rehearsal and role play are so important during the "Fix It" phase, and the developmental reason for this.
- What single accomplishment by the child is necessary during the "Fix It" phase.
- Three criteria for ending the "Fix It" phase.
- The intent of the "Smile" phase and the developmental reason for having this phase.

APPENDIX A: A HISTORY OF LSI AND ITS FIELD VALIDATION

As we prepare to move into the 21st century, we are able to look back with the clarity that time and distance bring to see the extensive growth and field validation that have occurred over the past 40 years with LSI. It is evident that:

1. LSI has been the only specific verbal intervention strategy to emerge for use by educators and others who work directly with students in crisis.

2. Results of studies of a steadily expanding range and variety of field application for LSI have been published over the past 40 years, primarily using qualitative research methodology.

3. Fundamental concepts and strategies have remained essentially intact since LSI was first developed in the fifties, while variations for specific applications have been expanded considerably and reported in detail.

4. LSI has been included as an intervention strategy in most college textbooks for working with students who are socially, emotionally, or behaviorally disturbed, yet there has not been a textbook published, until this book, solely to teach LSI.

In the early fifties, Fritz Redl and David Wineman introduced the concept of the "marginal interview," later to be renamed "Life Space Interview" (Redl, 1959a, 1966) and now, "Life Space Intervention" (LSI). Redl defined a student's life space as "direct life experience in connection with the issues that become the interview focus" (Redl, 1966, pp. 40–41). The term *life space* was used to imply greater proximity to a student's natural environment than is usual in more formal clinical settings. The term *marginal interview* had been used to distinguish the process from a formal psychiatric interview or clinical psychotherapy, which were, at that time, the only forms of intervention for troubled children and youth other than play therapy. Special education had not yet been developed for such students. There were few residential treatment programs; community mental health had not yet become a nationwide service concept as an alternative to mental hospitals where children and youth were typically housed on adult wards; and a juvenile justice system was just beginning.

The work of Redl and Wineman centered on the study of behavioral controls of aggressive, delinquent children and youth. They believed that behavioral control not only was the essential issue for the students they worked with but also was "the daily job of the most normal child in the pursuit of everyday life" (Redl & Wineman, 1952, p. 27). It was their contention that the problems of delinquent and aggressive students were magnified and intensified pictures of some of the troubles every child has to go through in learning self-control. They were convinced that the difference between the normal and the severely disturbed child was in the ability (or lack of ability) to handle the task of self-control. They observed that normal students could handle adverse circumstances by themselves if given adequate support, while disturbed students could not make use of genuine support from adults. It was their belief that strategies for effective teaching of behavior controls were an equally important pursuit for the parent and teacher of any child. LSI was one result of their extensive experimentation and study about ways to teach behavior controls while providing adequate support for personality development.

The roots of LSI go back to the early 1900s in Europe with the publication of a monumental book entitled *Wayward Youth* by August Aichorn (1935), a Viennese educator and psychoanalyst who directed a school for delinquent boys. In his work, Aichorn translated psychoanalytic concepts into operating principles that were useful in treating delinquents. This book had a major influence on the field of mental health and attracted international recognition by describing a psychodynamic approach to understanding and treating aggressive youths. The importance of the book is still recognized, and it is on the recommended reading list for all serious students in the field of mental health and delinquency.

One of Aichorn's most famous students, who also served as a teacher in his school, was Fritz Redl. Trained as a psychoanalyst by Anna Freud, Redl developed a commitment to treating delinquents. In 1936, he left Vienna and moved to New York City and then, in 1940, to Detroit where he continued to expand his innovative approach with aggressive, delinquent, and disturbed youth in three major projects: The Detroit Group Project, The Detroit Project Group Summer Camp, and Pioneer House. The boys in these programs had generally poor controls over themselves and their aggressive impulses. They were frequently involved in personal and group crisis in which they either refused to participate or attempted to destroy group activity and cohesiveness. Redl's clinical research concerns were "to understand why children's controls break down, how some of them defend themselves so successfully against the adults in their lives, and what can be done to prevent and treat such childhood disorganization" (Redl & Wineman, 1957, p. 13).

To conduct these studies, Redl joined clinical insights and experience with David Wineman, a social worker from Wayne State University. Together, Redl and Wineman planned, directed, and described in exacting detail the characteristics, conditions, and interventions developed during the life of these projects. They used carefully structured, descriptive, and anecdotal procedures to record the behavior of many hundreds of children with whom they worked in these various settings. Their population was "literally thousands of 'behavioral incidents,' comparable in nature, sampled on the basis of the involvedness of certain personality-parts, with the adult's clinical attitude kept relatively constant" (Redl & Wineman, 1957, p. 14).

These clinical attitudes Redl refers to are psychodynamic principles and a milieu concept around which the programs were organized (Redl, 1959b). The basic LSI was formulated here. Redl's three fundamental clinical premises continue to have considerable application today, as general program guides for those who work with troubled students in various settings.

1. Maintain an emotionally healthy environment. A therapeutic environment provides a psychological climate of support and protection from traumatic handling by any personnel; some interpersonal gratifications that are not contingent on behavior; and a degree of tolerance for deviancy and regression with safeguards to protect against escalation.

2. Plan program activities for psychological support. Provide for gratifying play free from goals associated with highly competitive games, arts and crafts, or other gadgets that require a level of performance that is difficult to achieve.

3. Use LSI for therapeutic management of life events. Use LSI to highlight the implications of what the student's behavior is doing, what it means in value demands, and the consequences when "behavior spurts into the life scene with great velocity and intensity" (Redl & Wineman, 1957, p. 37).

As part of this research, detailed descriptions of behavior of individual students and groups were compiled. The table at the end of this appendix summarizes the 22 major characteristics Redl and Wineman reported.

Redl and Wineman believed that adults who work directly with students during a crisis have the greatest opportunity to intervene in therapeutic ways and to make lasting change. They emphasized the importance of the peer group, believing that the spontaneous nature of group life for delinquents is so volatile that the adults who work with them frequently have to intervene immediately and therapeutically during a crisis. There is not time to wait until a caseworker or therapist can get to the scene. This conviction is made quite clear in two publications, *Children Who Hate* (Redl & Wineman, 1951) and *Controls from Within* (Redl & Wineman, 1952), later republished in a single volume, *The Aggressive Child*, in 1957. These writings provide the original field validation for LSI. LSI had become a major crisis intervention strategy; furthermore, it was a new way to take traditional clinical concepts of psychotherapy and apply them in action settings (Dittman & Kitchener, 1959).

The work of Redl and Wineman was radical and innovative for the times. It made a significant impact on the entire field of mental health, residential treatment, and special education for children and youth in crisis. It was the milestone contribution that expanded mental health concepts for those who work directly with troubled students in their natural settings on a daily basis. Redl also was convinced that while LSI was developed to help troubled children and youth, it can be used successfully with normal children who do not need therapy or special programs, but who are temporarily overwhelmed by intense, conflicting feelings like rage, fear, shame, anger, guilt; or who are overwhelmed by unusual life events such as death, divorce, illness, accident, failure, or abuse. During these times, normal children behave more like emotionally disturbed children and need intervention of a supportive and skilled adult trained in LSI.

While Redl and Wineman were directing the Pioneer House Project, 60 miles northwest of Detroit William Morse established the University of Michigan's Fresh Air Camp (FAC) for emotionally disturbed boys. LSI was adopted as a treatment strategy. Wineman had joined the staff in 1950, and Redl served as a major consultant. For the next several years, FAC became the mecca for milieu treatment, providing interdisciplinary training for psychiatry, social work, psychology, and special education (Morse & Wineman, 1957). Contemporary practices for using LSI in therapeutic camping also have their origins in this work (Morse & Small, 1959).

In 1953, Redl was appointed chief of the Child Research Branch of the National Institute of Mental Health (NIMH). During this time LSI was further refined and field-tested when Redl instituted another field validation of the Pioneer House Project. Nicholas J. Long, who had been trained by Morse, accepted an offer by Redl in 1956 to become director of this NIMH Residential Program. The project involved six of the most aggressive boys on the East Coast of the United States. With ample clinical and research staff for this project, Redl continued his systematic study of various techniques and strategies of psychotherapy, program structure, group processes, LSI, and remedial education. This was an exciting and expanding time for those who worked in mental health for children and youth.

During the sixties, major conceptual advances were made for providing therapeutic interventions other than traditional mental health services for emotionally disturbed and troubled youth. In this period, LSI was extended beyond clinical settings into educational environments where it could be used by social workers, special educators, and principals (Long, Stoeffler, Krause, & Jung, 1961; Vernick, 1963). In 1961, Long and Newman published a monograph, *The Teacher's Handling of Children in Conflict*, one of the first attempts to translate many of Redl's principles, including LSI, into specific educational techniques for classroom teachers. Redl (1963) also edited a collection of reports about these applications of LSI for the *American Journal of Orthopsychiatry*, including a paper by Long (1963) reporting on issues in teaching LSI to graduate students. In 1964, Newman and Keith published a monograph entitled *The School-Centered Life Space Interview*. This publication included a series of expanded applications of LSI in school settings (Bernstein, 1963; Redl, 1963). In 1964, Morse, Cutler, and Fink published their national research findings on the first large-scale study of public school classes for the emotionally disturbed, highlighting the great variation of theories, strategies, and techniques developing in the field. From this study, it was clear that more structure and field validation of methods of intervention, including LSI, were needed.

Among the responses to this need for further field validation was a series of publications reporting on new uses for LSI (Faulk & Faulk, 1965; Morse, 1965). In 1965, Long published a monograph, *Direct Help to the Classroom Teacher*, emphasizing the role of the school psychologist as a consultant for crisis intervention and behavior management. That publication was followed by *Conflict in the Classroom* by Long, Morse, and Newman (1965), the first major textbook merging theory and practice from the fields of mental health and education. This textbook helped define the psychoeducational approach to special education and continued in use for two decades as a major resource for training teachers of emotionally and behaviorally disturbed students to use LSI as a psychoeducational strategy.

The significance of the social environment on the behavior of aggressive and disturbed students received increasing attention in the latter part of the sixties. Redl expanded his previous work to reflect this new emphasis for LSI in a book entitled *When We Deal with Children* (1966). Newman (1967) continued expanding her applications of LSI in public school settings. Trieschman, Whittaker, and Brendtro (1969) published *The Other 23 Hours*, stressing the importance of controlling the milieu when designing and implementing an intervention program. They emphasized the necessity for structuring everything that happens to, with, and for pupils in that environment. LSI was among the procedures they advocated.

The seventies brought several important refinements to the basics of the psychoeducational approach and to LSI as a major intervention strategy. In 1971, Morse published his concept of the Crisis Teacher and an operational outline for use of LSI by a Crisis Teacher. This work developed from Morse's consultation in the public schools and day treatment centers in the Ann Arbor, Michigan, area. These publications provided additional substance and structure to LSI as a specific intervention strategy and the specific skills needed by an adult to deal effectively with crisis intervention in the public schools.

During this period, Mary M. Wood established the Rutland Psychoeducational Program in Georgia for seriously emotionally disturbed children and youth. This statewide network of services has reached 10,000 seriously disturbed students annually since 1975 (Swan, Wood, & Jordon, 1990). Wood and colleagues published the first therapeutic curriculum guide, *Developmental Therapy* (Wood, 1975, 1986a, 1986b; Wood, Swan, & Newman, 1981), an applied synthesis of developmental theories. By expanding the psychoeducational approach within a developmental framework, Wood was able to combine mental health and education practices with existing theories about the development of learned behavior, personality, values, motivation, and anxiety. Wood's Developmental Therapy curriculum expanded the idea that LSI can be used in a therapeutic intervention program to meet developmental objectives concerned with social communication and socialization in groups, as well as with behavioral, emotional, and cognitive development (Bachrach, Mosley, Swindle, & Wood, 1978; Williams & Wood, 1977; Wood, 1975, 1979).

Wood and associates conducted two separate field research studies and 23 replications of the Developmental Therapy model over a 5-year period. Their findings on the progress of students and on teachers' mastery of skills, including LSI, were submitted to the National Institute of Education Joint Dissemination Review Panel in 1975 and again in 1981, receiving validation both times as a data-based, field-validated program, which subsequently was disseminated as an exemplary curriculum by the National Diffusion Network for over a decade (Kaufman, Paget, & Wood, 1981; National Dissemination Study Group, 1985).

The eighties produced a series of studies and publications about LSI focusing on effective ways to use LSI for particular types of problems and adapting it to various age groups and program settings (DeMagistris & Imber, 1980). In 1981, Long and Fagen produced an entire issue of *The Pointer* on LSI. This publication compiled the field-based experiences of many leaders in the field who, by now, were using LSI and including it in their college "methods" classes (e.g., Bloom, Dembinski, Fagen, Heuchert, Long, Merritt, Morgan, Morse, Sanders, Tompkins, and Werner). In this publication, Wood and Weller (1981) described how they adapted LSI to the developmental characteristics of students and emphasized the instructional objectives in the social-emotional-behavioral domains that can be achieved through LSI.

The work of Wood and associates through the Developmental Therapy Institute at the University of Georgia continued to include LSI as a major verbal intervention strategy for teaching social communication and value-based problem solving, adapted to the particular developmental characteristics of students in various age groups—preschool, elementary and middle school, and secondary school. Their work also focused on teaching these intervention skills to both professionals and paraprofessionals (Wood, 1981, 1982, 1986a, 1986b; Wood, Combs, & Walters, 1986; Wood, Swan, & Newman, 1981).

In 1987, Naslund published a longitudinal, data-based study entitled "Life Space Interviewing: A Psychoeducational Intervention Model for Teaching Pupil Insights and Measuring Program Effectiveness" (Naslund, 1985, 1987). She kept detailed records on 1,404 LSIs—every LSI conducted for an academic year with 28 severely disturbed students at Rose School in Washington, DC. This school, established by Nicholas Long, was a community-based psychoeducational program combining mental health and special education services for severely emotionally and behaviorally disturbed students between ages 6 and 13. LSI was a major intervention strategy used in the program. Naslund's study examined changes that took place during an academic year while using LSI, looking specifically at (a) frequency with which a student is sent to the "crisis room" for LSI, (b) reasons for being sent, (c) frequency of use of physical restraint prior to LSI, and (d) types of LSI used.

Naslund's results revealed that students who were developmentally older, those who had been in the program longer, and those who had above average IQ showed the greatest decrease in their use of the crisis room for LSI, over time. Her results also showed that peer-related crisis was most frequent (364 incidents); while classroom work was the least-used reason for referring a student (41 incidents). Other significant findings included an increase in classroom work, a decrease in loss of self-control, and an increase in self-initiated requests for follow-up LSIs, which suggests that students, by the end of the year, were utilizing LSI (talking about a problem with an adult) as a way to manage their own stresses.

Results of Naslund's study also provided valuable information about the extent to which physical intervention, preliminary to LSI, was needed. Four out of five students in her study could be managed on a verbal level without physical restraint during crisis intervention. She also reports, "One out of every 14 was severely out of control and required restraint for over 15 minutes" (Naslund, 1985, p. 108). The use of physical restraint declined significantly during the year.

Among the five types of LSI used during Naslund's study period, those designed to organize reality and correct distortions in perceptions ("Reality Rub-In") were used most frequently (170 times). However, over time, this type of LSI decreased while LSIs designed to teach new social skills ("New Tools Salesmanship") increased. This is consistent with the expectation that as students drop dysfunctional behavior and see social events and interpersonal interactions more realistically, they need additional LSIs to learn more effective social skills for handling stress and crisis.

Looking back at the use of LSI over the past decades, we find that it has been included as an intervention strategy in almost every college textbook and many research studies of behavior management strategies for teachers of emotionally disturbed and behaviorally disordered students (Apter & Conoley, 1984; Bommarito, 1977; Brown, McDowell, & Smith, 1981; Heuchert, 1983; Kaufmann & Lewis, 1974; Kerr & Nelson, 1989; Knoblock, 1983; McDowell, Adamson, & Wood, 1981; Morse, 1985; Newcomer, 1980; Paul & Epanchin, 1982; Reinert, 1980; Rich, 1982; Shea & Bauer, 1987; Swanson & Reinert, 1979).

As we enter the nineties, attention is drawn again to the concerns of 50 years ago—delinquent youth. The level of violence and alienation now is almost out of control. Students and the adults who work with them in educational settings, the juvenile justice system, alternative school programs, group homes, substance abuse programs, and detention centers are confronted with brutality, crisis, and aggression daily. Major institutions of business, industry, education, government, and church are seeking ways to restore values in the regulation of behavior.

We have seen this need and believe that LSI is a powerful tool for teaching responsibility for self-regulated behavior and values. Using the work of Kohlberg and associates, we have drawn again on the benchmark work of others to enrich LSI applications by going back to theory. The research and theories of Coles (1986), Kohlberg (1981, 1984), Lickona (1976, 1983), Selman (1980), Selman, Beardless, Schultz, Krupa, and Podorefsky (1986), and Turiel (1983) provide a wealth of new knowledge about values and moral development, the growth of interpersonal understanding and social perspective-taking by children and youth. This new knowledge provides an added dimension to our understanding of how to make LSI powerful and effective for students in crisis. The expanded LSI puts emphasis on behavior regulated by insight and a value system. This is a big step for today's disturbed and alienated youth. But unless we make that step with them, it will not happen!

CHARACTERISTIC DISTURBANCES FOUND IN THE AGGRESSIVE, DELINQUENT BOYS FOR WHOM LSI WAS INITIALLY DEVELOPED

1. Low frustration tolerance.

2. Reaction to insecurity, anxiety, or fear with rapid, extreme measures of panic, flight, or ferocious attacks and diffuse destruction.

3. Low resistance to temptations to behave unacceptably from situations, gadgets, or the contagious behavior of others.

4. Vulnerability to high levels of excitement stimulated by group excitement.

5. Responsiveness to inner urges and impulses without restrictions or sublimation.

6. Lack of capacity for ''responsible care'' of possessions, even those most valued.

7. Resistance to change.

8. Inability to prevent or deal with past emotional memories flooding the entire system and dominating a present life event.

9. Fewer guilt feelings when the conscience should be working and inability to take corrective action to assuage their guilt.

10. Inability to see their own contribution in a chain of events.

11. Inability to switch to inner control when external controls on behavior are withdrawn.

12. Exaggerated demanding and rejecting reactions to trusted adults who offer acceptance and relationship.

13. Inability to recall and revive memories of previous satisfactions, achievements, or pleasures when a present life event proves unsatisfactory.

14. Lack of realism about rules and regulations.

15. Difficulty with concepts of time, postponement of an event, or future consequences.

16. A paradoxical insensitivity to interpersonal reactions with a heightened sensitivity to peer group attitudes, principles, and expectations.

17. Inability to learn from past experience about future situations, whether pleasant or unpleasant.

18. Impaired ability to draw realistic inferences from what happens to others.

19. Exaggerated reactions to failure, success, and mistakes.

20. Impaired ability to react constructively to competitive challenges of sports and games.

21. Difficulty adjusting to group roles that demand both sharing and restricting personal urges to control or dominate others.

22. Failure to use rational problem solving in times of social crisis.

(From *The Aggressive Child* by Fritz Redl and David Wineman, 1957, pp. 74–140.)

APPENDIX B: CHECKLIST FOR RATING YOUR OWN LSI SKILLS

Note: Checklist may be reproduced.

Date: _____ Student: _____

Place: _____ Length of LSI: _____

Was physical intervention needed before beginning LSI? ____

What was the central issue? _____

_____

_____

If Emotional First Aid was used, check type:

☐ Drain off emotions

☐ Support student engulfed in emotion

☐ Maintain communication

☐ Regulate social behavior

☐ Act as umpire

Check type of therapeutic goal(s) (if more than one, number in order of occurrence):

☐ Organize reality

☐ Confront behavior

☐ Build self-esteem

☐ Teach new skills

☐ Expose exploitation

Check the LSI steps you were able to complete during this LSI:

☐ 1. Focus on the Incident

☐ 2. Students in Crisis Need to Talk

☐ 3. Find the Central Issue and Select a Therapeutic Goal

☐ 4. Choose a Solution Based on Values

☐ 5. Plan for Success

☐ 6. Get Ready to Resume the Activity

From the following list, check the skills you believe you were able to use effectively in this LSI. If you care to rate yourself on a scale, use 1 = ineffective through 5 = highly effective.

NONVERBAL BODY LANGUAGE

☐ Convey support and alliance through body posture.   1 2 3 4 5

☐ Use eye contact or the opposite as needed to provide "space."   1 2 3 4 5

☐ Vary voice quality and volume as needed.   1 2 3 4 5

☐ Use physical stance to convey needed adult role and relationship.   1 2 3 4 5

☐ Avoid excessive touch.   1 2 3 4 5

☐ Maintain physical proximity or distance as needed.   1 2 3 4 5

☐ Stand or sit as needed to convey control.   1 2 3 4 5

☐ Convey interest, support, or other messages through facial expressions.   1 2 3 4 5

VERBAL STYLE

☐ Use concrete words for clarity.   1 2 3 4 5

☐ Use imagery to motivate.   1 2 3 4 5

☐ Convey ideas clearly.   1 2 3 4 5

☐ Maximize student's talk.   1 2 3 4 5

☐ Minimize own talk.   1 2 3 4 5

☐ Use a time line to help student organize events. 1 2 3 4 5

☐ Assist student in clarifying an issue. 1 2 3 4 5

☐ Assist student in seeing cause-and-effect relationships. 1 2 3 4 5

☐ Use reflection effectively. 1 2 3 4 5

☐ Use interpretation effectively. 1 2 3 4 5

☐ Decode accurately. 1 2 3 4 5

☐ Use third-person form to generalize. 1 2 3 4 5

☐ Limit use of references to yourself. 1 2 3 4 5

☐ Limit use of negative statements. 1 2 3 4 5

RELATIONSHIP

☐ Use active listening. 1 2 3 4 5

☐ Communicate respect for student. 1 2 3 4 5

☐ Communicate interest. 1 2 3 4 5

☐ Communicate calm self-control. 1 2 3 4 5

☐ Convey confidence and optimism. 1 2 3 4 5

☐ Convey focus and competence. 1 2 3 4 5

☐ Avoid intruding into student's private "space." 1 2 3 4 5

☐ Avoid value judgments. 1 2 3 4 5

☐ Avoid role reversals. 1 2 3 4 5

☐ Avoid counteraggression. 1 2 3 4 5

APPENDIX C: CHARTS FOR THE SIX LSI STEPS

These charts may be reproduced for studying the details in the six LSI steps. When spread out horizontally in a sequence, they illustrate the dynamic flow of LSI from beginning to end.

Step 1: Focus on the Incident

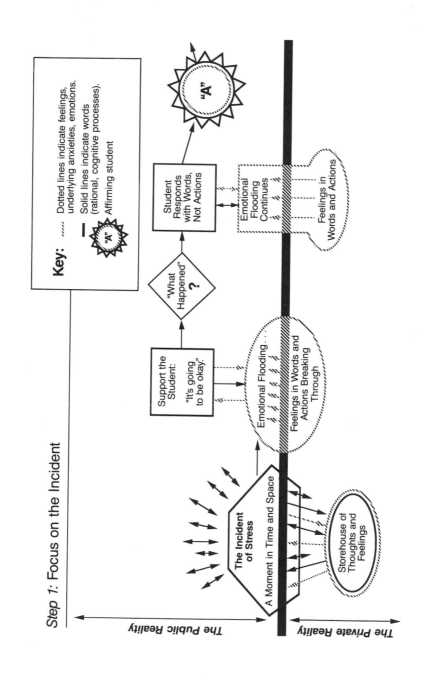

Step 2: Students in Crisis Need to Talk

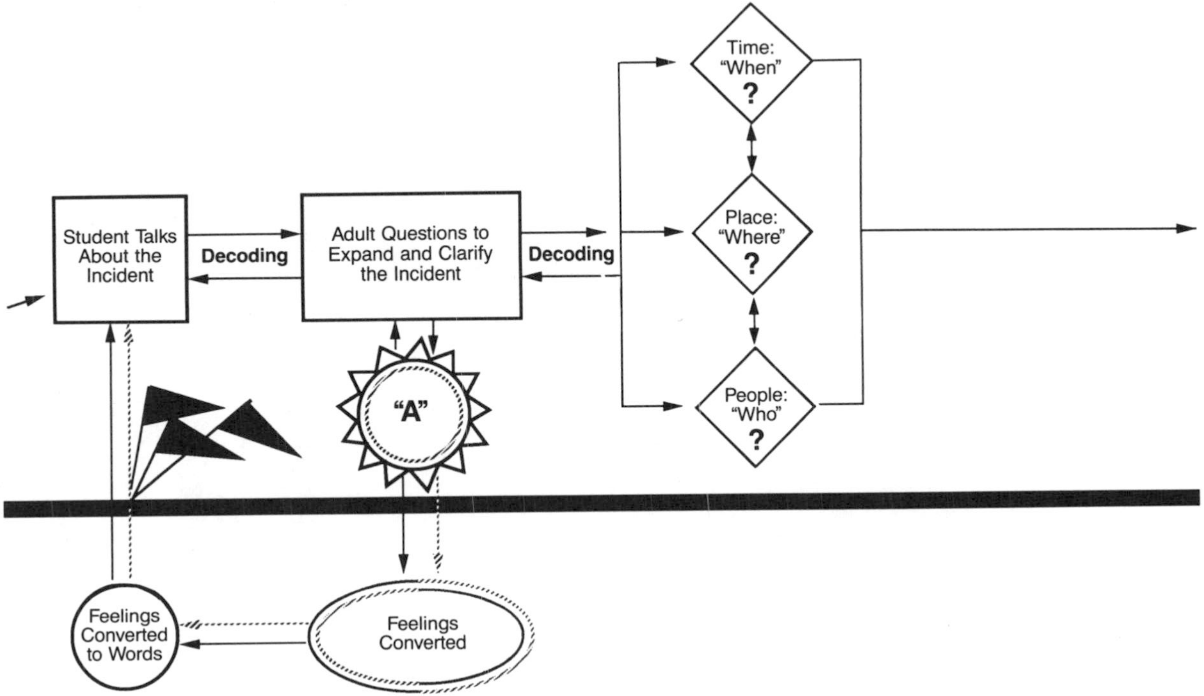

Step 3: Find the Central Issue and Select a Therapeutic Goal

Central Issue: The Incident?

What is the Central Issue?

State the Issue

Motivation to Change? ↔ Insight?

Which Therapeutic Goal is Needed?

Goals
1. Organize reality.
2. Confront behavior.
3. Build values for self-control.
4. Teach new skills.
5. Expose exploitation.

"A"

Feelings Converted

Central Issue: Underlying Anxiety?
• Aloneness?
• Inadequacy?
• Guilt?
• Conflict?
• Identity?

Step 4: Choose a Solution Based on Values

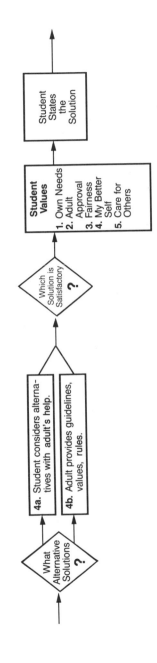

Step 5: Plan for Success

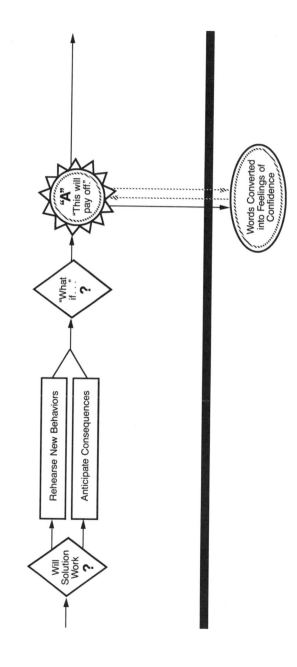

Step 6: Get Ready to Resume the Activity

How to End and Return to Group **?**

"What is the group doing now" **?**

"What will you do" **?**

"What will the others do" **?**

"A" Student Empowered

Therapeutic Goal Accomplished: *Student's Insight and Self-Esteem Enhanced*

Storehouse of Better Thoughts and Feelings

Aichorn, A. (1935). *Wayward youth.* New York: Viking.

Apter, S. J., & Conoley, J. C. (1984). *Child behavior disorders and emotional disturbance.* Englewood Cliffs, NJ: Prentice-Hall.

Bachrach, A. W., Mosley, A. R., Swindle, F. L., & Wood, M. M. (1978). *Developmental therapy for young children with autistic characteristics.* Austin, TX: PRO-ED.

Bernstein, M. (1963). Life space interview in the school setting. In R. Newman & M. Keith (Eds.), *The school-centered life space interview* (pp. 35–44). Washington, DC: School Research Project, Washington School of Psychiatry.

Bettelheim, B. (1977). *The uses of enchantment.* New York: Vintage.

Bloom, R. B. (1981). The reality rub-in interview with emotionally disturbed adolescents. *The Pointer, 25,* 22–25.

Bommarito, J. W. (1977). *Prevention and clinical management of troubled children.* Washington, DC: University Press of America.

Brown, G., McDowell, R. L., & Smith, J. (1981). *Educating adolescents with behavior disorders.* Columbus, OH: Merrill.

Children's Defense Fund. (1986). *Vanishing dreams: The growing economic plight of America's families.* Chicago: Center for Labor Market Studies, Northwestern University.

Coles, R. (1967). *Children of crisis.* Boston: Little, Brown.

Coles, R. (1986). *The moral life of children.* Boston: Atlantic Monthly Press.

DeMagistris, R. J., & Imber, S. C. (1980). The effects of life space interviewing on academic and social performance of behaviorally disordered children. *Behavior Disorders, 6,* 12–25.

Dembinski, R. J. (1981). The opening gambit: How students avoid the LSI. *The Pointer, 25,* 12–15.

Dittman, A., & Kitchener, R. (1959). Life space interviewing and individual play therapy: A comparison of techniques. *American Journal of Orthopsychiatry, 29,* 19–26.

Earls, F. (1986). Developmental perspective on psychosocial stress in childhood. In M. W. Yogman & T. B. Brazelton (Eds.), *In support of families* (pp. 29–41). Cambridge, MA: Harvard University Press.

Elder, G. H., Liker, J. K., & Cross, C. E. (1984). Parent-child behavior in the great depression: Life course and intergenerational influences. In P. B. Baltes & O. Brim, Jr. (Eds.), *Life span development and behavior* (Vol. 6, pp. 109–158). New York: Academic Press.

Fagen, S. A. (1981). Conducting an LSI: A process model. *The Pointer, 25*, 9–11.

Faulk, U., & Faulk, G. (1965). Use of social worker and the life space interview with institutionalized children in the public school. *Child Study Center Bulletin, 1*, 27–31.

Heuchert, C. M. (1983). Can teachers change behavior? Try interviews. *Academic Therapy, 18*, 321–328.

Heuchert, C. M., & Long, N. J. (1981). A brief history of life space interviewing. *The Pointer, 25*, 5–8.

Kagan, J. (1986). Stress on and in the family. In M. W. Yogman & T. B. Brazelton (Eds.), *In support of families* (pp. 42–54). Cambridge, MA: Harvard University Press.

Kaufman, A. S., Paget, K. D., & Wood, M. M. (1981). Effectiveness of developmental therapy for severely emotionally disturbed children. In F. Wood (Ed.), *Perspectives for a new decade: Education's responsibility for seriously disturbed and behaviorally disordered children and youth* (pp. 176–188). Reston, VA: Council for Exceptional Children.

Kaufmann, J., & Lewis, C. (1974). *Teaching children with behavior disorders.* Columbus, OH: Merrill.

Kerr, M. M., & Nelson, C. M. (1989). *Strategies for managing behavior problems in the classroom* (2nd ed.). Columbus, OH: Merrill.

Knoblock, P. (1983). *Teaching emotionally disturbed children.* Boston: Houghton Mifflin.

Kohlberg, L. (1981). *Essays on moral development, Vol. 1: The philosophy of moral development.* San Francisco: Harper & Row.

Kohlberg, L. (1984). *Essays on moral development, Vol. 2: The psychology of moral development.* San Francisco: Harper & Row.

Lickona, T. (Ed.). (1976). *Moral development and behavior.* New York: Holt, Rinehart & Winston.

Lickona, T. (1983). *Raising good children.* New York: Bantam.

Long, N. J. (1963). Some problems teaching life space interviewing techniques to graduate students in education in a large class at Indiana University. *American Journal of Orthopsychiatry, 33*, 723–727.

Long, N. J. (1965). *Direct help to the classroom teacher.* Washington, DC: School Research Project, Washington School of Psychiatry.

Long, N. J., Morse, W. C., & Newman, R. G. (Eds.). (1965). *Conflict in the classroom.* Belmont, CA: Wadsworth.

Long, N. J., & Newman, R. G. (1961). The teacher's handling of children in conflict. *Bulletin of the School of Education, Indiana University, 37*(4).

Long, N. J., Stoeffler, V., Krause, K., & Jung, C. (1961). Life space management of behavioral crisis. *Social Work, 6,* 38–45.

McDowell, R. L., Adamson, G. W., & Wood, F. H. (Eds.). (1981). *Emotional disturbance.* New York: Little, Brown.

Merritt, C. A. (1981). Bandaids for the bumps. *The Pointer, 25,* 16–19.

Morgan, R. (1981). Group life space interviewing. *The Pointer, 25,* 37–41.

Morse, W. C. (1965). The mental hygiene viewpoint on school discipline. *The High School Journal, 48,* 396–401.

Morse, W. C. (1971). The crisis or helping teacher. In N. J. Long, W. C. Morse, & R. G. Newman (Eds.), *Conflict in the classroom* (pp. 485–490). Belmont, CA: Wadsworth.

Morse, W. C. (1981). LSI tomorrow. *The Pointer, 25,* 67–70.

Morse, W. C. (1985). *The education and treatment of socioemotionally impaired children and youth.* Syracuse, NY: Syracuse University Press.

Morse, W. C., Cutler, R. L., & Fink, A. H. (1964). *Public school classes for the emotionally handicapped.* Washington, DC: Council for Exceptional Children.

Morse, W. C., & Small, E. (1959). Group life space interviewing in a therapeutic camp. *American Journal of Orthopsychiatry, 29,* 27–44.

Morse, W. C., & Wineman, D. (1957). Group interviewing in a camp for disturbed boys. *Journal of Social Issues, 13,* 23–31.

Naslund, S. R. (1985). Life space interviewing: A descriptive analysis of crisis intervention with emotionally disturbed children in a special education school. *Dissertation Abstracts International, 45/08A,* p. 2487; DER 84-25243.

Naslund, S. R. (1987). Life space interviewing: A psychoeducational interviewing model for teaching pupil insights and measuring program effectiveness. *The Pointer, 31,* 12–15.

National Dissemination Study Group. (1985). *Educational programs that work* (11th ed.). Longmont, CO: Sopris West.

Newcomer, P. L. (1980). *Understanding and teaching emotionally disturbed children.* Boston: Allyn & Bacon.

Newman, R. G. (1967). *Psychological consultation in the schools.* New York: Basic Books.

Newman, R., & Keith, M. (Eds.). (1963). *The school-centered life space interview.* Washington, DC: School Research Program, Washington School of Psychiatry.

Paul, J. L., & Epanchin, B. C. (1982). *Emotional disturbance in children.* Columbus, OH: Merrill.

Raspberry, W. (1989). New York kids see violence every day of lives. *Athens Daily News,* May 23.

Redl, F. (1959a). Strategy and techniques of the life space interview. *American Journal of Orthopsychiatry, 29,* 1–18.

Redl, F. (1959b). The concept of a therapeutic milieu. *American Journal of Orthopsychiatry, 29,* 721–736.

Redl, F. (1963). The life space interview in the school setting. *American Journal of Orthopsychiatry, 33,* 717–719.

Redl, F. (1966). *When we deal with children*. New York: The Free Press.

Redl, F., & Wineman, D. (1951). *Children who hate*. Glencoe, IL: The Free Press.

Redl, F., & Wineman, D. (1952). *Controls from within*. Glencoe, IL: The Free Press.

Redl, F., & Wineman, D. (1957). *The aggressive child*. Glencoe, IL: The Free Press.

Reinert, H. R. (1980). *Children in conflict* (2nd ed.). St. Louis: Mosby.

Rich, H. L. (1982). *Disturbed students: Characteristics and educational strategies*. Austin, TX: PRO-ED.

Rogers, C. R. (1965). The interpersonal relationship: The core of guidance. In Mosier (Ed.), *Guidance: An examination*. New York: Harcourt Brace & World.

Rogers, C. R. (1967). The therapeutic relationship: Recent theory and research. In C. H. Patterson (Ed.), *The counselor in the school: Selected readings* (pp. 228–240). New York: McGraw-Hill.

Sanders, L. S. (1981). New tool salesmanship interview. *The Pointer, 25,* 32–33.

Selman, R. L. (1980). *The growth of interpersonal understanding*. New York: Academic Press.

Selman, R. L., Beardless, W., Schultz, L. H., Krupa, M., & Podorefsky, D. (1986). Assessing adolescent interpersonal negotiation strategies: Toward the integration of structural and functional models. *Developmental Psychology, 22,* 450–459.

Shea, T., & Bauer, A. M. (1987). *Teaching children and youth with behavior disorders* (2nd ed.). Englewood Cliffs, NJ: Prentice-Hall.

Swan, W. W., Wood, M. M., & Jordon, J. (1990). Building a statewide program of mental health and special education services for children and youth. In G. K. Farley & S. G. Zimet (Eds.), *Day treatment for emotionally disturbed children* (Vol. 2, chap. 11). New York: Plenum.

Swanson, H. L., & Reinert, H. R. (1979). *Teaching strategies for children in conflict*. St. Louis: Mosby.

Thomas, A., & Chess, S. (1984). Genesis and evaluation of behavior disorders: From infancy to early adult life. *American Journal of Psychiatry, 141,* 1–9.

Tompkins, J. R. (1981). Symptom estrangement interview. *The Pointer, 25,* 26–28.

Trieschman, A. E., Whittaker, J. K., & Brendtro, L. K. (1969). *The other 23 hours*. Chicago: Aldine.

Turiel, E. (1983). *The development of social knowledge*. Cambridge, England: Cambridge University Press.

Vernick, J. (1963). The use of the life space interview on a medical ward. *Social Casework, 44,* 465–469.

Warnock, G. J. (1971). *The object of morality*. London: Methuen.

Werner, A. (1981). Massaging numb values interview. *The Pointer, 25,* 29–31.

Williams, G. H., & Wood, M. M. (1977). *Developmental art therapy.* Austin, TX: PRO-ED.

Wood, M. M. (1975). *Developmental therapy.* Baltimore: University Park Press.

Wood, M. M. (Ed.). (1979). *The developmental therapy objectives* (3rd ed.). Austin, TX: PRO-ED.

Wood, M. M. (Ed.). (1981). *Developmental therapy sourcebooks: Vol. 1: Music, movement, and physical skills. Vol. 2: Fantasy and make-believe.* Austin, TX: PRO-ED.

Wood, M. M. (1982). Developmental therapy: A model for therapeutic intervention in the schools. In T. B. Gutkin & C. R. Reynolds (Eds.), *A handbook for school psychology* (pp. 609–629). New York: Wiley.

Wood, M. M. (1986a). *Developmental therapy in the classroom.* Austin, TX: PRO-ED.

Wood, M. M. (1986b). Developmental therapy. In C. R. Reynolds, L. Mann (Eds.), *Encyclopedia of special education* (Vol. 3, pp. 499–500). New York: Wiley.

Wood, M. M., Combs, M. C., & Walters, L. H. (1986). Use of staff development by teachers and aides with emotionally disturbed and behavior disordered students. *Teacher Education and Special Education, 9,* 104–112.

Wood, M. M., Swan, W. W., & Newman, V. (1982). Developmental therapy for the severely emotionally disturbed and autistic. In R. L. McDowell, G. W. Adamson, & F. H. Wood (Eds.), *Teaching emotionally disturbed children* (pp. 264–299). New York: Little, Brown.

Wood, M. M., & Weller, D. (1981). How come it's different with some children? A developmental approach to life space interviewing. *The Pointer, 25,* 61–66.

Author Index

Subject Index

attributes of those conducting successful LSIs, 14–16
beginning when student not ready, 168–70
charts pertaining to, *9, 112, 116, 125, 135, 141, 147,* 320–325
checklist of skills for, 159–61, 160t, 315–17
compared to psychotherapy, 6
counteraggression by adult during, 174–75
decoding and, 59, 62, 66
digressions during, 170–72
essential strategies for, 108–109
examples of, 16–29, 196–204, 212–19, 227–39, 249–59, 268–75
failure to build dialogue from student's responses, 176–77
failure to consider negative solutions and consequences, 183
failure to state central issue simply and concisely, 181–82
field validation of, 309–12
goals of, x, 11–14, 82, 101–102, 109
as helping process, 157
historical development of, xi, 6–7, 303–12
importance of, 298–99
invasion of student's private space, 177–78
jumping to solutions prematurely, 182–83
pitfalls of, 168–83
points to remember concerning, 153–54, 186–87
prerequisite skills needed by students, 11, 12t
purposes of, 5–7, 106
reactions of peer group following, 150–51

Redl's clinical premises concerning, 305–306
reflection of negatives during, 173–74
regression and, 149–50
resistance to, 6, 149–50, 169–70
role reversal attempts by students, 179–81
series of, 109
significance of name change from Life Space Interview, xi–xii, 7
silence and, 168–70
Step 1: focus on incident, 8, *9,* 110–15, *112, 320*
Step 2: students' need to talk, 8–9, *9,* 115–23, *116, 321*
Step 3: central issue and therapeutic goal, *9, 9,* 124–33, *125, 322*
Step 4: solution based on values, 9–10, *9,* 133–40, *135, 323*
Step 5: plan for success, *9,* 10, 140–46, *141, 324*
Step 6: resume activity, *9,* 10, 146–53, *147, 325*
steps in, 7–10, *9,* 109–53, *112, 116, 125, 135, 141, 147*
trivialization of complex feelings during, 175–76
use with a group, 106–107
values and, 88, 90
verbal nature of, ix, xi–xii, 106
Life Space Interview, xi, 6–7, 303–307
Listening, 120, 132
LSI. *See* Life Space Intervention (LSI)

Make-believe situations, 286
Manipulation, 55, 77
"Manipulation of the Boundaries of the Self," 130–31, 132, 261–75

Rose School, 310
Rutland Psychoeducational
    Program, 309

Schools
    early use of LSI in, 307–11
    stress involved in, 38–39
Secondary interpretation of
    feelings, 63–64
Self-control, 129–30, 132, 221–39
Self-esteem, 13
Self-regulation, 3–4, 84, 94, 304
Separation, 91
Shared skills, 79–80
Silence, 168–70
Social blindness, 191
Social skills
    adult's task in teaching new
        social skills, 242–43
    example of LSI pertaining to,
        249–59
    examples of students who can
        benefit from, 244–48, 311
    teaching of, 130, 132, 241–42
Storytelling, 286
Stress. *See also* Conflict Cycle;
    Crisis
    crisis and, 33
    examples of, ix–x
    in students' lives, 36–39
Students. *See also* Develop-
        mentally delayed students
    and adult authority, 74–75
    adult strategies that influence
        behavior of, 75–81
    and changing roles of adults,
        75, 76t
    concerns during crisis, 73–74
    Conflict Cycle and, 105–106
    consequences and, 80–81
    decoding ability, 59
    developmental anxieties of,
        52–58, 53, 66
    emotional dependence of,
        90–93, 94
    example of student values,
        97–98

Existential Crisis of, 82–86, 88,
    89, 94, 166–67
feelings and anxieties of,
    40–41
friendship with adults, 77–79
gratification and, 14–15
motivation of, 76–77, 86–94
and postexistential phase,
    85–86
prerequisite skills for LSI, 11,
    12t
protection by adults, 14
reactions to crisis, 5–6, 51
relationship and, 15
responsibility and, 15, 82–94
self-regulation and, 3–4, 84
shared skills and, 79–80
stress in lives of, 36–39
and therapeutic goal of
    building values to
    strengthen self-control,
    224–27
and therapeutic goal of
    confrontation of unaccept-
    able behavior, 208–12
and therapeutic goal of
    exposure of exploitation,
    265–68
and therapeutic goal of
    organization of reality,
    193–96
and therapeutic goal of
    teaching new social skills,
    244–48
values of, 87–90, 89, 93–94
violence and, ix–x, 37, 311
Sublimation, 43t, 44
Substitution, 43t, 44
Surface interpretation of
    feelings, 61–62
"Symptom Estrangement," 129,
    132, 205–19

Therapeutic goals
    of abbreviated LSI, 288

The Authors

Mary Margaret Wood is professor emerita of special education at the University of Georgia and director of the Developmental Therapy Institute in Athens, Georgia. She is widely recognized as one of the early leaders in work with troubled children because of her sustained advocacy efforts, outstanding program development, and over 25 years of direct work with children and youth who have social, emotional, and behavioral handicaps. Dr. Wood is founder of the statewide  Georgia Psychoeducational Network, 24 programs serving about 9,000 troubled students annually. She also formulated the first curriculum adaptations of developmental theory for special education with severely disturbed students.

Dr. Wood received her undergraduate education at Goucher College in Baltimore, Maryland, and her doctorate in special education from the University of Georgia in 1963. She is author and editor of a number of books and articles dealing with the developmental approach to teaching social and emotional skills. In 1970, while on the faculty at the University of Georgia, she founded and served as the first director of the Rutland Psychoeducational Program in the public schools in Athens, Georgia. There she led a team of professionals from mental health and education to develop and field-test the Developmental Therapy Curriculum. During this time, she also established the first graduate training program in Georgia for teachers of disturbed children and youth. Both the Developmental Therapy Curriculum model and the parallel teacher preparation program have received national recognition as exemplary educational programs.

Dr. Wood served as policy advisor to U.S. Secretary of Education Elliot Richardson on the National Child Development Advisory Council, advised the U.S. Department of Education Special Education Programs and the Council for Exceptional Children, was a consultant to several national projects including Head Start and Mr. Roger's Neighborhood, advised several Georgia governors on mental health services for children, and lectured on many university and college campuses.

Working with the staff at the Rutland Center and the Developmental Therapy Institute, Dr. Wood and her colleagues have worked with thousands of children, educators, and mental health professionals in 30 states and in other nations. Developmental Therapy has been translated into Spanish, Chinese, Portuguese, and German. Called Peggy by colleagues and friends, Dr. Wood has touched many lives through her influence as a role model, her written works, and her ideas about effective interventions.

Dr. Wood and her husband, Norman, an economist, have six children and six grandchildren and live in Athens, Georgia.

Nicholas J. Long is professor of special education at American University in Washington, DC, and is recognized nationally as one of the leaders in reeducating emotionally disturbed children and youth. He received his doctorate in child development and educational psychology from the University of Michigan in 1956 and has been a licensed psychologist since 1974. He has taught at the University of Michigan, Indiana University, George Washington University, and  Georgetown University in addition to teaching at American University for the past 20 years, and he has published extensively.

Dr. Long has been directly involved with troubled children and youth for over 30 years as clinician, educator, and administrator in private and public schools, day treatment programs, and residential facilities. From 1955 to 1956 he served as first principal at Children's Psychiatric Hospital in Ann Arbor, Michigan, and for the following 2 years he served as chief of the Children's

Residential Treatment Center at the National Institute of Mental Health for Fritz Redl's study on aggressive youth. From 1962 to 1972 he served as executive director of Hillcrest Children's Psychiatric Center in Washington, DC. From 1972 to 1990 Dr. Long developed and directed a model psychoeducational treatment program called The Rose School, serving severely disturbed students excluded from the Washington, DC, public schools.

As director of The Rose School, Dr. Long expanded his psychoeducational approach and demonstrated an effective way for professionals from mental health and education to work collaboratively. He also established a federally funded, field-based graduate training program in emotional disturbance, which received national recognition and awards for its unique contributions to urban educational problems. In addition he has served on many state advisory boards and has been the child psychologist for *National Geographic World* magazine for children since 1974.

Many of today's leaders in fields related to the reeducation of troubled children and youth received their advance training under Dr. Long's guidance. They came away sharing his commitment to children and youth, his enthusiasm for seeing the potential in each student, and his dedication to standards of excellence in professional practice.

Dr. Long and his wife, Jody, a school psychologist, have five children and live in Chevy Chase, Maryland.